D1070059

Sparse Matrix Computations

Academic Press Rapid Manuscript Reproduction

Proceedings of the Symposium on Sparse Matrix Computations
at Argonne National Laboratory
on September 9–11, 1975

Sparse Matrix Computations

Edited by

JAMES R. BUNCH

University of California, San Diego

and

DONALD J. ROSE

Harvard University

Academic Press Inc. New York San Francisco London 1976

A Subsidiary of Harcourt Brace Jovanovich, Publishers

ACADEMIC PRESS, INC.
111 Fifth Avenue, New York, New York 10003

United Kingdom Edition published by
ACADEMIC PRESS, INC. (LONDON) LTD.
24/28 Oval Road, London NW1

Library of Congress Cataloging in Publication Data

Symposium on Sparse Matrix Computations, Argonne
National Laboratory, 1975.
Sparse matrix computations.

Bibliography: p.
Includes index.
1. Matrices–Data processing–Congresses.
2. Mathematical optimization–Data processing–Congresses.
Data processing–Congresses. I. Bunch, James R.
II. Rose, Donald J. III. United States. Argonne
National Laboratory, Lemont, Ill. IV. Title.
QA188.S9 1975 512.9′43 76-7964
ISBN 0–12–141050–1

PRINTED IN THE UNITED STATES OF AMERICA

Contents

Contributors

Randolph E. Bank, Department of Mathematics, University of Chicago, Chicago, Illinois 60637.

Åke Björck, Department of Mathematics, Linköping University, Linköping, Sweden.

William L. Briggs, Applied Mathematics, Aiken Computation Laboratory, Harvard University, Cambridge, Massachusetts 02138.

James R. Bunch, Department of Mathematics, University of California, San Diego, La Jolla, California 92093.

B. L. Buzbee, Los Alamos Scientific Laboratory, Los Alamos, New Mexico 87544.

Alan K. Cline, Departments of Computer Sciences and of Mathematics and Center for Numerical Analysis, University of Texas at Austin 78712.

Paul Concus, Lawrence Berkeley Laboratory, University of California, Berkeley, California 94720.

Jim Douglas, Jr., Department of Mathematics, University of Chicago, Chicago, Illinois 60637.

Todd Dupont, Department of Mathematics, University of Chicago, Chicago, Illinois 60637

S. C. Eisenstat, Department of Computer Science, Yale University, New Haven, Connecticut 06520.

W. Morven Gentleman, Department of Computer Science, University of Waterloo, Waterloo, Ontario, Canada.

Alan George, Department of Computer Science, University of Waterloo, Waterloo, Ontario, Canada

Philip E. Gill, Division of Numerical Analysis and Computing, National Physical Laboratory, Teddington, Middlesex, England.

D. Goldfarb, Department of Computer Science, City College–CUNY, New York, New York 10031.

Gene H. Golub, Department of Computer Science, Stanford University, Stanford, California 94305

Fred Gustavson, IBM T. J. Watson Research Center, Yorktown Heights, New York 10598.

Gary Hachtel, IBM, Office Products Division, Boulder, Colorado 80302.

Joel E. Hirsh, Center for Earth and Planetary Physics, Harvard University, Cambridge, Massachusetts 02138.

Thomas D. Howell, Department of Computer Science, Cornell University, Ithaca, New York 14853.

W. Kahan, Department of Electrical Engineering and Computer Science, University of California, Berkeley, California 94720.

T. L. Magnanti, Department of Mathematics, Sloan School, MIT, Cambridge, Massachusetts 02139.

Walter Murray, Division of Numerical Analysis and Computing, National Physical Laboratory, Teddington, Middlesex, England.

Dianne P. O'Leary, Department of Mathematics, University of Michigan, Ann Arbor, Michigan 48104.

B. N. Parlett, Computer Science Division, University of California, Berkeley, California 94720.

George W. Platzman, Department of Geophysical Sciences, University of Chicago, Chicago, Illinois 60637.

T. A. Porsching, Department of Mathematics, University of Pittsburgh, Pittsburgh, Pennsylvania 15260.

Donald J. Rose, Applied Mathematics, Aiken Computation Laboratory, Center for Research in Computing Technology, Harvard University, Cambridge, Massachusetts 02138.

Alberto Sangiovanni-Vincentelli, Department of Electrical Engineering and Computer Sciences, University of California, Berkeley, California 94720.

Michael A. Saunders, Systems Optimization Laboratory, Department of Operations Research, Stanford University, Stanford, California 94305.

M. H. Schultz, Department of Computer Science, Yale University, New Haven, Connecticut 06520.

A. H. Sherman, Department of Computer Science, University of Illinois, Urbana, Illinois 61801.

G. W. Stewart, Department of Computer Science, University of Maryland, College Park, Maryland 20742.

Robert Endre Tarjan, Computer Science Department, Stanford University, Stanford, California 94305.

Richard S. Varga, Department of Mathematics, Kent State University, Kent, Ohio 44242.

George F. Whitten, Applied Mathematics, Aiken Computation Laboratory, Center for Research in Computing Technology, Harvard University, Cambridge, Massachusetts 02138.

P. T. Woo, Chevron Oil Field Research Company, La Habra, California 90631.

Preface

The papers in this volume were presented at the Symposium on Sparse Matrix Computations held at Argonne National Laboratory in September 1975. The editors, the organizing committee, and the sponsors sought to present a symposium that would give a broadly based view of the field as represented by current work in the subject. Hence the papers contain contributions in several areas of matrix computations, and include some of the most active research in numerical linear algebra.

We have organized the papers into general categories which deal, respectively, with sparse elimination, sparse eigenvalue calculations, optimization, mathematical software for sparse matrix computations, partial differential equations, and applications involving sparse matrix technology. We would like to emphasize, however, the considerable overlap between these categories; some papers could and should be considered in two or more of the general categories mentioned.

In a more general context this volume presents research in applied numerical analysis but with considerable influence from computer science. In particular most of the papers deal explicitly (or implicitly) with the design, analysis, implementation, and application of computer algorithms. Such an emphasis includes not only the establishment of space and time complexity bounds, but also an attempt to understand the algorithms and the computing environment in enough detail to make design decisions for effective mathematical software to be used as tools in science and engineering.

The editors wish to thank the other members of the organizing committee (Robert Brayton, Gene Golub, John Hopcroft, Martin Schultz, and Brian Smith) for their help in organizing the technical program and James C. T. Pool and the staff of the Applied Mathematics Division at Argonne National Laboratory for their excellent job in providing the arrangements. We would also like to thank the Energy Research and Development Administration, the Office of Naval Research, and Argonne National Laboratory for sponsoring the symposium.

J. R. Bunch
D. J. Rose

Sparse Matrix Computations

I

Design and Analysis of Elimination Algorithms

GRAPH THEORY AND GAUSSIAN ELIMINATION

ROBERT ENDRE TARJAN
Computer Science Department
Stanford University
Stanford, California 94305

Abstract

This paper surveys graph-theoretic ideas which apply to the problem of solving a sparse system of linear equations by Gaussian elminination. Included are a discussion of bandwidth, profile, and general sparse elimination schemes, and of two graph-theoretic partitioning methods. Algorithms based on these ideas are presented.

1. Introduction

Consider the system $xM = c$, where M is a nonsingular real-valued n by n matrix, x is a one by n vector of variables, and c is a one by n vector of constants. We wish to solve this system of equations for x. In many applications, M is a large sparse matrix; that is, M has many zero elements. If the system is solved using Gaussian elimination or some other direct method, many of the zeros in M may become non-zero. To make the solution process efficient, we would like to avoid having to explicitly examine the zeros of M, and to keep the number of non-zero elements small.

We can model the zero - non-zero structure of M by a directed graph and study the effect of a solution method on this graph. This graph-theoretic analysis has several important benefits, including the following.

(1) For some sparse matrices, a graph-theoretic representation is a good one, allowing efficient access of non-zero matrix elements.

(2) We can devise a good solution procedure for an entire class of matrices (those with the same zero - non-zero structure) at a time. If several matrices with the same zero - non-zero structure occur in an application, then spending extra time initially to devise a good solution procedure may result in later savings as the procedure is reused.

This research was supported in part by National Science Foundation grant DCR72-03752 A02.

(3) The approach illuminates the applicability of solution methods for linear systems to other kinds of graph problems such as those arising in global flow analysis and operations research.

This paper surveys several graph-theoretic aspects of the solution of linear systems. We consider several graph-theoretic methods for choosing a good ordering scheme for Gaussian elimination. These include bandwidth minimization, profile minimization, and general sparse techniques. We also discuss graph-theoretic block methods based on the strongly-connected components and dominators of the underlying graph. Finally, we discuss the problem of choosing a set of pivot positions for Gaussian elimination.

The paper contains seven sections. Section 2 introduces necessary graph-theoretic notation. Section 3 discusses representation of a system of linear equations as a graph, a decomposition method which uses strongly connected components, and a graphical version of Gaussian elimination. Section 4 discusses methods for choosing a pivot order. Section 5 discusses a decomposition method using dominators. Section 6 discusses selection of a set of pivot positions. Section 7 contains further remarks. Results in Sections 3, 4, and 6 are not new, though some are as yet unavailable in print. Section 5 contains new results.

2. Graph-Theoretic Notation

A directed graph $G = (V,E)$ is a finite set V of $n = |V|$ elements called vertices and a finite set $E \subseteq V \times V$ of $m = |E|$ vertex pairs called edges. An edge of the form (i,i) is a loop. If $(i,j) \in E$ if and only if $(j,i) \in E$, we say G is symmetric. (A symmetric directed graph corresponds to the undirected graph given by making $\{i,j\}$ an undirected edge if $(i,j) \in E$. We prefer to use symmetric directed graphs instead of undirected graphs since they correspond more closely to the computer representation of graphs.)

A graph $G' = (V',E')$ is a subgraph of G if $V' \subseteq V$ and $E' \subseteq E$. If $V'' \subseteq V$ and $G(V'') = (V'',E(V''))$ where $E(V'') = \{(i,j) \in E \mid i,j \in V''\}$, then $G(V'')$ is the subgraph of G induced by the vertex set V''. Similarly, if $E'' \subseteq E$ and $G(E'') = (V(E''),E'')$ where $V(E'') = \{i \in V \mid \exists (i,j) \in E''$ or $\exists (j,i) \in E''\}$, then $G(E'')$ is the subgraph of E induced by the edge set E''.

A sequence of edges $p = (v_1,v_2),\ldots,(v_k,v_{k+1})$ is a path from v_1 to v_{k+1}. By convention there is a path of no edges from every vertex to itself. If $v_1 = v_{k+1}$ the path is a cycle. Every cycle contains at least one edge. The path is said to contain vertices $v_1, v_2, \ldots, v_{k+1}$ and edges $(v_1, v_2), \ldots, (v_k, v_{k+1})$ and to avoid all other vertices and edges. If $v_1, v_2, \ldots, v_{k+1}$ are distinct, except possibly $v_1 = v_{k+1}$, p is simple.

If there is a path from a vertex v to a vertex w, v is reachable from w. If every vertex in a graph G is reachable from every other vertex, G is strongly connected. The maximal strongly connected subgraphs of a graph G are vertex-disjoint and are called its strongly connected components. If u, v, w are distinct vertices of a graph G such that every path from u to w contains v, then v is a dominator of w with respect to u. If G contains no three distinct vertices u, v, w such that v dominates w with respect to u, G is strongly biconnected. The maximal strongly biconnected subgraphs of G are edge disjoint (except for loops) and are called the strongly biconnected components of G.

A (directed, rooted) tree T is a graph with a distinguished vertex r such that there is a unique path from r to any vertex. If v is on the path from r to w, we write $v \overset{*}{\rightarrow} w$ and say v is an ancestor of w and w is a descendant of v. If (v,w) is a tree edge, we write $v \rightarrow w$ and say v is the parent of w and w is a child of v. If $v \overset{*}{\rightarrow} w$ and $v \neq w$, we write $v \overset{+}{\rightarrow} w$ and say v is a proper ancestor of w and w is a proper descendant of T.

If $G = (V,E)$ is any graph, the symmetric (or undirected) extension of G is the graph $G' = (V, \{(i,j) \mid (i,j) \in E \text{ or } (j,i) \in E\})$. If T is a tree, its symmetric extension is called a symmetric (or undirected) tree. If $G = (V,E)$ is any graph, its reversal is the graph $G^R = (V, \{(j,i) \mid (i,j) \in E\})$.

For a graph $G = (V,E)$ an ordering α of V is bijection $\alpha: \{1,2,\ldots,n\} \leftrightarrow V$. $G_\alpha = (V,E,\alpha)$ is an ordered graph.

3. Gaussian Elimination on a Graph

Let $xM = c$ be a set of n linear equations, where $M = (m_{ij})$ is an n by n non-singular matrix. We can represent the zero - non-zero structure of the matrix M by an ordered graph $G_\alpha = (V,E,\alpha)$, where $V = \{1,2,\ldots,n\}$, $E = \{(i,j) \mid m_{ij} \neq 0 \text{ or } i = j\}$, and $\alpha(i) = i$ for $1 \leq i \leq n$. The unordered graph $G = (V,E)$ corresponds to the set of matrices PMP^T , where P is a permutation matrix.

We can represent the system $xM = c$ by assigning to vertex i the value $c(i) = -c_i$ and the variable $x(i) = x_i$ and assigning to edge (i,j) the value $m(i,j) = m_{ij}$ if $k \neq j$, $m(i,j) = m_{ij}+1$ if $i = j$. The system $xM = c$ becomes

$$Q = \left\{ \sum_{(i,j)\in E} x(i)m(i,j) + c(i) = x(j) \mid 1 \leq j \leq n \right\} .$$

Henceforth we consider the system of equations defined graph-theoretically in this way. (The variable $x(j)$ appears on the right side of the j-th equation for reasons to be discussed later.)

Corresponding to any subgraph $G' = (V',E')$ of G is a system of equations

$$Q' = \left\{ \sum_{(i,j)\in E'} x(i)m(i,j) + c(i) = x(j) \mid j\in V' \right\} .$$

We shall discuss solving the system Q by Gaussian elimination. First, it is useful to consider a way of decomposing Q into subsystems Q' such that the solution to the subsystems gives the solution to the whole system. Let $G_1 = (V_1,E_1),\ldots,G_k = (V_k,E_k)$ be the strongly connected components of the graph G . These components can be ordered so that if (v,w) is an edge of G with $v\in V_i$ and $w\in V_j$, then $i \geq j$. Such an ordering is a topological sorting [26] of the components. Given the components in a topologically sorted order, the following method solves the system Q .

```
SOLVE: for i := k step -1 until 1 do begin
         solve the system Q_i;
         for (v,w) ∈ E such that v∈V_i , w∈V_j with j > i do
             c(w) := c(w) + x(v)·m(v,w);
       end SOLVE;
```

This scheme is well-known and its validity is easy to check. The strongly connected components of G correspond to the irreducible blocks of the matrix M [43].

We can find the strongly connected components of a graph G and topologically sort them in $O(n+m)$ time using depth-first search [40]. The running time of SOLVE is thus $O(n+m)$ plus the time to solve the subsystems Q_i if the graph G is represented as a set of adjacency lists [40]. Reference [6] contains a more detailed complexity analysis.

One special case of SOLVE is important. If each strongly connected component of G consists of a single vertex (i.e., G is acyclic except for loops), each subsystem Q_i is a single equation $x(i)a(i) + c(i) = x(i)$. Solving such an equation requires one subtraction and one division: $x(i) = c(i)[1-a(i)]^{-1}$. In this case SOLVE requires $O(n+m)$ total. This special case is the final step, called back-solving, of the Gaussian elimination method.

The first step of Gaussian elimination consists of the following algorithm.

```
ELIMINATE: for j := 1 until n-1 do
    for (j,k) ∈ E with k > j do begin
    a:  c(k) := c(k) + c(j)·[1-m(j,j)]^-1·m(j,k);
        for (i,j) ∈ E with i > j do begin
            if (i,k) ∉ E then add (i,k)
                with value m(i,k) = 0 to E;
            m(i,k) := m(i,k) + m(i,j)·[1-m(j,j)]^-1·m(j,k);
end end ELIMINATE;
```

It is well-known and easy to verify that when ELIMINATE terminates, the solution to the original equation set Q can be found by applying SOLVE to the graph $G' = (V,E')$ defined by $E' = \{(i,j) \in E \cup F \mid i \geq j\}$, where F is the set of added edges (i,k), called fill-in edges, created by ELIMINATE. The values on edges when ELIMINATE terminates give an LU decomposition of M [14].

ELIMINATE requires $O(n+|E \cup F|)$ numeric storage and

$$O\left(n + \sum_{\substack{(i,j) \in E \cup F \\ i>j}} \left(1 + \sum_{\substack{(j,k) \in E \cup F \\ k>j}} 1\right)\right)$$

arithmetic operations. SOLVE requires $O\left(n + \sum_{\substack{(i,j) \in E \cup F \\ i \geq j}} 1\right)$ time once

ELIMINATE is applied. Solving the system Q for a new set of constants requires $O(n+|E\cup F|)$ time given the LU decomposition computed by ELIMINATE. For a more detailed complexity analysis, see [6].

Implementing ELIMINATE to achieve the bounds above for total storage and total operation count is not simple. Two methods of implementation suggest themselves.

(1) Representation of $m(i,j)$ using a hash table [25].

(2) Representation of $m(i,j)$ using adjacency lists for G [40].

It is straightforward to implement ELIMINATE using a hash table to store edge values. This representation will achieve the desired storage bound in the worst case and the desired operation bound on the average (but not in the worst case). Because the hash table must be stored, the storage requirements will exceed the storage necessary for adjacency lists, but the average running time is apt to be faster than using adjacency lists.

Careful implementation of ELIMINATE using adjacency lists allows us to achieve the desired storage and time bounds in the worst case. Gustavson [19] discusses many of the ideas important in such an implementation. We use a two-pass method. First, we compute the set F of fill-in edges. An algorithm described in [34] is adequate for this step. Next we use the following modification of ELIMINATE for the LU decomposition. We assume that, for each vertex k, a list B(k) of vertices j such that $(j,k)\in E\cup F$ is available, and that these vertices j are in order by number, smallest to largest, in the list B(k). Associated with each entry $j\in B(k)$ is the value $m(j,k)$. The procedure below carries out the computation column-by-column. This method of elimination is sometimes called the Crout method or the Doolittle method [14].

```
CELIMINATE: begin
    for i := 1 until n do array(i) := 0;
    for k := 2 until n do begin
        for j ∈ B(k) do array(j) := m(j,k);
        for j ∈ B(k) with j < k do begin
            c(k) := c(k)+c(j)·[1-m(j,j)]^-1·m(j,k);
    b:  for i ∈ B(j) with i > j do
            array(i) := array(i)+m(i,j)·[1-m(j,j)]^-1 m(j,k);
    end;
    for (j,k) ∈ E ∪ F do m(j,k) := array(j);
end end CELIMINATE;
```

Variable array is used here to make the computation in Step b easy. It is easy to see that this procedure works correctly and achieves the desired storage bound and operation count. The correctness of CELIMINATE depends on the fact that the entries in each list $B(k)$ are in order by number. This representation seems to require that the fill-in F be precomputed.

So far, little is known about the efficiency of using adjacency lists versus using a hash table. Most likely, the hash table method uses less time, and the adjacency list method uses less space. See [8,19,23] for details concerning implementation of Gaussian elimination using adjacency lists.

The time and storage requirements of ELIMINATE depend only on the structure of G and on the ordering α. By reordering the vertices of G, we may greatly improve the efficiency of ELIMINATE. The next section discusses the problem of choosing a good ordering. Because of the complexity of implementing ELIMINATE for sparse graphs, various researchers have studied special methods which handle certain types of sparse graphs. Two such methods, the bandwidth method, and the profile method, are discussed in the next section, in addition to the general sparse method.

Symmetry plays an important role in the solution process. If the matrix M is symmetric (i.e., $m_{ij} = m_{ji}$), it is possible to save a factor of two in storage and computing time by using the symmetry [19]. If the matrix M is structurally symmetric (i.e., G is symmetric), it is much easier to compute the fill-in and other properties of the elimination order [33]. In some applications it may be useful to make G symmetric by adding an edge (j,i) for each edge (i,j). This may simplify the implementation of ELIMINATE and decrease the time necessary to find a good elimination ordering. These savings must be balanced against the time and storage costs for handling the added edges.

If one of the pivot elements $m(j,j)$ equals one, the j-th iteration of the main loop in ELIMINATE cannot be carried out. Furthermore if any of the $m(j,j)$ are close to one, the method is numerically unstable [14]. For certain types of matrices, however, ELIMINATE is guaranteed to work and to be numerically stable. These include the diagonally dominant matrices and the symmetric positive definite matrices [14]. Henceforth we shall not worry about numeric stability but shall assume that ELMINATE using any vertex ordering will produce an acceptable answer. In practice, however, it is important to verify stability.

4. Elimination Schemes

One method used to avoid the complexity of implementing ELIMINATE for general sparse graphs is the bandwidth method. If α is an ordering of the vertices of G, we define the bandwidth b of G to be $\max_{(i,j)\in E} |\alpha(i)-\alpha(j)|$. The bandwidth method finds a band of width $2b+1$ about the main diagonal outside of which all entries are zero, and performs Gaussian elimination within the band. The bandwidth version of Gaussian elimination appears below.

```
BELIMINATE: for j := 1 until n-1 do
     for k := j+1 until j+b do begin
          c(k) := c(k) + c(j)·[1-m(j,j)]^-1·m(j,k);
          for i := j+1 until j+b do
               m(i,k) := m(i,k) + m(i,j)·[1-m(j,j)]^-1·m(j,k);
end BELIMINATE;
```

Bandwidth elimination requires $O(bn)$ storage using array storage and $O(b^2n)$ time. The difficulty with the bandwidth method is finding an ordering which produces a small bandwidth. A graph for which there is an ordering such that all edges within the bandwidth are present is called a dense bandwidth graph. It is easy to test in $O(n+m)$ time whether a graph G is a dense bandwidth graph. If it is, the ordering which makes G a dense bandwidth graph is easy to compute.

A graph with an ordering which produces bandwidth one is tridiagonal [14]. (Edges within the bandwidth may be missing, so a tridiagonal graph need not be a dense bandwidth graph.) It is easy to test in $O(n+m)$ time whether a graph is tridiagonal. Garey and Johnson [16] have devised an $O(n+m)$ time method to find a bandwidth two ordering if one exists. We know of no efficient method to test for bandwidth three.

Various heuristics exist for finding orderings with small bandwidth. A breadth-first search method proposed by Cuthill and McKee [11] works well on some examples.

Unfortunately, the problem of determining whether a given graph G has an ordering which produces a bandwidth of given size b or less belongs to a class of problems called NP-complete. The NP-complete problems have the following properties.

(1) If *any* NP-complete problem has a polynomial-time algorithm, then *all* NP-complete problems have polynomial-time algorithms.

(2) If <u>any</u> NP-complete problem has a polynomial-time algorithm, then any problem solvable non-deterministically in polynomial time has a deterministic polynomial-time algorithm.

Such well-studied problems as the travelling salesman problem, the tautology problem of propositional calculus, and the maximum clique problem are NP-complete. It seems unlikely that any NP-complete algorithm has a polynomial-time algorithm. Papadimitriou [29] first proved the minimum bandwidth problem NP-complete; Garey and Johnson [16] proved the problem NP-complete even for trees! This negative result reduces the appeal of the bandwidth scheme except for problems for which a good choice of ordering is explicit or implicit in the problem description.

An extension of the bandwidth method is the <u>profile method</u>. If α is an ordering of the vertices of G, the <u>profile</u> $b(j)$ of vertex j is $\max\{\alpha(j)-\alpha(i) \mid (i,j)\in E$ or $(j,k)\in E$ and $\alpha(j) > \alpha(i)\}$. The profile method assumes that all entries are within an envelope of varying width about the main diagonal. For implementation of the profile method, see [38]. Profile elimination requires $O\left(\sum_{j=1}^{n} b(j)\right)$ storage and $O\left(\sum_{j=1}^{n} b(j)^2\right)$ time. As with the bandwidth method, there is still the problem of finding an ordering with small profile.

A graph G for which there is an ordering such that all edges within the profile are present is called a <u>dense profile graph</u>. That is, $G = (V,E)$ is a dense profile graph if and only if G is symmetric and there is an ordering α of the vertices such that if $(i,j)\in E$ with $\alpha(i) < \alpha(j)$, and k satisfies $\alpha(i) \leq \alpha(k) \leq \alpha(j)$, then $(k,j)\in E$.

There is a nice characterization of dense profile graphs which has apparently not appeared in print before. We call a graph $G = (V,E)$ an <u>interval graph</u> if there is a mapping I of the vertices of G into sets of consecutive integers such that $(i,j)\in E$ if and only if $I(i) \cap I(j) \neq \emptyset$.

<u>Theorem 1</u>. G is a dense profile graph if and only if G is an interval graph.

<u>Proof</u>. Suppose $G = (V,E)$ is dense profile with appropriate ordering α. For each vertex $v\in V$, let $I(v) = \{\alpha(w) \mid (w,v)\in E$ and $\alpha(w) \leq \alpha(v)\}$. By the dense profile property, each set $I(v)$ is a set of consecutive integers. Suppose $(i,j)\in E$ with $\alpha(i) < \alpha(j)$. Then $\alpha(i) \in I(i) \cap I(j)$.

Suppose $\alpha(k) \in I(i) \cap I(j)$. Then $(k,i),(k,j) \in E$, $\alpha(k) \leq \alpha(i)$, $\alpha(k) \leq \alpha(j)$. Without loss of generality, suppose $\alpha(i) \leq \alpha(j)$. Then by the dense profile property, $(i,j) \in E$. Thus the intervals $I(v)$ faithfully represent the edges of G .

Conversely, suppose G is an interval graph with appropriate intervals $I(v)$. G is symmetric since $I(i) \cap I(j) = I(j) \cap I(i)$. Let $\alpha(j)$ be an ordering such that $\alpha(j) \leq \alpha(i)$ implies the largest integer in $I(\alpha(j))$ is no greater than the largest integer in $I(\alpha(i))$. Let $(i,j) \in E$ with $\alpha(i) < \alpha(j)$ and suppose $\alpha(i) \leq \alpha(k) \leq \alpha(j)$. Then $I(\alpha(i)) \cap I(\alpha(j)) \neq \emptyset$, so $I(\alpha(j))$ contains all integers between the largest integer in $I(\alpha(i))$ and the largest integer in $I(\alpha(j))$. This set includes the largest integer in $I(\alpha(k))$. Thus $I(\alpha(j)) \cap I(\alpha(k)) \neq \emptyset$, and $(k,j) \in E$. □

Lueker and Booth [28] have devised an $O(n+m)$ -time test for the interval graph property. The test is constructive, so an appropriate ordering for a dense profile graph can be found in $O(n+m)$ time.

The breadth-first search method of Cuthill and McKee produces small profile on some examples. A reverse breadth-first search based on the Cuthill-McKee method does as well or better [27]. Little is known theoretically about the behavior of such heuristics. The problem of finding an ordering to minimize $\sum_{i=1}^{n} b(j)$ (or $\sum_{i=1}^{n} b(j)^2$) has not yet been proved NP-complete. For results on the NP-completeness of a similar problem, see [15]. See [10] for further discussion of bandwidth, profile, and related ordering schemes.

It is easy to generalize the definitions of bandwidth and profile to allow different envelopes on either side of the diagonal. See [10]. In view of the difficulty of finding good orderings for minimizing symmetric bandwidth and profile, we do not pursue this idea further.

Several facts reduce the appeal of the bandwidth and profile schemes except on problems for which a good choice of ordering is explicit or implicit in the problem description. First, it is not easy to find a good ordering. Second, and more important, the bandwidth and profile schemes may be overly pessimistic in that they may examine many matrix elements which are in fact zero. This will happen with sparse graphs having large bandwidth or profile. A practical example is the square k by k grid graph, which arises in finite difference solutions to partial differential equations

[43]. Any bandwidth or profile method for this problem requires $O(k^3)$ storage and $O(k^4)$ time [22,30], whereas the nested dissection method [17,36], a special type of general sparse ordering, requires only $O(k^2 \log k)$ storage and $O(k^3)$ time.

We consider now the general sparse method. A graph is a <u>perfect elimination graph</u> if there is an ordering which produces no fill-in. We can test for the perfect elimination property in $O(nm)$ time [34]. This property is computationally at least as hard as testing a directed graph for transitivity, so improving the time bound beyond $O(nm)$ would be a significant result. Given any ordering, we can compute its fill-in in $O(nm)$ time [34]. Such an algorithm is useful if we wish to precompute the fill-in before performing the numeric calculations. Computing the fill-in is at least as hard as computing the transitive closure of a directed graph [34].

The problem of finding an ordering which minimizes the size of the fill-in is NP-complete [34]. However, a related problem has a polynomial time algorithm. We call a set of fill-in edges F <u>minimal</u> if no ordering produces a fill-in $F' \subset F$. If α is an ordering which produces fill-in F, α is a <u>minimal</u> ordering. Minimal orderings are not necessarily close to minimum, but given any ordering we can improve it to a minimal one in $O(n^4)$ time [34].

These problems are easier for symmetric graphs. We can test a symmetric graph for the perfect elimination property in $O(n+m)$ time, compute the fill-in of any ordering in $O(n+m)$ time, and find a minimal ordering in $O(nm)$ time [35]. These algorithms, especially the one to compute fill-in, may have important practical uses.

In view of the NP-completeness results, we cannot hope to solve the general problem of efficiently implementing sparse Gaussian elimination. We can only try to solve the problem for special cases. Approaches include the following.

(1) Develop and study heuristics for producing orderings with small fill-in. Several heuristics have been proposed, including the minimum degree and minimum fill-in heuristics [31,32]. These methods seem to work well in practice, but nothing is known about their theoretical behavior.

(2) Develop good ordering schemes for special types of graphs. A successful example of this approach is the nested dissection method [17,36].

(3) Develop methods which avoid the necessity of computing all the fill-in. In some cases values on fill-in edges can be stored implicitly rather than explicitly, resulting in a savings of time and storage.

We consider in the next section a method which combines ideas (3) and (4).

Another possible approach would be to study the average behavior of elimination methods. This approach is not a good one, however, for two reasons. (1) Most graphs which occur in practical problems are highly non-random in their structure. (2) Erdös and Even [13] have shown that "most" symmetric graphs with order $n \log n$ edges have a fill-in of order n^2 (most graphs with less than order $n \log n$ edges are not connected). Thus a dense matrix method is as good (to within a constant factor) as any sparse method, on random graphs which are not too sparse.

5. A Decomposition Method Using Dominators

This section presents a decomposition method for solving systems of linear equations which is more powerful than the decomposition into strongly connected components discussed in Section 3. The idea of the method is as follows. Suppose $G = (V,E)$ is a directed graph and there exists a triple u, v, w of distinct vertices such that v dominates w with respect to u. We can partition V into $V = \{v\} \cup V_1 \cup V_2$ such that V_1 contains u and all vertices reachable by a path from u which avoids v. Let $G_1 = (\{v\} \cup V_1, E(\{v\} \cup V_1))$, $G_2 = (\{v\} \cup V_2, E(\{v\} \cup V_2))$. Suppose we are given a set of equations defined on G. We solve the set by the following method.

Step 1: For each vertex $w \in V_2$, solve for $x(w)$ in terms of $x(v)$ using the system of equations defined on G_2. That is, represent $x(w)$ as $x(w) = x(v) \cdot a(v,w) + b(v,w)$ for some real values $a(v,w)$, $b(v,w)$.

Step 2: Replace each edge (x,y) with $x \in V_2$, $y \in \{v\} \cup V_1$, by an edge (v,y) with value $m(v,y) = 0$, if such an edge does not exist already. Set $m(v,y) := a(v,x) \cdot m(x,y) + m(v,y)$. Set $c(y) := b(v,x) \cdot m(x,y) + c(y)$.

Step 3: In the new graph G', solve the system of equations defined on G'_1.

Step 4: Using the equations found in Step 1, solve for the values of the variables $x(w)$, $w \in V_2$.

This method solves the system of equations defined on graph G by solving the two smaller systems defined on G_2 and G_1' and combining the solutions. It is equivalent to carrying out Gaussian elimination on G in an order so that all the vertices in V_2 are ordered first, followed by vertex v, followed by all the vertices in V_1 . For each edge (x,y) with $x \in V_2$, $y \in \{v\} \cup V_1$, this elimination order may create a large number of fill-in edges (x',y) with $x' \in V_2$. None of these fill-in edges are really necessary to the computation; only the corresponding fill-in edge (v,y) is necessary. By computing the value of this edge directly, we avoid computing many of the fill-in edges and thus save time and storage space.

We generalize this scheme as follows. Henceforth we assume $G = (V,E)$ is strongly connected. Let r be some fixed, distinguished vertex of G. If v dominates w with respect to r and no vertex dominates w with respect to v , we say v is the <u>immediate dominator</u> of w (with respect to r). We denote this relationship by $v = \underline{\text{idom}}(w)$.

<u>Theorem 2</u> [1]. Each vertex $w \neq r$ has a unique immediate dominator. The rooted tree $T = (V, \{(\underline{\text{idom}}(v),v) \mid v \neq r\})$, called the <u>dominator tree</u> of G, has the property that, for every vertex w, its dominators with respect to r are exactly its ancestors in T.

Our solution method works as follows.

Step 1: Choose a fixed vertex r of G. Compute the corresponding dominator tree T.

Step 2: Working from the leaves of T to the root, solve for each variable $x(v)$ in terms of $x(\underline{\text{idom}}(v))$.

Step 3: Solve for $x(r)$ and for all other variables $x(v)$ by backsolving using the equations computed in Step 2.

Step 2 will compute, for each variable $x(v)$, a pair of numbers $a(v)$ and $b(v)$ such that $x(v) = x(\underline{\text{idom}}(v)) \cdot a(v) + b(v)$. As we work through the tree in Step 2, we must compose such affine functions. We will assume the existence of two primitive instructions for this purpose. Given two ordered pairs (a,b) and (c,d) , let $(a,b) \cdot (c,d) = (ac , bc+d)$ (this operation corresponds to forming the composition of the affine functions $ax+b$ and $cy+d$).

The two operations will construct T and place ordered pairs of real numbers on its edges. Initially T has no edges constructed. The operation LINK($\underline{idom}(v)$, v , (a,b)) adds the edge $(\underline{idom}(v),v)$, with associated value $c(\underline{idom}(v),v) = (a,b)$ to T . The operation EVAL(v) returns the ordered triple (u,x,y) such that $(x,y) = c(e_1)\cdot c(e_2)\cdot \ldots \cdot c(e_k)$, where $e_1,e_2,\ldots,e_k$ is the longest path to vertex v in the part of T so far constructed by LINK instructions, and this path starts at vertex u . (If v has no entering edge yet constructed, EVAL(v) returns the triple $(v,1,0)$; the pair $(1,0)$ corresponds to the identity function.)

Now we give the details of the algorithm.

Step 1: Choose a fixed vertex r of G . Compute the corresponding dominator tree T of G . Number the vertices of T from 1 to n in postorder. For each vertex v , let $s(v)$ be the set of children of v in T .

Step 2:
```
for u := 1 until n do begin
    initialize E(u) = ∅;
    for v∈s(u) do
        for each edge (w,v) of G do begin
            (z,a,b) := EVAL(w);
            if (z,v) is not an edge of E(u) then
                add (z,v) with value
                    m(z,v) = m(w,v)·a to E(w)
            else m(z,v) := m(w,v)·a + m(z,v);
            c(v) := m(w,v)·b + c(v);
        end;
    find the strongly connected components of the
        graph G(u) = ({v} ∪ s(u),E(u)) and
        topologically sort them;
    solve the system of equations
```

$$Q(u) = \left\{ \sum_{(w,v)\in E(u)} m(w,v)\cdot x(w)+c(v) = x(v) \;\middle|\; v\in s(u) \right\}$$

```
            to give an equation x(v) = a(v)·x(u) + b(v)
            for each v∈s(u), by using Gaussian
            elimination and the strongly connected
            components decomposition as discussed in
            Section 3;
        for v∈s(u) do LINK(u , v , (a(v),b(v)) ;

Step 3:  for each edge (w,n) of G do begin
        (z,a,b) := EVAL(w);
        m(n,n) := m(w,n)·a + m(n,n);
        c(n) := m(w,n)·b + c(n);
        x(n) := c(n)·[1-m(n,n)]^-1;
    end;
    for i := n-1 step -1 until 1 do
        x(i) := x(idom(i))·a(i) + c(i);
```

This method uses Gaussian elimination on the strongly connected components of the graphs $G(u)$ and combines the solutions to give the solution to the entire problem. The time to combine solutions is almost-linear in the size of G; thus if the method breaks the graph into several parts it is certainly faster than Gaussian elimination applied to the whole graph.

More precisely, the running time of Step 1 is $O(m\,\alpha(m,n))$ [41], where $\alpha(m,n)$ is a very slowly growing function related to a functional inverse of Ackermann's function. Step 1 requires $O(m)$ storage. Step 2 requires $O(m\,\alpha(m,n))$ time and $O(m)$ storage for the LINK and EVAL instructions [41]. Step 2 requires $O(m)$ time and storage except for the Gaussian elimination steps and the LINK and EVAL instructions. Step 3 requires $O(n)$ time and storage. Thus the entire algorithm requires $O(m\,\alpha(m,n))$ time and $O(m)$ storage exclusive of the Gaussian elimination steps.

If each strongly connected component of every graph $G(u)$ consists of a single vertex, then the algorithm runs in $O(m\,\alpha(m,n))$ time total. A graph G for which this happens is called a reducible graph [2,9] (not to be confused with a reducible matrix). Though reducible graphs do not seem to arise in numerical problems, they often arise in global optimization of computer code, to which the ideas in this paper also apply. Thus this decomposition method may have considerable practical value. Indeed, similar methods for reducible graphs have been extensively studied by computer scientists [2,9,18,21,24,42].

If no root r can be found for which G breaks into several pieces using this decomposition scheme, the same idea can be applied to the reverse of G. The algorithm must be changed somewhat, but the idea is similar. In fact, a more general algorithm which divides G into strongly biconnected components and solves a set of equations on each component can be developed. The trouble with such an algorithm is that at present no efficient method exists for dividing a graph into strongly biconnected components. Research is in progress in this area.

6. Selection of a Set of Pivot Positions

When considering orderings for Gaussian elimination in Section 4, we restricted our attention to simultaneous row and column permutations, represented by a renumbering of vertices in the graph representing the system of equations. Thus we always used the positions on the main diagonal as pivot positions. In numeric problems, there is no reason to restrict our attention to such reorderings, however. We can easily allow independent row and column permutations, and thus use an arbitrary transversal of the matrix as a set of pivot positions (a matrix transversal is a set of n matrix elements, no two in the same row or column).

There are two reasons for selecting a transversal other than the main diagonal.

(1) To improve the stability of Gaussian elimination.

(2) To improve the resource requirements of Gaussian elimination.

The well-known partial and complete pivoting methods [14] choose a transversal to improve stability. They choose a set of matrix elements of large absolute value as pivots. These methods depend on the actual numeric entries and not on the zero - non-zero structure of the matrix.

If we do not know the actual entries of the matrix, but only its zero - non-zero structure, then any transversal consisting of non-zero elements is as good as any other for purposes of stability. Such a transversal may be found in $O(n^{1/2} m)$ time by using a bipartite matching algorithm of Hopcroft and Karp [44]. Dulmage and Mendelsohn [12] extensively discuss this and related problems. Essentially no research has been done on the problem of picking a non-zero transversal which minimizes resource requirements. One theorem is known however.

Theorem 2. Let M be any matrix. Let Q be any permutation matrix such that MQ has a non-zero main diagonal. Let $G(Q)$ be the directed graph corresponding to MQ. Then the vertex partition induced by the strongly connected components of $G(Q)$ is independent of Q.

This theorem follows from results of Dulmage and Mendelsohn [12]. Howell has given a nice proof [45]. The theorem implies that the strong component decomposition method discussed in Section 3 produces the same number of components independent of the transversal chosen, though the components themselves may be different.

Ignoring questions of stability, there is no reason not to choose a transversal some of whose elements are initially zero and only become non-zero as the elimination proceeds. Such a choice may result in substantial computational savings. Bank and Rose [4] have provided a practical example of this idea. Though their method is numerically unstable, it can be modified to make it stable without degrading its efficiency too much [5].

In summary, the problem of choosing the best set of pivot positions, for stability or efficiency or both, is very poorly understood. The results of Bank and Rose indicate that allowing only transversals which are initially non-zero is too restrictive. It is likely that the problem is too hard for a general solution, and the most promising areas for research seem to be the development of heuristics and special-case algorithms.

7. Remarks

Though we have assumed throughout this discussion that the matrix M consists of numbers, there is no reason to do so. The techniques of linear algebra, such as Gaussian elimination, apply to other algebraic structures having two operations $+$ and $\cdot$. Thus the methods discussed in this paper can be used to compute path sets in labelled graphs [3,37] (a problem of automata theory), find shortest paths and other kinds of optimal paths in directed graphs [7], and to do global flow analysis of computer code [2,9,18,24, 42]. The algorithms remain the same; only the interpretation changes.

We must assume the existence, for any a, of an element a^* such that, for all b, $a^* \cdot b$ is a solution to the equation $x = a \cdot x + b$. For numbers, $a^* = [1-a]^{-1}$ exists whenever $a \neq 1$, and Gaussian elimination requires non-unit pivots.

References

[1] A. V. Aho and J. D. Ullman, The Theory of Parsing, Translation, and Compiling, Vol. II: Compiling, Prentice-Hall, Englewood Clifss, N.J. (1972).

[2] F. E. Allen, "Control flow analysis," SIGPLAN Notices 5 (1970), 1-19.

[3] R. C. Backhouse and B. A. Carré, "Regular algebra applied to pathfinding problems," J. Inst. Maths. Applics., 15 (1975), 161-186.

[4] R. E. Bank and D. J. Rose, "An $O(n^2)$ method for solving constant coefficient boundary value problems in two dimensions," SIAM J. Numer. Anal., to appear.

[5] R. E. Bank and D. J. Rose, "Marching algorithms for elliptic boundary va ue problems I: the constant coefficient case," SIAM J. Numer. Anal., submitted.

[6] J. R. Bunch and D. J. Rose, "Partitioning, tearing, and modification of sparse linear systems," J. Math. Anal. Appl., 48 (1974), 574-593.

[7] B. A. Carré, "An algebra for network routing problems," J. Inst. Maths. Applics., 7 (1971), 273-294.

[8] A. Chang, "Application of sparse matrix methods in electric power systems," Sparse Matrix Proceedings, R. A. Willoughby, ed., IBM Research, Yorktown Heights, N.Y. (1968), 113-122.

[9] J. Cocke, "Global common subexpression elimination," SIGPLAN Notices, 5 (1970), 20-24.

[10] E. Cuthill, "Several strategies for reducing the bandwidth of matrices," Sparse Matrices and Their Applications, D. Rose and R. Willoughby, eds., Plenum Press, N. Y. (1972), 157-166.

[11] E. Cuthill and J. McKee, "Reducing the bandwidth of sparse symmetric matrices," Proc. ACM National Conf. (1969), 157-172.

[12] A. Dulmage and N. Mendelsohn, "Graphs and Matrices," Graph Theory and Theoretical Physics, F. Harary, ed., Academic Press, N. Y. (1967), 161-227.

[13] S. Even, private communication (1974).

[14] G. E. Forsythe and C. B. Moler, Computer Solution of Linear Algebraic Equations, Prentice-Hall, Englewood Cliffs, N.J. (1967).

[15] M. R. Garey, D. S. Johnson, and L. Stockmeyer, "Some simplified NP-Complete problems," Proc. Sixth Annual ACM Symp. on Theory of Computing (1974), 47-63.

[16] M. R. Garey and D. S. Johnson, private communication (1975).

[17] J. A. George, "Nested dissection of a regular finite element mesh," SIAM J. Numer. Anal., 10 (1973), 345-363.
[18] S. Graham and M. Wegman, "A fast and usually linear algorithm for global flow analysis," Conf. Record of the Second ACM Symp. on Principles of Prog. Lang. (1975), 22-34.
[19] F. G. Gustavson, "Some basic techniques for solving sparse systems of linear equations," Sparse Matrices and Their Applications, D. Rose and R. Willoughby, eds., Plenum Press, N.Y. (1972), 41-52.
[20] F. G. Gustavson, W. Liniger, and R. Willoughby, "Symbolic generation of an optimal Crout algorithm for sparse systems of linear equations," J. ACM, 17 (1970), 87-109.
[21] M. S. Hecht and J. D. Ullman, "Characterizations of reducible flow graphs," J. ACM, 21 (1974), 367-375.
[22] A. J. Hoffman, M. S. Martin and D. J. Rose, "Complexity bounds for regular finite difference and finite element grids," SIAM J. Numer. Anal., 9 (1961), 364-369.
[23] A. Jennings, "A compact storage scheme for the solution of symmetric linear simultaneous equations," Comput. J. 9 (1966), 281-285.
[24] K. W. Kennedy, "Node listings applied to data flow analysis," Conf. Record of the Second ACM Symp. on Principles of Prog. Lang. (1973), 10-21.
[25] D. Knuth, The Art of Computer Programming, Vol. 3: Sorting and Searching, Addison-Wesley, Reading, Mass. (1973), 506-549.
[26] D. Knuth, The Art of Computer Programming, Vol. 1: Fundamental Algorithms, Addison-Wesley, Reading, Mass. (1968), 258-265.
[27] W. H. Liu and A. H. Sherman, "Comparative analysis of the Cuthill-McKee and reverse Cuthill-McKee ordering algorithms for sparse matrices," SIAM J. Numer. Anal., to appear.
[28] G. S. Lueker and K. S. Booth, "Linear algorithms to recognize interval graphs and test for the consecutive ones property," Proc. Seventh Annual ACM Symp. on Theory of Computing (1975), 255-265.
[29] C. H. Papadimitriou, "The NP-completeness of the bandwidth minimization problem," Computing, to appear.
[30] D. J. Rose, private communication (1975).
[31] D. J. Rose, "Triangulated graphs and the elimination process," J. Math. Anal. Appl., 32 (1970), 597-609.

[32] D. J. Rose, "A graph-theoretic study of the numerical solution of sparse positive definite systems of linear equations," Graph Theory and Computing, R. Read, ed., Academic Press, N.Y., (1973), 183-217.

[33] D. J. Rose and R. E. Tarjan, "Algorithmic aspects of vertex elimination," Proc. Seventh Annual ACM Symp. on Theory of Computing, (1975), 245-254.

[34] D. J. Rose and R. E. Tarjan, "Algorithmic aspects of vertex elimination on directed graphs," to appear.

[35] D. J. Rose, R. E. Tarjan, and G. S. Lueker, "Algorithmic aspects of vertex elimination on graphs," SIAM J. Comput., to appear.

[36] D. J. Rose and G. F. Whitten, "Automatic nested dissection," Proc. ACM Conference (1974), 82-88.

[37] A. Salomaa, Theory of Automata, Pergamon Press, Oxford, England (1969), 120-123.

[38] A. H. Sherman, "Subroutines for envelope solution of sparse linear equations," Research Report No. 35, Dept. of Computer Science, Yale University (1974).

[39] A. H. Sherman, Ph.D. thesis, Yale University (1975).

[40] R. E. Tarjan, "Depth-first search and linear graph algorithms," SIAM J. Comput., 1 (1972), 146-160.

[41] R. E. Tarjan, "Applications of path compression on balanced trees," to appear.

[42] J. D. Ullman, "A fast algorithm for the elimination of common subexpressions," Acta Informatica, 2 (1973), 191-213.

[43] R. S. Varga, Matrix Iterative Analysis, Prentice-Hall, Englewood Cliffs, N. J. (1962).

[44] J. E. Hopcroft and R. M. Karp, "An $n^{5/2}$ algorithm for maximum matching in bipartite graphs," SIAM J. Comput. 2 (1973), 225-231.

[45] T. D. Howell, "Partitioning Using PAQ," this Proceedings.

PARTITIONING USING PAQ[+]

Thomas D. Howell

Cornell University

Ithaca, New York

ABSTRACT

The so-called PAQ problem is concerned with the solution of sparse systems of linear equations $Ax = b$ using the transformation $PAQy = Pb$, $x = Qy$. An algorithm is given for choosing P and Q to partition the matrix A into its irreducible components. A theorem on which this algorithm is based has long been known, yet no simple, easily understood proof appears in the literature. Such a proof is given here. Remarks are made concerning some unsolved problems related to the PAQ problem.

1. INTRODUCTION

Let A be a large, sparse, nonsingular, nonsymmetric matrix. Consider solving the system of linear equations $Ax = b$ by Gaussian elimination. It is often advantageous to permute the equations and variables and solve instead, $(PAQ)y = Pb$, $x = Qy$, where P and Q are permutation matrices. The permutations P and Q may then be chosen to minimize the cost of performing Gaussian elimination on PAQ. This cost may be measured by the number of storage locations or the number of arithmetic operations used, or a combination of the two.

[+]This research was supported in part by the Office of Naval Research Grant N00014-67-A-0077-0021, by a National Science Foundation Graduate Fellowship and by Argonne National Laboratory.

The choice of P and Q may be viewed as a two step process. The first step is to partition the matrix A. This is accomplished by finding P_0 and and Q_0 such that P_0AQ_0 is block upper triangular:

$$P_0AQ_0 = \begin{bmatrix} A_{11} & A_{12} & \cdots & A_{1k} \\ 0 & A_{22} & \cdots & A_{2k} \\ \cdot & & \cdot \;\cdot & \cdot \\ \cdot & & \cdot \;\cdot & \cdot \\ \cdot & & \cdot \;\cdot & \cdot \\ 0 & \cdots & 0 & A_{kk} \end{bmatrix} . \qquad (1)$$

Each A_{ii} is a nonsingular matrix which can not be further partitioned.

The second step is to find $\bar{P}_i$ and $\bar{Q}_i$, $1\le i\le k$, which permute the equations and variables of the i-th block in a way which minimizes the cost of performing elimination on $\bar{P}_iA_{ii}\bar{Q}_i$. Then,

$$PAQ = P_k \cdots P_1P_0AQ_0Q_1 \cdots Q_k = \begin{bmatrix} \bar{A}_{11} & \bar{A}_{12} & \cdots & \bar{A}_{1k} \\ 0 & \bar{A}_{22} & \cdots & \bar{A}_{2k} \\ \cdot & & \cdot \;\cdot & \cdot \\ \cdot & & \cdot \;\cdot & \cdot \\ \cdot & & \cdot \;\cdot & \cdot \\ 0 & \cdots & 0 & \bar{A}_{kk} \end{bmatrix} ,$$

where P_i and Q_i are $\bar{P}_i$ and $\bar{Q}_i$ extended to the size of A.

The partitioning step may be accomplished as follows:

1. Find a permutation matrix R so that the main diagonal of RA is nonzero.
2. Form the directed graph G(RA) whose adjacency matrix is RA.

3. Find the strong components $C_1, \ldots, C_k$ of G(RA) and order them so that for $i<j$ no path exists from C_j to C_i. Let Q_0^t be a permutation which corresponds to this ordering of vertices.

4. Now, $Q_0^t RAQ_0 \equiv P_0AQ_0$ is block upper triangular with diagonal blocks $A_{11}, \ldots, A_{kk}$ corresponding to components $C_1, \ldots, C_k$ of G(RA).

This algorithm appears in the literature at least as early as 1962, [1], [2]. It is based upon a theorem which states that any permutation used in step 1 to make the main diagonal nonzero gives rise to the same strong component vertex partition in step 3, although the components themselves may be different. Johnson, Dulmage and Mendelsohn [1] prove a related theorem which has the above theorem as a consequence. Their proof, which is given in graph theoretic terms, is difficult to follow and to apply to the partitioning of sparse matrices. The idea for a simpler proof of this theorem is given by Steward [2]. His argument is only intuitive; he does not state a theorem or give a formal proof. The main result of this paper is to give a new, short, simple proof of the theorem mentioned above.

2. PRELIMINARY REMARKS

The reason partitioning is important is that it allows a large system of equations to be replaced by several smaller systems. Suppose P_0 and Q_0 have been found so that P_0AQ_0 is in the form of equation (1). Let y and P_0b be partitioned into $y_1, \ldots, y_k$ and $(P_0b)_1, \ldots, (P_0b)_k$ to correspond to the partitioning of P_0AQ_0 in equation (1). The system $(P_0AQ_0)y = P_0b$, $x = Q_0y$ may now be solved by block back substitution:

$$A_{kk}\, y_k = (P_0b)_k$$

$$\vdots$$

$$A_{ii}\, y_i = (P_0b)_i - \sum_{j=i+1}^{k} A_{ij}\, y_j$$

$$\vdots$$

$$A_{11}\, y_1 = (P_0 b)_1 - \sum_{j=2}^{k} A_{1j}\, y_j$$

$$x = Q_0\, y$$

Now we need only apply Gaussian elimination to the diagonal blocks A_{ii}, $1 \leq i \leq k$. This is certainly no more costly than doing elimination on the entire matrix, and can be much less costly if the partitioning is fine.

Let $R = QP$. Then $PAQ = Q^t RAQ$. The permutation matrices R and Q then have the follwing interpretations: R permutes the rows of A so that the elements which will be used as "pivot elements" for Gaussian elimination lie on the main diagonal of RA. The symmetric permutation $Q^t(RA)Q$ then determines the order in which these pivot elements will be used. Once R and Q have been chosen, Gaussian elimination is applied to the diagonal blocks of matrix $Q^t RAQ$ with no further pivoting. For this reason we will almost always consider only those permutations R for which the main diagonal of RA (hence also of $Q^t RAQ$) is entirely nonzero. It would be possible to leave some zeroes on the main diagonal of RA and trust the eleimination process to "fill them in" (change them to nonzero) before they are used as pivot elements. We will see in Section 5 that this can only cause the partition generated to be coarser than that which would be generated using a nonzero main diagonal.

The second step in the choice of P and Q is very difficult in general. It requires finding $\bar{P}_i$ and $\bar{Q}_i$ to minimize the cost of Gaussian elimination on the diagonal block A_{ii}. No algorithm is known which finds the optimum permutations without essentially trying all possible permutations. This problem will be discussed further in Section 6.

As mentioned above, the cost of Gaussian elimination may be measured in a variety of ways. A simple, commonly used cost function is called "fill-in". The fill-in is the number of new nonzero elements created during the elimination process. These elements are said to have been "filled in". Fill-in determines the number of additional storage locations required to perform Gaussian elimination on A. This cost function will be used for the remainder of this paper.

3. DEFINITIONS

Definition 1: A diagonal of the nxn matrix A is any set of n elements containing one element from each row and column of A. The set $\{a_{ii} | 1 \leq i \leq n\}$ is the main diagonal of A.

Because A is nonsingular, its determinant is not zero, and therefore at least one diagonal of A has all of its elements nonzero.

Definition 2: Let π be a nonzero diagonal of A. The directed graph associated with A and π is $G(A,\pi)$ which is defined as follows: Let R and S be permutation matrices such that the elements of π form the main diagonal of RAS. Then $G(A,\pi) = (V,E)$ where $V = \{v_i | 1 \leq i \leq n\}$ and $E = \{(v_i,v_j) | i \neq j$ and $(RAS)_{ij} \neq 0\}$. When π_0 is the main diagonal of A, $G(A,\pi_0)$ will also be written simply as $G(A)$.

It remains to show that $G(A,\pi)$ is well defined, i.e., that $G(A,\pi)$ is unaffected by the choice of R and S. Note that all graphs used in this paper are directed graphs.

Theorem 1: Let $G_1(A,\pi)$ and $G_2(A,\pi)$ be the graphs described by Definition 2 using pairs of permutations (R_1,S_1) and (R_2,S_2) respectively. Then $G_1(A,\pi)$ and $G_2(A,\pi)$ are isomorphic.

Proof: Let Π be the nxn permutation matrix whose 1's are in the positions corresponding to the locations of the elements of π in A. Since (R_1,S_1) and (R_2,S_2) move π to the main diagonal, $R_1\Pi S_1 = R_2\Pi S_2 = I$. Then $S_1 = \Pi^t R_1^t$, $S_2 = \Pi^t R_2^t$. Let $C = R_2 R_1^t$. Then,

$$\begin{aligned} C(R_1AS_1)C^t &= R_2R_1^tR_1AS_1R_1R_2^t \\ &= R_2A\Pi^tR_2^t \\ &= R_2AS_2 \end{aligned}$$

The matrices R_1AS_1 and R_2AS_2 are the same up to a renumbering of the rows and columns by the permutation C, hence $G_1(A,\pi) \cong G_2(A,\pi)$ with isomorphism given by C.

A consequence of Theorem 1 is that with no loss of generality we may always move the pivot elements to the main diagonal using row permutations only, i.e., using $R = \Pi^t$, $S = I$.

Definition 3: A directed graph is strongly connected if there is a path from each vertex to each other vertex. A directed graph which is not strongly connected can be divided in a unique way into maximal strongly connected subgraphs which are called strong components.

These strong components $C_1, \ldots, C_k$ can be ordered in such a way that for $i<j$ no path exists from vertices in C_j to vertices in C_i. An efficient algorithm for finding the strong components in this order has been given by Tarjan [3]. This algorithm can also be found in [4], as can definitions of other graph theoretic terms used here with which the reader may not be familiar.

Definition 4: If $PAQ = \begin{bmatrix} A_{11} & A_{12} \\ 0 & A_{22} \end{bmatrix}$ where A_{11} and A_{22} are square and of order at least one, and P and Q are permutation matrices, then A is bireducible. Otherwise A is fully indecomposable. If

$$PAQ = \begin{bmatrix} A_{11} & A_{12} & \cdots & A_{1k} \\ 0 & A_{22} & \cdots & A_{2k} \\ \cdot & \cdot & \cdot & \cdot \\ \cdot & & \cdot & \cdot \\ \cdot & & & \cdot \\ 0 \ldots & & 0 & A_{kk} \end{bmatrix},$$

and each A_{ii} is fully indecomposable, then the submatrices A_{ii}, $1 \leq i \leq k$, are the indecomposable components of A.

Definition 5: Let A be an nxn matrix with nonzero main diagonal. Let G(A) be its graph. We say that A (or G(A)) has property C if the following condition holds:

Matrix statement - Every subset of $k < n$ rows of A has nonzero entries in at least $k+1$ columns.

Graph statement - Every subset of $k < n$ vertices of G(A) has at least one directed edge from a vertex in the subset to a vertex outside the subset.

4. MAIN THEOREM

Theorem 2: Let A be an nxn matrix with nonzero main diagonal and G(A) be its graph. The following three statements are equivalent:

(1) A has property C.
(2) A is fully indecomposable.
(3) G(A) is strongly connected.

Proof: not (1) => not (2) : Let S be a subset of $k < n$ rows of A on which property C fails. In S are k elements of A's main diagonal. These nonzero elements are in k different columns. Since property C fails the other columns of S must be zero. Let P be a permutation matrix which orders rows in S last. Then

$$PAP^t = \begin{bmatrix} A_{11} & A_{12} \\ 0 & A_{22} \end{bmatrix} \} \text{ rows in S },$$

and A is bireducible.

not (2) => not(1) : Let $PAQ = \begin{bmatrix} A_{11} & A_{12} \\ 0 & A_{22} \end{bmatrix}$.

Then property C fails on the rows of A_{22}, hence also on the rows of A from which they are formed.

not (1) => not(3) : Let S be a subset of rows of A on which property C fails. Then no path exists in G(A) from vertices corresponding to rows in S to vertices outside this set, hence G(A) is not strongly connected.

not (3) => not(1) : If G(A) is not strongly connected it must have a proper strong component

C_k with no outward edges. The rows of A corresponding to vertices in C_k have nonzero entries only in columns corresponding to vertices in C_k, violating property C.

Corollary: Let R and S be permutation matrices such that RA and SA both have nonzero main diagonal. Then G(RA) is strongly connected if and only if G(SA) is strongly connected.

Proof: By Theorem 2 property C is equivalent to strong connectedness. Property C depends only on the set of rows in the matrix, not on their order. The matrices RA and SA have the same set of rows.

The essence of the main theorem is contained in Theorem 2 and its corollary. It remains only to apply these results to matrices whose graphs have several strong components.

Theorem 3: Every nonzero diagonal π of A induces the same strong component vertex partition in $G(A,\pi)$.

Proof: The proof is by induction on the number of strong components of $G(A,\pi)$. Let k be this number, and let the components be $C_1,\ldots,C_k$ where no path exists from C_j to C_i with $i<j$. The corollary to Theorem 2 takes care of the case k=1. For the inductive step assume the theorem holds whenever $G(A,\pi')$ has k-1 components for some nonzero diagonal π'. Let X be the set of indices of the vertices in C_k, the last strong component. Let $\bar{X}$ be the rest of the indices: $X \cup \bar{X} = \{1,\ldots,n\}$. Let A_X and $A_{\bar{X}}$ be square submatrices of A consisting of the rows and columns whose indices are in X and $\bar{X}$ respectively. Let ρ be another diagonal of A. Let π_X, $\pi_{\bar{X}}$, ρ_X and $\rho_{\bar{X}}$ be the elements of π and ρ in A_X and $A_{\bar{X}}$ respectively. Consider the rows of A whose indices lie in X. Because C_k has no outward edges in $G(A,\pi)$, the only nonzero entries in these rows are in A_X. Since ρ_X contains one nonzero entry from each of these rows, ρ_X forms a diagonal of A_X. It follows that $\rho_{\bar{X}}$ forms a diagonal of $A_{\bar{X}}$. The inductive hypothesis implies that $G(A_{\bar{X}},\rho_{\bar{X}})$ has the same strong component vertex partition as $G(A_{\bar{X}},\pi_{\bar{X}})$. The corollary to Theorem 2 implies that $G(A_X,\rho_X)$ is strongly connected. The rows of A containing A_X are zero outside of A_X, so $G(A_X,\rho_X)$ re-

mains as a separate strong component of $G(A,\rho)$. Therefore $G(A,\pi)$ and $G(A,\rho)$ have the same strong component vertex partition, and the theorem follows by induction.

5. TIME BOUNDS

Let A be a nonsingular nxn matrix with $f(n)$ nonzero elements ($n \leq f(n) \leq n^2$). The partitioning algorithm can be rewritten as follows:

1. Find a nonzero diagonal π and permutation R moving π to the main diagonal.
2. Form $G(A,\pi)$.
3. Find strong components of $G(A,\pi)$ in an order such that for $i<j$, no path exists from C_j to C_i. Let the permutation Q^t order row k before row ℓ whenever $V_k \in C_i$, $V_\ell \in C_j$, and $i<j$.
4. Form $PAQ \equiv Q^t RAQ$ which is block upper triangular.

Steps 2 and 4 are straight forward and take time and storage bounded by a constant multiple of $f(n)$. Step 3 can be accomplished using the algorithm of Tarjan also in time and storage proportional to $f(n)$. The fastest known algorithm to do step 1 in general is due to Hopcroft and Karp [5] and requires time proportional to $n^{2.5}$. This may be too costly to be feasible for very large values of n, however for many specific problems a diagonal may be found more easily. In many cases A has a ready-made nonzero main diagonal.

Remark: Suppose A has a zero on its main diagonal which will be filled in by the elimination process before it is used as a divisor. One could still construct G(A) and do the partitioning as before. By Theorem 3 this graph would have the same strong component vertex partition as the one generated as follows:

1. Color the zero on the main diagonal red.
2. Find R so that RA has nonzero main diagonal and construct G(RA) as before.
3. Add an edge corresponding to the red zero which is now somewhere off the main diagonal of RA.

This added edge may or may not change the strong component structure of G(RA). It is possible for it to merge two or more components of G(RA) into a single component.

6. DIRECTIONS FOR FURTHER RESEARCH

Consider for a moment the case in which A is symmetric and positive definite. Then A has a ready-made nonzero diagonal. As long as this main diagonal is used, which means that only reorderings of the form Q^tAQ are considered, symmetry and positive definiteness are preserved. Because Q^tAQ is positive definite, pivoting is not necessary to insure the numerical stability of Gaussian elimination. This leaves us free to choose Q to minimize the cost of doing the elimination on Q^tAQ. This case has been studied extensively by Rose and others [6], [7], [8].

When A is not symmetric, positive definite there is no reason to restrict reorderings to the form Q^tAQ, so we consider PAQ (or Q^tRAQ). In contrast to the previous case, the ready-made nonzero main diagonal and the guaranteed numerical stability have been lost, but the freedom to choose a nonzero main diagonal (via R) has been gained. The partitioning problem using PAQ is fairly easy, and a reasonably good solution has been described above. Theorem 3 shows that for partitioning it does not matter which nonzero diagonal is chosen. This leaves us free to choose both P and Q for each indecomposable component of A in a way which attempts to solve two more difficult problems which we have ignored up to now.

One problem is numerical stability. In the case of dense matrices the main purpose of pivoting is to insure stability. In the sparse, positive definite, symmetric case pivoting is used to minimize the cost of Gaussian elimination. In the sparse, nonsymmetric case one would like to use the freedom to choose both P and Q in order to achieve both goals: stability and low cost. This paper has ignored the first goal. By distinguishing only between zero and nonzero numbers we have factored out the problem of numerical stability. Any practical algorithm for the PAQ problem will have to face this problem.

The second difficult problem is concerned with the goal of low cost. How should $\bar{P}_i$ and $\bar{Q}_i$ be chosen to minimize the cost of doing elimination on $\bar{P}_i A_{ii} \bar{Q}_i$ where A_{ii} is an indecomposable component of A? Rose and Tarjan [9] have shown that after $R_i = Q_i P_i$ has been chosen so that $R_i A_{ii}$ has a nonzero main diagonal, the problem of choosing Q_i to minimize the cost (measured in terms of fill-in) of doing elimination on $\bar{Q}_i^t R_i \bar{A}_{ii} \bar{Q}_i \equiv \bar{P}_i A_{ii} \bar{Q}_i$ is NP-complete. For a discussion of the meaning of NP-complete see [4], [10], or [11]. The problem of choosing both $\bar{P}_i$ and $\bar{Q}_i$ to minimize cost seems at least as hard as choosing $\bar{Q}_i$ given $\bar{R}_i$; it would be interesting to learn whether this problem is also NP-complete. The problem is hard enough that practical algorithms probably will have to use heuristics to help find "good" permutations $\bar{P}_i$ and $\bar{Q}_i$ rather than insisting on finding the best ones. One strategy is to fix $\bar{R}_i$, then choose $\bar{Q}_i$ using heuristics developed for the symmetric $Q^t AQ$ problem. Two such heuristics are "minimum degree" and "minimum deficiency", and are discussed in [6].

7. EXAMPLES

The first example will illustrate the entire process of solving Ax = b using the method described above.

$n = 5$

$$Ax = b \qquad \begin{bmatrix} 2 & 0 & 4 & 0 & 0 \\ 1 & 1 & 0 & 0 & 3 \\ 0 & 0 & 3 & 4 & 0 \\ 2 & 2 & 0 & 1 & 5 \\ 0 & 0 & 1 & 1 & 0 \end{bmatrix} \qquad x = \begin{bmatrix} 0 \\ 2 \\ 2 \\ 2 \\ 1 \end{bmatrix}$$

1. Choose any nonzero diagonal, say $\pi = (a_{11}, a_{25}, a_{34}, a_{42}, a_{53})$. Let R move it to the main diagonal.

$$R = \begin{bmatrix} 1 & 0 & 0 & 0 & 0 \\ 0 & 0 & 0 & 1 & 0 \\ 0 & 0 & 0 & 0 & 1 \\ 0 & 0 & 1 & 0 & 0 \\ 0 & 1 & 0 & 0 & 0 \end{bmatrix}$$

$$RAx = Rb \qquad \begin{bmatrix} 2 & 0 & 4 & 0 & 0 \\ 2 & 2 & 0 & 1 & 5 \\ 0 & 0 & 1 & 1 & 0 \\ 0 & 0 & 3 & 4 & 0 \\ 1 & 1 & 0 & 0 & 3 \end{bmatrix} \qquad x = \begin{bmatrix} 0 \\ 2 \\ 1 \\ 2 \\ 2 \end{bmatrix}$$

2. Form $G(A,\pi) = G(RA)$

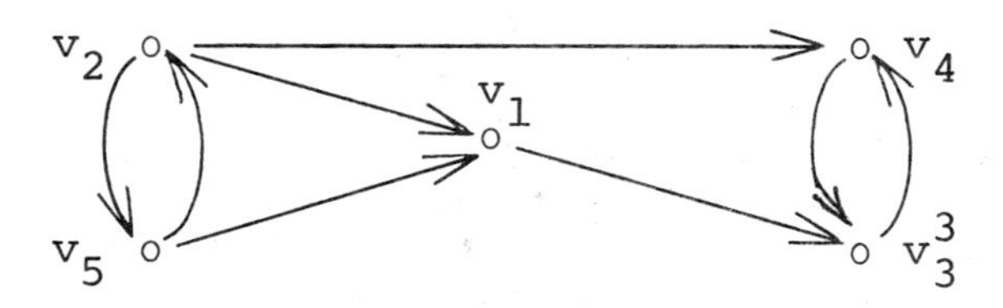

3. Find strong components in order. Let Q^t reflect this order.

$$\{v_2, v_5\}, \ \{v_1\}, \ \{v_3, v_4\}$$

$$Q^t = \begin{bmatrix} 0 & 1 & 0 & 0 & 0 \\ 0 & 0 & 0 & 0 & 1 \\ 1 & 0 & 0 & 0 & 0 \\ 0 & 0 & 1 & 0 & 0 \\ 0 & 0 & 0 & 1 & 0 \end{bmatrix} \qquad Q = \begin{bmatrix} 0 & 0 & 1 & 0 & 0 \\ 1 & 0 & 0 & 0 & 0 \\ 0 & 0 & 0 & 1 & 0 \\ 0 & 0 & 0 & 0 & 1 \\ 0 & 1 & 0 & 0 & 0 \end{bmatrix}$$

4. Form $PAQ = Q^t RAQ$.

$$PAQy = Pb \qquad \left[\begin{array}{cc|c|cc} 2 & 5 & 2 & 0 & 1 \\ 1 & 3 & 1 & 0 & 0 \\ \hline 0 & 0 & 2 & 4 & 0 \\ \hline 0 & 0 & 0 & 1 & 1 \\ 0 & 0 & 0 & 3 & 4 \end{array}\right] y$$

$$= \begin{bmatrix} A_{11} & A_{12} & A_{13} \\ 0 & A_{22} & A_{23} \\ 0 & 0 & A_{33} \end{bmatrix} \begin{bmatrix} y_1 \\ y_2 \\ y_3 \end{bmatrix} = \left[\begin{array}{c} 2 \\ 2 \\ \hline 0 \\ \hline 1 \\ 2 \end{array}\right]$$

5. Block back substitution.

$$A_{33} y_3 = (Pb)_3$$

$$\begin{bmatrix} 1 & 1 \\ 3 & 4 \end{bmatrix} y_3 = \begin{bmatrix} 1 \\ 2 \end{bmatrix} \quad \Rightarrow \quad y_3 = \begin{bmatrix} 2 \\ -1 \end{bmatrix}$$

$$A_{22}y_2 = (Pb)_2 - A_{23}y_3$$

$$\begin{bmatrix} 2 \end{bmatrix} y_2 = \begin{bmatrix} -8 \end{bmatrix} \Rightarrow \quad y_2 = \begin{bmatrix} -4 \end{bmatrix}$$

$$A_{11}y_1 = (Pb)_1 - A_{12}y_2 - A_{13}y_3$$

$$\begin{bmatrix} 2 & 5 \\ 1 & 3 \end{bmatrix} y_1 = \begin{bmatrix} 11 \\ 6 \end{bmatrix} \Rightarrow \quad y_1 = \begin{bmatrix} 3 \\ 1 \end{bmatrix}$$

$$y = \begin{bmatrix} y_1 \\ y_2 \\ y_3 \end{bmatrix} = \begin{bmatrix} 3 \\ 1 \\ -4 \\ 2 \\ -1 \end{bmatrix} \quad x = Qy = \begin{bmatrix} -4 \\ 3 \\ 2 \\ -1 \\ 1 \end{bmatrix}$$

Remark: If we had used the main diagonal of A without applying the permutation R to make it nonzero, the resulting graph would have been strongly connected, and no partitioning would have resulted.

The second example shows that the choice of a nonzero main diagonal can affect the cost of Gaussian elimination on a component of A. Suppose,

$$A_{ii} = \begin{bmatrix} x & x & 0 & x \\ 0 & x & x & 0 \\ 0 & x & x & x \\ x & 0 & 0 & x \end{bmatrix} \quad x = \text{arbitrary nonzero entry}$$

The choice $\bar{R}_i = I$ leads to fill-in of at least one, regardless of the choice of $\bar{Q}_i$. The choice,

$$\bar{R}_i = \begin{bmatrix} 0 & 0 & 0 & 1 \\ 0 & 1 & 0 & 0 \\ 0 & 0 & 1 & 0 \\ 1 & 0 & 0 & 0 \end{bmatrix} ,$$

leads to no fill-in if $\bar{Q}_i$ is chosen to be I, for example.

In the preceeding example A_{ii} was nonsymmetric. It has been conjectured that when the zero-nonzero structure of A_{ii} is symmetric with nonzero main diagonal the cost cannot be reduced by using a diagonal such that the zero-nonzero structure of $\bar{R}_i A_{ii}$ is nonsymmetric instead of one for which it is symmetric. While not disproving this conjecture, a final example will show that the situation is not as simple as might be imagined. Consider:

$$A = R_0A = \begin{bmatrix} x & 0 & x & x & x \\ 0 & x & x & x & x \\ x & x & x & 0 & 0 \\ x & x & 0 & x & 0 \\ x & x & 0 & 0 & x \end{bmatrix} \quad R_1A = \begin{bmatrix} x & x & x & 0 & 0 \\ 0 & x & x & x & x \\ x & 0 & x & x & x \\ x & x & 0 & x & 0 \\ x & x & 0 & 0 & x \end{bmatrix} \quad R_2A = \begin{bmatrix} x & x & x & 0 & 0 \\ x & x & 0 & x & 0 \\ x & 0 & x & x & x \\ 0 & x & x & x & x \\ x & x & 0 & 0 & x \end{bmatrix}$$

m	R_0A	R_1A	R_2A
2	36	12	0
3	0	48	48
4	36	12	48
5	0	24	24
6	48	24	0
Avg. cost	4.2	4.0	3.8
Min. cost	2	2	3

Each entry of the table gives the number of the 120 possible permutations Q such that the cost for Q^tR_iAQ is m. The cost is still measured by fill-in.

The average cost over all permutations Q can be reduced below that of R_0A by using a nonsymmetric permutation R_1A, even though the minimum cost over all Q is unaffected. The average cost can be reduced still further by using another nonsymmetric permutation R_2A, yet in this case the minimum cost actually rises!

8. CONCLUSION

An algorithm has been given for choosing permutation matrices P_0 and Q_0 to partition a sparse matrix A into its indecomposable components. A new proof of a theorem underlying this algorithm has been given which is simpler and clearer than the original proof. Bounds on the running time of the algorithm are stated and an example is given illustrating the algorithm. Areas for further research are indicated, and some remarks are made concerning the conjecture that for symmetric matrices reorderings of the form Q^tAQ can do as well as those of the form PAQ.

9. ACKNOWLEDGEMENT

I wish to thank Professor Donald J. Rose for many helpful discussions and comments.

REFERENCES

[1] Johnson, D.M., A.L. Dulmage, and N.S. Mendelsohn, "Connectivity and Reducibility of Graphs", Can. J. Math., 14 (1962), pp. 529-439.

[2] Steward, D.V., "On an Approach to Techniques for the Analysis of the Structure of Large Systems of Equations", SIAM Review, 4 (1962), pp. 321-342.

[3] Tarjan, R.E., "Depth First Search and Linear Graph Algorithms", SIAM J. Computing, 1 (1972), pp. 146-160.

[4] Aho, A.V., J.E. Hopcroft, and J.D. Ullman, The Design and Analysis of Computer Algorithms, Addison-Wesley, Reading, Mass. (1974), pp. 189-195.

[5] Hopcroft, J.E. and R.M. Karp, "An $n^{5/2}$ algorithm for Maximum Matchings in Bipartite Graphs", SIAM J. Computing, 2 (1972).

[6] Rose, D.J., "A Graph-theoretic Study of the Numerical Solution of Sparse Positive Definite Systems of Linear Equations", Graph Theory and Computing, R. Read, ed., Academic Press, N.Y. (1973), pp. 183-217.

[7] Bunch, J.R. and D.J. Rose, "Partitioning, Tearing and Modification of Sparse Linear Systems", J. Math. Anal. and Applications, 48 (1974), pp. 574-593.

[8] Rose, D.J., and R.E. Tarjan, "Algorithmic Aspects of Vertex Elimination", Proceedings Seventh Annual ACM Symposium on Theory of Computing (1975), pp. 245-254.

[9] Rose, D.J. and R.E. Tarjan, "Algorithmic Aspects of Vertex Elimination on Directed Graphs", joint Harvard-Stanford Technical Report.

[10] Cook, S., "The Complexity of Theorem-proving Procedures", Proceedings Third Annual ACM Symposium on Theory of Computing (1975), pp. 245-254.

[11] Karp, R.M., "Reducibility Among Combinatorial Problems", Complexity of Computer Computations, R.E. Miller and J.W. Thatcher, eds., Plenum Press, N.Y. (1972), pp. 85-104.

BLOCK METHODS FOR SOLVING SPARSE LINEAR SYSTEMS

James R. Bunch

Department of Mathematics
University of California, San Diego
La Jolla, California 92093

Abstract. Graph-theoretic techniques for analyzing the solution of large sparse systems of linear equations by partitioning and using block methods are developed. Questions concerning stability of block methods are discussed.

1. Introduction.

Let M be a nonsingular (sparse) matrix of order n. Suppose M has been partitioned so that

$$M = \left[\begin{array}{c|c|c|c} A_{11} & A_{12} & \cdots & A_{1k} \\ \hline A_{21} & A_{22} & & \\ \hline \vdots & & \ddots & \\ \hline A_{k1} & & & A_{kk} \end{array}\right] \equiv A , \tag{1.1}$$

where A_{ii} is of order n_i and $\sum_{i=1}^{k} n_i = n$. Or, in general, we have permuted M into A and then partitioned A as above.

Wilkinson [16] has said that a matrix should be considered sparse whenever it has become worthwhile (to whomsoever) to take advantage of the zeros in the matrix. Conversely, a matrix may be considered dense (by whomsoever) whenever it is not worthwhile (to him) to take advantage of the zeros in the matrix.

The special case when $k = 2$ and $n_1 = 2$ is discussed in Section 2 and the analysis is applied to the diagonal pivoting decomposition for symmetric indefinite matrices.

Graph techniques for block methods are discussed in Section 3 (non-symmetric case) and 5 (symmetric case) and questions of stability of block methods are discussed in Sections 4 (non-symmetric) and 6 (symmetric).

2. Graph Techniques: 2×2 Blocks.

Rose [13] associated an undirected graph with a symmetric (sparse) matrix and interpreted the point LDL^t factorization (point symmetric elimination) of the matrix as an elimination process on the undirected graph. Bunch and Rose [7] generalized this, associating a directed graph with a nonsymmetric matrix and interpreting the point LU factorization of the matrix as an elimination process on the associated directed graph.

In the general case (non-symmetric), an $n \times n$ matrix M with nonzero diagonal is associated with a directed graph $G(M) = (X, \mathcal{C})$, where $X = \{x_1, \cdots, x_n\}$ is a set of n vertices and $\mathcal{C} \subseteq X \times X$ is a set of arcs such that $(x_i, x_j) \in \mathcal{C}$ (there is an arc from x_i to x_j in G) if and only if $m_{ij} \neq 0$ and $i \neq j$.

Let $M = \left[\begin{array}{c|c} p & r^t \\ \hline c & B \end{array}\right]$, where p is 1×1, r and c are $(n-1) \times 1$ and B is $(n-1) \times (n-1)$. Then the first step of the point LU factorization of M gives

$$M = \left[\begin{array}{c|c} 1 & 0 \\ \hline c/p & I \end{array}\right] \left[\begin{array}{c|c} p & r^t \\ \hline 0 & B-cr^t/p \end{array}\right].$$

We can define a corresponding process on $G(M)$. Let x_1 be the vertex associated with the first row and column of M. The elimination graph G_{x_1} is obtained from G by deleting x_1 and its arcs and adding an arc from y to z (when none is already there) if there is an arc from y to x_1 and an arc from x_1 to z. Thus, $G_{x_1} = (Y, \mathcal{B})$, where

$Y = \{x_2, \cdots, x_n\}$ and $(y, z) \in \mathcal{B}$ iff $(y, z) \in \mathcal{G}$ or (y, x_1), $(x_1, z) \in \mathcal{G}$.

If the subtraction of cr^t/p from B has not created any zeros in B where there were nonzeros before, then $G(B - cr^t/p) \equiv G_{x_1}$. We shall ignore the accidental creation of zeros during the elimination process; this is reasonable (cf. [11]) since the testing for the creation of a zero at each step would be too costly. Ignoring the accidental creation of zeros, we then know the zero-nonzero structure of the reduced matrix $B - cr^t/p$ from the elimination graph G_{x_1}.

If M is symmetric, then (x_i, x_j) is an arc iff (x_j, x_i) is an arc. We shall consider the two as identical and call (x_i, x_j) an edge (an unordered pair). The first step of the point LDL^t factorization of M (symmetric elimination) then gives

$$M = \left[\begin{array}{c|c} p & c^t \\ \hline c & B \end{array}\right] = \left[\begin{array}{c|c} 1 & 0 \\ \hline c/p & 1 \end{array}\right] \left[\begin{array}{c|c} p & 0 \\ \hline 0 & B - cc^t/p \end{array}\right] \left[\begin{array}{c|c} 1 & c^t/p \\ \hline 0 & I \end{array}\right]$$

Once again, ignoring accidental creation of zeros, $G(B - cc^t/p) \equiv G_{x_1}$.

Let us now consider a symmetric matrix $M = \left[\begin{array}{c|c} P & C^t \\ \hline C & B \end{array}\right]$, where P is 2×2, C is $(n-2) \times 2$, B is $(n-2) \times (n-2)$, and $\det P \neq 0$. The block LDL^t decomposition of M is given by

$$M = \left[\begin{array}{c|c} I_2 & \\ \hline CP^{-1} & I_{n-2} \end{array}\right] \left[\begin{array}{c|c} P & 0 \\ \hline 0 & B - CP^{-1}C^t \end{array}\right] \left[\begin{array}{c|c} I_2 & P^{-1}C^t \\ \hline 0 & I_{n-2} \end{array}\right].$$

Let

$$P = \begin{bmatrix} p_{11} & p_{21} \\ p_{21} & p_{22} \end{bmatrix},$$

$$C = \begin{bmatrix} c_{11} & c_{12} \\ \vdots & \vdots \\ c_{n-2,1} & c_{n-2,2} \end{bmatrix}, \quad \text{and} \quad B = \begin{bmatrix} b_{11} & \cdots & b_{n-2,1} \\ \vdots & \ddots & \\ b_{n-2,1} & & b_{n-2,n-2} \end{bmatrix}.$$

Then

$$(CP^{-1})_{k1} = \frac{c_{k1}\, p_{22} - c_{k2}\, p_{21}}{p_{11}\, p_{22} - p_{21}^2}$$

and

$$(CP^{-1})_{k2} = \frac{c_{k2}\, p_{11} - c_{k1}\, p_{21}}{p_{11}\, p_{22} - p_{21}^2}.$$

If $p_{11}, p_{22}, p_{21} \neq 0$, then $(CP^{-1})_{k1} = 0$ iff ($c_{k1} = 0$ and $c_{k2} = 0$) or ($c_{k1}\, p_{22} = c_{k2}\, p_{21}$), and $(CP^{-1})_{k2} = 0$ iff ($c_{k1} = 0$ and $c_{k2} = 0$) or ($c_{k2}\, p_{11} = c_{k1}\, p_{21}$).

Ignoring accidental creation of zeros once again (i.e. $c_{k1}\, p_{22} = c_{k2}\, p_{21}$ or $c_{k2}\, p_{11} = c_{k1}\, p_{21}$), $(CP^{-1})_{k1} = 0$ (or $(CP^{-1})_{k2} = 0$) iff $c_{k1} = 0$ and $c_{k2} = 0$. Conversely, $(CP^{-1})_{k1} \neq 0$ (and $(CP^{-1})_{k2} \neq 0$) iff $c_{k1} \neq 0$ or $c_{k2} \neq 0$. The (i, j) element of the reduced matrix $B - CP^{-1}C^t$ is $b_{ij} - (CP^{-1})_{i1}\, c_{j1} - (CP^{-1})_{i2}\, c_{j2}$. Once again, ignoring accidental creation of zeros, $(B - CP^{-1}C^t)_{ij} \neq 0$ iff $b_{ij} \neq 0$ or ($(CP^{-1})_{i1} \neq 0$ and $c_{j1} \neq 0$) or ($(CP^{-1})_{i2} \neq 0$ and $c_{j2} \neq 0$). But $(CP^{-1})_{i1} \neq 0$ implies that $c_{i1} \neq 0$ or $c_{i2} \neq 0$ and $(CP^{-1})_{i2} \neq 0$ implies that $c_{i1} \neq 0$ or $c_{i2} \neq 0$. Thus, a fill-in occurs iff ($c_{i1} \neq 0$ or $c_{i2} \neq 0$) and ($c_{j1} \neq 0$ or $c_{j2} \neq 0$).

Let us describe an analogous process on the graph $G(M) = (X, \mathcal{E})$ of M. We shall construct a new graph G(M, P) corresponding to the partitioning of M into P, C and B.

Since P and C correspond to the vertices x_1 and x_2, we shall replace them by a single "block" vertex y. If $c_{k1} \neq 0$ or $c_{k2} \neq 0$, then there will be an edge between x_{k+2} and y in the graph G(M, P). Thus $G(M, P) = (Y, \mathcal{E})$, where $Y = \{y, x_3, \ldots x_n\}$ while $(x_i, x_j) \in \mathcal{E}$ iff $(x_i, x_j) \in \mathcal{G}$ for $3 \leq i, j \leq n$ and $(y, x_k) \in \mathcal{E}$ iff $(x_1, x_k) \in \mathcal{G}$ or $(x_2, x_k) \in \mathcal{G}$ for $3 \leq k \leq n$.

Note that C and CP^{-1} do not necessarily have the same zero-nonzero structure since $c_{k1} \neq 0$ and $c_{k2} = 0$ implies $(CP^{-1})_{k1} \neq 0$ and $(CP^{-1})_{k2} \neq 0$, and similarly $c_{k1} = 0$ and $c_{k2} \neq 0$ implies $(CP^{-1})_{k1} \neq 0$ and $(CP^{-1})_{k2} \neq 0$. However, ignoring accidental creation of zeros, CP^{-1} has at least as many nonzeros and at most twice as many nonzeros as C has.

Consider the following example. Let

$$M = \left[\begin{array}{c|c} P & C^t \\ \hline C & B \end{array}\right] ,$$

where

$$P = \left[\begin{array}{c|c} p_{11} & p_{21} \\ \hline p_{21} & p_{22} \end{array}\right] ,$$

$$C = \begin{bmatrix} \alpha & 0 \\ 0 & 0 \\ 0 & \beta \\ \gamma & \delta \\ 0 & 0 \\ 0 & \epsilon \\ \varphi & 0 \end{bmatrix} , \quad B = \begin{bmatrix} m_{33} & m_{43} & 0 & m_{63} & 0 & m_{83} & 0 \\ m_{43} & m_{44} & 0 & m_{64} & 0 & 0 & m_{94} \\ 0 & 0 & m_{55} & 0 & 0 & 0 & m_{95} \\ m_{63} & m_{64} & 0 & m_{66} & 0 & m_{86} & 0 \\ 0 & 0 & 0 & 0 & m_{77} & 0 & m_{97} \\ m_{83} & 0 & 0 & m_{86} & 0 & m_{88} & 0 \\ 0 & m_{94} & m_{95} & 0 & m_{97} & 0 & m_{99} \end{bmatrix}$$

where the designated elements are considered nonzero. Then

G(M):

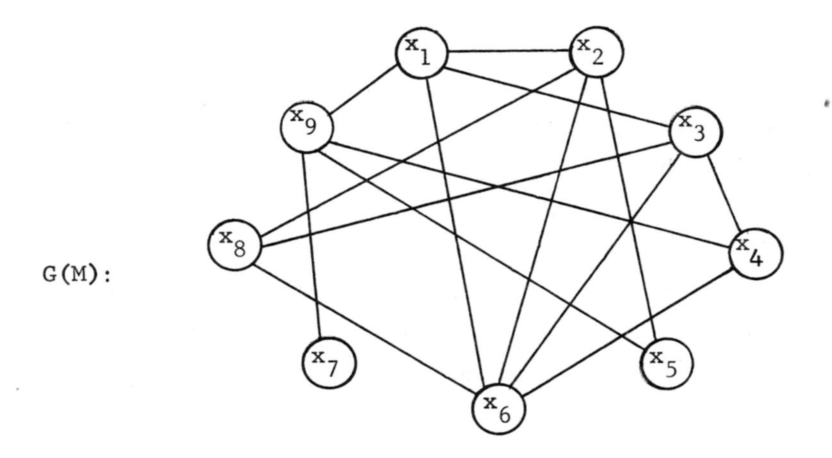

and

G(M, P):

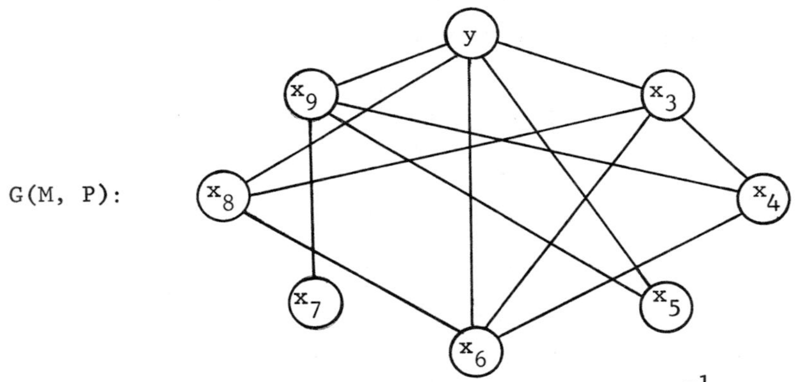

The graph G(M, P) shows that the rows of CP^{-1} corresponding to x_3, x_5, x_6, x_8 and x_9 (i.e. the first, third, fourth, sixth and seventh rows) will be nonzero and the rows corresponding to x_4 and x_7 (i.e. the second and fifth rows) will be zero. Further, we see from G(M, P) that $CP^{-1}C^t$ will have nonzeros in the entries corresponding to (x_3, x_5), (x_3, x_6), (x_3, x_8), (x_3, x_9), (x_5, x_6), (x_5, x_8), (x_5, x_9), (x_6, x_8), (x_6, x_9) and (x_8, x_9).

The elimination graph G_y is:

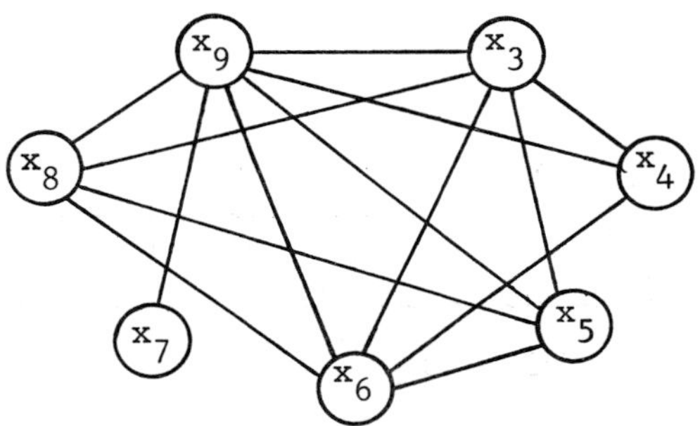

and since we are ignoring accidental creation of zeros, this is exactly $G(B - CP^{-1}C^t)$.

Verifying this, the zero-nonzero structures are

$$CP^{-1} = \begin{bmatrix} X & X \\ 0 & 0 \\ X & X \\ X & X \\ 0 & 0 \\ X & X \\ X & X \end{bmatrix}, \quad CP^{-1}C^t = \begin{bmatrix} X & 0 & X & X & 0 & X & X \\ 0 & 0 & 0 & 0 & 0 & 0 & 0 \\ X & 0 & X & X & 0 & X & X \\ X & 0 & X & X & 0 & X & X \\ 0 & 0 & 0 & 0 & 0 & 0 & 0 \\ X & 0 & X & X & 0 & X & X \\ X & 0 & X & X & 0 & X & X \end{bmatrix},$$

and

$$B-CP^{-1}C^t = \begin{bmatrix} X & X & X & X & 0 & X & X \\ X & X & 0 & X & 0 & 0 & X \\ X & 0 & X & X & 0 & X & X \\ X & X & X & X & 0 & X & X \\ 0 & 0 & 0 & 0 & X & 0 & X \\ X & 0 & X & X & 0 & X & X \\ X & X & X & X & X & X & X \end{bmatrix}.$$

Now, suppose that we had used two 1×1 pivots instead of one 2×2 pivot. Then

$$M = \left[\begin{array}{cc|ccccccc} p_{11} & p_{21} & \alpha & 0 & 0 & \gamma & 0 & 0 & \varphi \\ \hline p_{21} & p_{22} & 0 & 0 & \beta & \delta & 0 & \epsilon & 0 \\ \hline \alpha & 0 & m_{33} & m_{43} & 0 & m_{63} & 0 & m_{83} & 0 \\ 0 & 0 & m_{43} & m_{44} & 0 & m_{64} & 0 & 0 & m_{94} \\ 0 & \beta & 0 & 0 & m_{55} & 0 & 0 & 0 & m_{95} \\ \gamma & \delta & m_{63} & m_{64} & 0 & m_{66} & 0 & m_{86} & 0 \\ 0 & 0 & 0 & 0 & 0 & 0 & m_{77} & 0 & m_{97} \\ 0 & \epsilon & m_{83} & 0 & 0 & m_{86} & 0 & m_{88} & 0 \\ \varphi & 0 & 0 & m_{94} & m_{95} & 0 & m_{97} & 0 & m_{99} \end{array}\right]$$

So the elimination graph G_{x_1} obtained after eliminating x_1 is

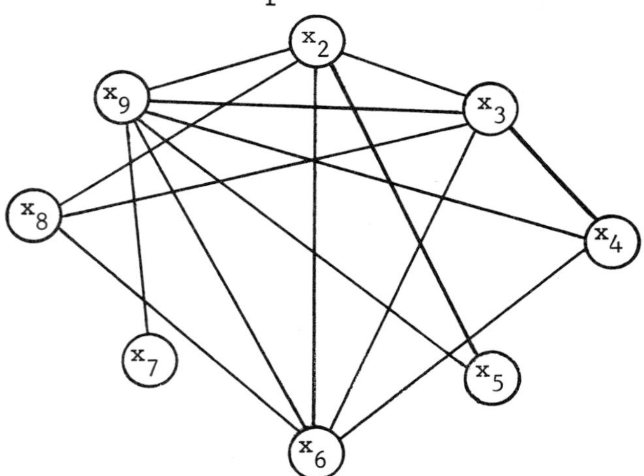

and fill-in occurred at (x_2, x_3), (x_2, x_9), (x_3, x_9) and (x_6, x_9).

In fact, $M = L_1 D_1 L_1^t$, where

$$L_1 = \left[\begin{array}{c|c} 1 & \\ \hline p_{21}/p_{11} & \\ \alpha/p_{11} & \\ 0 & \\ 0 & I_8 \\ \gamma/p_{11} & \\ 0 & \\ 0 & \\ \varphi/p_{11} & \end{array}\right],$$

$$D_1 = \left[\begin{array}{c|c|ccccccc} p_{11} & 0 & 0 & 0 & 0 & 0 & 0 & 0 & 0 \\ \hline 0 & \tilde{p}_{22} & \theta & 0 & \beta & \tilde{\delta} & 0 & \epsilon & \eta \\ \hline 0 & \theta & \times & \times & 0 & \times & 0 & \times & \times \\ 0 & 0 & \times & \times & 0 & \times & 0 & 0 & \times \\ 0 & \beta & 0 & 0 & \times & 0 & 0 & 0 & \times \\ 0 & \tilde{\delta} & \times & \times & 0 & \times & 0 & \times & \times \\ 0 & 0 & 0 & 0 & 0 & 0 & \times & 0 & \times \\ 0 & \epsilon & \times & 0 & 0 & \times & 0 & \times & 0 \\ 0 & \eta & \times & \times & \times & \times & \times & 0 & \times \end{array}\right],$$

$\tilde{p}_{22} = p_{22} - p_{21}^2/p_{11}$, $\theta = -\alpha p_{21}/p_{11}$, $\tilde{\delta} = \delta - \gamma p_{21}/p_{11}$, $\eta = -\varphi p_{21}/p_{11}$, and $\times$ denotes other nonzero elements. We shall assume $\tilde{p}_{22} \neq 0$.

If x_2 is now eliminated, the elimination graph G_{x_1,x_2} becomes:

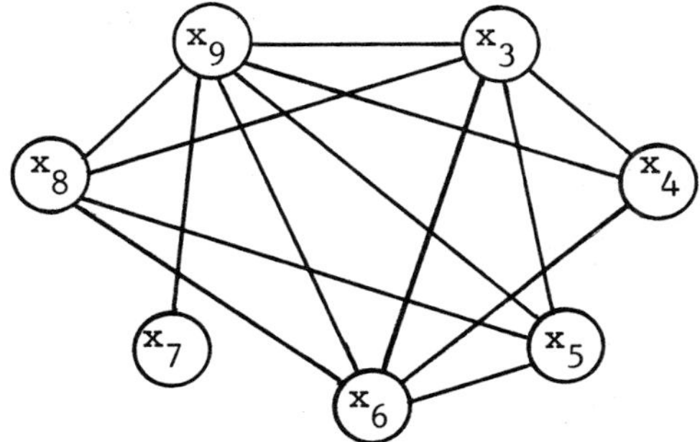

and fill-in occurred at (x_3, x_5), (x_5, x_6), (x_5, x_8), and (x_8, x_9).

Note that $G_{x_1,x_2} \equiv G_y$. (This is to be expected since the two reduced matrices are identical.) Now, $D_1 = L_2 D_2 L_2^t$, where

$$L_2 = \left[\begin{array}{c|c|ccccccc} 1 & 0 & 0 & 0 & 0 & 0 & 0 & 0 & 0 \\ \hline 0 & 1 & 0 & 0 & 0 & 0 & 0 & 0 & 0 \\ \hline 0 & \theta/\tilde{p}_{22} & & & & & & & \\ 0 & 0 & & & & & & & \\ 0 & \beta/\tilde{p}_{22} & & & & & & & \\ 0 & \tilde{\delta}/\tilde{p}_{22} & & & & I_7 & & & \\ 0 & 0 & & & & & & & \\ 0 & \epsilon/\tilde{p}_{22} & & & & & & & \\ 0 & \eta/\tilde{p}_{22} & & & & & & & \end{array}\right]$$

and D_2 has the zero-nonzero structure

$$\left[\begin{array}{c|c|ccccccc} x & 0 & 0 & 0 & 0 & 0 & 0 & 0 & 0 \\ \hline 0 & x & 0 & 0 & 0 & 0 & 0 & 0 & 0 \\ \hline 0 & 0 & x & x & x & x & 0 & x & x \\ 0 & 0 & x & x & 0 & x & 0 & 0 & x \\ 0 & 0 & x & 0 & x & x & 0 & x & x \\ 0 & 0 & x & x & x & x & 0 & x & x \\ 0 & 0 & 0 & 0 & 0 & 0 & x & 0 & x \\ 0 & 0 & x & 0 & x & x & 0 & x & x \\ 0 & 0 & x & x & x & x & x & x & x \end{array}\right].$$

Thus, $M = (L_1L_2)D_2(L_1L_2)^t$ and $L_1L_2 =$

$$\left[\begin{array}{c|c|ccccccc} 1 & 0 & 0 & 0 & 0 & 0 & 0 & 0 & 0 \\ \hline p_{21}/p_{11} & 1 & 0 & 0 & 0 & 0 & 0 & 0 & 0 \\ \hline \alpha/p_{11} & \theta/\tilde{p}_{22} & & & & & & & \\ 0 & 0 & & & & & & & \\ 0 & \beta/\tilde{p}_{22} & & & & & & & \\ \gamma/p_{11} & \tilde{\delta}/\tilde{p}_{22} & & & & I_7 & & & \\ 0 & 0 & & & & & & & \\ 0 & \epsilon/\tilde{p}_{22} & & & & & & & \\ \varphi/p_{11} & \eta/\tilde{p}_{22} & & & & & & & \end{array}\right]$$

L_1L_2 has 9 multipliers while CP^{-1} has 10. But $P \oplus (B-CP^{-1}C^t)$ requires 1 more storage than does D_2. Thus, using a 2 x 2 pivot here requires 2 more storage locations than does the use of two 1 x 1 pivots.

If $P = \begin{bmatrix} p_{11} & p_{21} \\ p_{21} & p_{22} \end{bmatrix}$ has zeros, then the situation is different.

If $p_{21} = 0$, so $P = \begin{bmatrix} p_{11} & 0 \\ 0 & p_{22} \end{bmatrix}$, then

$$CP^{-1} = \begin{bmatrix} \alpha/p_{11} & 0 \\ 0 & 0 \\ 0 & \beta/p_{22} \\ \gamma/p_{11} & \delta/p_{22} \\ 0 & 0 \\ 0 & \epsilon/p_{22} \\ \varphi/p_{11} & 0 \end{bmatrix} \quad \text{and} \quad L_1L_2 = \left[\begin{array}{cc|c} 1 & 0 & \\ 0 & 1 & 0 \\ \hline \multicolumn{2}{c|}{CP^{-1}} & I_7 \end{array}\right].$$

So the use of two 1 x 1 pivots or a 2 x 2 is identical.

If $p_{11} = 0 = p_{22}$, then the 1 x 1 pivots cannot be used, but the 2 x 2 can be. Then

$$P = \begin{bmatrix} 0 & p_{21} \\ p_{21} & 0 \end{bmatrix} \quad \text{and} \quad CP^{-1} = \begin{bmatrix} 0 & \alpha/p_{21} \\ 0 & 0 \\ \beta/p_{21} & 0 \\ \delta/p_{21} & \gamma/p_{21} \\ 0 & 0 \\ \epsilon/p_{21} & 0 \\ 0 & \varphi/p_{21} \end{bmatrix},$$

so the zero-nonzero structure of CP^{-1} is obtained by switching the columns of C.

In the tridiagonal method [1, 12], a symmetric matrix is reduced to tridiagonal form by (nonsingular) congruences, i.e. $QAQ^t = LTL^t$, where Q is a permutation matrix, L is unit lower triangular, and T is symmetric tridiagonal. However, Duff and Reid [8, pp. 7-8] have observed in several tests that the tridiagonal method fills-in much more than the diagonal pivoting method does.

If M is a general (i.e. non-symmetric) matrix of order n,

$$M = \left[\begin{array}{c|c} P & A \\ \hline C & B \end{array}\right], \text{ where } P = \begin{bmatrix} p_{11} & p_{12} \\ p_{21} & p_{22} \end{bmatrix}. \quad \text{Let } G(M) = (X, \mathcal{A}), \text{ where}$$

$X = \{x_1, \cdots, x_n\}$ and $(x_i, x_j) \in \mathcal{A}$ if and only if $m_{ij} \neq 0$ for $i \neq j$. Similarly, we shall construct a new graph $G(M, P)$ corresponding to the partitioning of M into P, A, B, and C. Since P corresponds to the vertices x_1 and x_2, we replace x_1 and x_2 by a single "block" vertex y. If c_{k1} or $c_{k2} \neq 0$, then there will be an arc from x_{k+2} to y in the graph $G(M, P)$; if a_{1k} or $a_{2k} \neq 0$, then there will be an arc from y to x_{k+2}. Thus $G(M, P) = (Y, \mathcal{E})$, where $Y = \{y, x_2, \cdots, x_n\}$ while $(x_i, x_j) \in \mathcal{E}$ if and only if $(x_i, x_j) \in \mathcal{A}$ for $3 \le i, j \le n$, $(y, x_k) \in \mathcal{E}$ if and only if $(x_1, x_k) \in \mathcal{A}$ or $(x_2, x_k) \in \mathcal{A}$ for $3 \le k \le n$, and $(x_k, y) \in \mathcal{E}$ if and only if $(x_k, x_1) \in \mathcal{A}$ or $(x_k, x_2) \in \mathcal{A}$.

The situation for CP^{-1} is exactly the same as in the symmetric case: ignoring accidental creation of zeros, CP^{-1} has at least as many nonzeros and at most twice as many nonzeros as C has. Similarly, $(CP^{-1}A)_{ij} \neq 0$ for $i \neq j$ if and only if there is a path of length 2 from x_{i+2} to x_{j+2} going through y, i.e. iff $(x_{i+2}, y) \in \mathcal{E}$ and $(y, x_{j+2}) \in \mathcal{E}$. Thus, the elimination graph G_y, which is identical to the graph of $B-CP^{-1}A$ ignoring accidental creation of zeros, is given by $(X_p, \mathcal{A}_p)$, where $X_p = \{x_3, \cdots, x_n\}$ and $(x_i, x_j) \in \mathcal{A}_p$ iff $(x_i, x_j) \in \mathcal{E}$ or $(x_i, y), (y, x_j) \in \mathcal{E}$ for $3 \le i, j \le n$.

3. Graph Techniques for Large Blocks: Non-symmetric Case.

Suppose an $n \times n$ non-symmetric matrix M has been partitioned so that

$$(3.1) \qquad M = \left[\begin{array}{c|c|c|c} A_{11} & A_{12} & \cdots & A_{1k} \\ \hline A_{21} & A_{22} & & \\ \hline \vdots & & \ddots & \\ \hline A_{k1} & & \cdots & A_{kk} \end{array}\right] \equiv A ,$$

where A_{ii} is of order n_i and $\sum_{i=1}^{k} n_i = n$. (Or, in general, $PMP^t = A$ or $PMQ = A$ where P and Q are permutation matrices.)

We could similarly make a graph construction of this situation for $n_i > 0$. If A_{11} is sparse and we wish to take advantage of the sparsity, then the construction is complicated. We shall restrict ourselves here to the case when $A_{11}, \cdots , A_{kk}$ are dense, but the A_{ij} for $i \neq j$ are sparse; this is a very reasonable situation in practice.

In general, assume that $PMP^t = A$ or $PMQ = A$, where P and Q are permutation matrices, and we have permuted M into the partitioned matrix A above, where $A_{11}, \cdots , A_{kk}$ are dense and the A_{ij} for $i \neq j$ are sparse.

Let $G(M) = (X, \mathcal{A})$ be the graph of M. We could replace the vertices $x_{q_i+1}, \cdots , x_{q_{i+1}}$, $q_i = \sum_{\ell=1}^{i-1} n_\ell$, associated with A_{ii} by a single "block" vertex y_i for $1 \leq i \leq k$ and the graph $(Y, \mathcal{E})$ associated with the partitioning has $Y = \{y_1, \cdots , y_k\}$ and $(y_i, y_j) \in \mathcal{E}$ iff there exist x_r, $q_i + 1 \leq r \leq q_{i+1}$, and x_s, $q_j + 1 \leq s \leq q_{j+1}$, such that $(x_r, x_s) \in \mathcal{A}$. We could then construct elimination graphs for this block elimination just as was done for point elimination: y_1 and its associated arcs are deleted and an arc from y_i to y_j is added (if it is not there already) iff there is an arc from y_i to y_1 and from y_1 to y_j. This will give the zero-nonzero structure of the blocks, and all theorems in [7] can be reformulated for block elimination as defined above. However, this graph construction allows us to take advantage of the sparsity in an off-diagonal block A_{ij}, $i \neq j$, only in the case that the block is null. Let us give a graph construction of block elimination which will allow us to take advantage of the sparsity of the off-diagonal blocks.

We will replace by a single "block" vertex y_1 the vertices $x_1, \cdots, x_{n_1}$ associated with A_{11} only. Then $G(M, A_{11}) = (Y_1, \mathcal{E}_1)$, where $Y_1 = \{y_1, x_{n_1+1}, \cdots, x_n\}$ while for $n_1 + 1 \leq r \leq n$, $(y_1, x_r) \in \mathcal{E}_1$ (or $(x_r, y) \in \mathcal{E}_1$) iff for some s, $1 \leq s \leq n_1 + 1$, $(x_s, x_r) \in \mathcal{G}$ (or $(x_r, x_s) \in \mathcal{E}_1$) and for $n_1 + 1 \leq i,j \leq n$, $(x_i, x_j) \in \mathcal{E}_1$ iff $(x_i, x_j) \in \mathcal{G}$. Wherever the block A_{i1} has a nonzero element in some row, the multiplier block $A_{i1}A_{11}^{-1}$ will have a full row there. The blocks in the reduced matrix will be of the form $A_{ij} - A_{i1}A_{11}^{-1}A_{1j}$. Let $Z \equiv A_{i1}A_{11}^{-1}A_{1j}$.

Of course, if $n_1 > 2$, we do not really want to invert A_{11} and form A_{11}^{-1} explicitly. Instead, we do the triangular factorization of $A_{11} = L_1U_1$. Then $Z = A_{i1}U_1^{-1}L_1^{-1}A_{1j}$. The computation of $Z = A_{i1}U_1^{-1}L_1^{-1}A_{1j}$ can be done in various ways [7, 10]:

$$(3.2)\quad Z = (A_{i1}U_1^{-1})(L_1^{-1}A_{1j}) \quad \text{by} \quad \begin{cases} \text{(a) solving } U_1^tX^t = A_{i1}^t \\ \text{(b) solving } L_1Y = A_{ij} \\ \text{(c) forming } Z = XY \end{cases},$$

$$(3.3)\quad Z = A_{i1}(U_1^{-1}(L_1^{-1}A_{1j})) \quad \text{by} \quad \begin{cases} \text{(a) solving } L_1Y = A_{1j} \\ \text{(b) solving } U_1X = Y \\ \text{(c) forming } Z = A_{i1}X \end{cases}$$

$$(3.4)\quad Z = ((A_{i1}U_1^{-1})L_1^{-1})A_{1j} \quad \text{by} \quad \begin{cases} \text{(a) solving } U_1^t X^t = A_{i1}^t \\ \text{(b) solving } L_1^t Y^t = X^t \\ \text{(c) forming } Z = YA_{1j} \end{cases} .$$

The factorizations $\mathfrak{F}_1$, $\mathfrak{F}_2$, and $\mathfrak{F}_3$ in [10] are identical to (3.2), (3.3), and (3.4), respectively, above. The operation counts, in general, will be different due to the various degrees of sparsity of A_{i1} and A_{1j}. Since we are assuming A_{11} is dense, L_1 and U_1 are dense.

In (3.2) we can sometimes save operations in (a) and (b) by taking advantage of the sparsity of A_{i1} and A_{1j}. (If the elements in the first row of A_{i1}^t and A_{1j} are nonzero, then no multiplications are saved, and (a) and (b) each require $\frac{1}{2}n_1^3$ multiplications.) In general, X and Y are dense, so forming $Z = XY$ requires n_1^3 multiplications. (The kth column of Y is nonzero after the first nonzero element in the kth column of A_{1j} is encountered.)

Similarly, in (3.3), multiplications can be saved in (a) by taking advantage of the sparsity of A_{1j}, however, if the first row of A_{1j} is nonzero, no multiplications will be saved and (a) will require $\frac{1}{2}n_1^3$ multiplications. Y will be nonzero in a column after the first nonzero element is encountered in the corresponding column of A_{1j}. Thus, in general, y is dense, and (b) will require $\frac{1}{2}n_1^3$ multiplications. However, in (c) we may save multiplications since A_{i1} is sparse. Thus, (3.3) may be better than (3.2).

The situation for (3.4) with respect to (3.2) is analogous with that in (3.3). Clearly, if A_{1j} has a nonzero first row, then (3.4) could be better than (3.3); and if A_{i1} has a nonzero first column, then (3.3) could be better than (3.4).

Deferred back solution [2, 7] is recommended for solving the right-hand side(s).

4. Stability of Non-symmetric Block Elimination.

If A_{11} is nonsingular but does not have a triangular factorization, then we can do pivoting; i.e., there exists a permutation matrix $\tilde{P}_1$ such that $\tilde{P}_1 A_{11} = L_1 U_1$. Then $Z = A_{i1} A_{11}^{-1} A_{1j} = A_{i1} U_1^{-1} L_1^{-1} \tilde{P}_1 A_{1j}$. We can compute this in a manner similar to (1), (2), (3) above since the additional work caused by the permutation matrix is trivial.

For (point) stability, we can use partial (or complete) pivoting to obtain the LU decomposition of $P_1 A_{11}$ (or $Q_1 A_{11} \tilde{Q}_1$), where P_1 is a permutation matrix (or Q_1, $\tilde{Q}_1$ are permutation matrices), and $Z = A_{i1} A_{11}^{-1} A_{1j} = A_{i1} U_1^{-1} L_1^{-1} P_1 A_{1j}$ (or $Z = A_{i1} \tilde{Q}_1 U_1^{-1} L_1^{-1} Q_1 A_{1j}$) can be computed in a manner similar to (3.2), (3.3), and (3.4) in Section 3. We would continue with partial (or complete) pivoting on the reduced block matrix. However, we also need block stability: that the norms of the reduced block matrices do not grow too large.

Of course, M being nonsingular does not imply that A_{11} is nonsingular. In fact, all the diagonal blocks could be singular. Consider

$$M = \left[\begin{array}{cc|cc} 0 & 0 & 0 & 4 \\ 0 & 0 & 3 & 0 \\ \hline 0 & 2 & 0 & 0 \\ 1 & 0 & 0 & 0 \end{array}\right] .$$

We would like a guarantee that this will not occur. Further, are there matrices for which we can be guaranteed that the norms of the reduced block matrices will not grow too large?

For point elimination such a class is provided by the diagonally dominant matrices. A matrix M is <u>(point) column diagonally dominant</u> if $|m_{ii}| \geq \sum_{j \neq 1} |m_{ji}|$ for $1 \leq i \leq n$; M is *(point) row diagonally dominant* if $|m_{ii}| \geq \sum_{\substack{j=1 \\ j\neq 1}}^{n} |m_{ij}|$ for $1 \leq i \leq n$.

If M is nonsingular and point diagonally dominant, then the LU decomposition of M exists, each reduced matrix is nonsingular and point diagonally dominant, and all the elements in all the reduced matrices are bounded by $2 \max_{i,j} |m_{ij}|$, independent of the order n of the matrix M [15, pp. 525-6].

Let M be partitioned as in (3.1). Then Feingold and Varga [9] say that M is *block (row) diagonally dominant* if A_{ii} is nonsingular and $\sum_{j\neq i} \|A_{ii}^{-1}\| \, \|A_{ij}\| \leq 1$ for $1 \leq i \leq n$. Similarly, we may say that M is *block column diagonally dominant* if A_{ii} is nonsingular and $\sum_{j\neq i} \|A_{ji}\| \, \|A_{ii}^{-1}\| \leq 1$ for $1 \leq i \leq n$. M is *block* (row or column) *strictly diagonally* dominant if there is strict inequality for each i.

George assumes point diagonal dominance when he uses block methods in [10]. However, what is needed is block diagonal dominance, and unfortunately, point diagonal dominance need not imply block diagonal dominance. Let

$$M = \left[\begin{array}{cc|cc} 4 & 3 & 1 & 0 \\ 1 & 7 & 4 & 1 \\ \hline 2 & -4 & 8 & 1 \\ 2 & 3 & 0 & 5 \end{array}\right] \equiv \left[\begin{array}{c|c} A & B \\ \hline C & D \end{array}\right].$$

M is point row diagonally dominant. But M is not block row diagonally dominant since $\|A^{-1}\|_\infty \|B\|_\infty = \frac{28}{5} > 1$ and $\|D^{-1}\|_\infty \|C\|_\infty = \frac{6}{5} > 1$ ($\|A^{-1}\|_1 \|B\|_1 = \frac{8}{5} > 1$ and $\|D^{-1}\|_1 \|C\|_1 = \frac{63}{40} > 1$).

Conversely, if M is block diagonally dominant, it need not be point diagonally dominant. Example [9, p. 1243]:

$$M = \left[\begin{array}{cc|cc} 0 & 1 & 0 & 0 \\ 2/3 & 0 & 1/3 & 0 \\ \hline 0 & 1/3 & 0 & 2/3 \\ 0 & 0 & 1 & 0 \end{array}\right]$$ is block column diagonally dominant,

but not point diagonally dominant.

Suppose M in (3.1) is a block tridiagonal matrix where the blocks A_{ii}, $1 \le i \le k$, and $A_{i+1,i}$, $A_{i,i+1}$, $1 \le i \le k-1$ are dense.

Varah [14] considers the block solution of such systems which arise from the approximation of certain time-dependent systems of linear partial differential equations in one space variable by an implicit finite difference scheme. Here, one has the appropriate block diagonal dominance.

5. Graph Techniques for Large Blocks: Symmetric Case.

Suppose an $n \times n$ symmetric matrix M has been partitioned so that

$$(5.1) \qquad M = \left[\begin{array}{c|c|c|c} A_{11} & A_{21} & \cdots & A_{k1} \\ \hline A_{21} & \ddots & & \\ \hline \vdots & & \ddots & \\ \hline A_{k1} & & & A_{kk} \end{array}\right] \equiv A ,$$

where A_{ii} is a dense matrix of order i, A_{ij} is sparse for $j \neq i$, and $n = \sum_{i=1}^{k} n_i$.

We can make graph constructions here as we did in Section 3 using undirected graphs. The triangular factorization of A_{11} can now be written as $A_{11} = L_1 D_1 L_1^t$, where L_1 is unit lower triangular and D_1 is diagonal.

Then $A_{ij} - A_{i1}A_{11}^{-1}A_{1j} = A_{ij} - A_{i1}L_1^{-t}D_1^{-1}L_1^{-1}A_{j1}$, and the computation of $Z = A_{i1}L_1^{-t}D_1^{-1}L_1^{-1}A_{j1}$ can be done several ways [7, 10]:

(5.2) $Z = (A_{i1}L_1^{-t})D_1^{-1}(L_1^{-1}A_{j1})$ by
- (a) solving $L_1X^t = A_{i1}^t$
- (b) solving $L_1Y = A_{j1}$
- (c) forming $Z = XD_1^{-1}Y$ by $(XD_1^{-1})Y$ or $X(D_1^{-1}Y)$

(5.3) $Z = A_{i1}(L_1^{-t}(D_1^{-1}(L_1^{-1}A_{j1})))$ by
- (a) solving $L_1Y = A_{j1}$
- (b) solving $D_1L_1X = Y$
- (c) forming $Z = A_{i1}X$

(5.4) $Z = ((A_{i1}L_1^{-t})D_1^{-1}L_1^{-1})A_{j1}$ by
- (a) solving $L_1X^t = A_{i1}^t$
- (b) solving $D_1L_1^tY^t = X^t$
- (c) forming $Z = YA_{j1}$

The remarks made in Section 3 also hold here.

6. Stability of Symmetric Block Elimination.

If M is positive definite, then so is each A_{ii}, and each block on the diagonal of each reduced matrix [15].

If M is indefinite, then the diagonal pivoting method [3, 4, 5, 6] or the tridiagonal method [1, 12] can be used to compute $Z = A_{i1}A_{11}^{-1}A_{1j}$ if A_{11} is not singular. However, non-singularity of each pivotal block during the process needs to be assured.

Acknowledgment. Partial support for this research was provided by AFOSR Grant 71-2006, by NSF Grant MPS75-06510 and by the Applied Mathematics Division, Argonne National Laboratory.

References

[1] Aasen, J.O., "On the reduction of a symmetric matrix to tridiagonal form", BIT, 11(1971), pp. 233-242.

[2] Bank, R.E., "Marching algorithms and block Gaussian elimination", this proceedings.

[3] Bunch, J.R., "Partial pivoting strategies for symmetric matrices",SIAM Numerical Analysis, 11(1974),pp.521-528.

[4] Bunch, J.R., and L.C. Kaufman, "Some stable methods for calculating inertia and solving symmetric linear systems", Univ. of Colorado C.S. Report #63, March, 1975.

[5] Bunch, J.R., L.C. Kaufman, and B.N. Parlett, "Decomposition of a symmetric matrix", University of Colorado C.S. Report #80, September,1975.

[6] Bunch, J.R., and B.N. Parlett, "Direct methods for solving symmetric indefinite systems of linear equations", SIAM Numerical Analysis, 8(1971), pp. 639-655.

[7] Bunch, J.R., and D.J. Rose. "Partitioning, tearing and modification of sparse linear systems", J. Math. Anal. Appl., 48(1974), pp. 574-593.

[8] Duff, I.S., and J.K. Reid, "A comparison of some methods for the solution of sparse overdetermined systems of linear equations", AERE Harwell Report C.S.S. 12, 1975.

[9] Feingold, D.G., and R.S. Varga, "Block diagonally dominant matrices and generalizations of the Gershgorin circle theorem", Pacific J. Math, 12(1962), pp. 1241-1250.

[10] George, A., "On block elimination for sparse linear systems", SIAM Numerical Analysis, 11(1974), pp.585-603.

[11] Gustavson, F.G., W. Liniger, and R. Willoughby, "Symbolic generation of an optimal Crout algorithm for sparse systems of linear equations", Journal ACM, 17(1970), pp. 87-109.

[12] Parlett, B.N., and J.K. Reid, "On the solution of a system of linear equations whose matrix is symmetric but not definite", BIT, 10 (1970), pp. 386-397.

[13] Rose, D.J., "A graph-theoretic study of the numerical solution of sparse positive definite systems of linear equations", Graph Theory and Computing, R. Read, editor, Academic Press, 1972, pp. 183-217.

[14] Varah, J.M., "On the solution of block-tridiagonal systems arising from certain finite-difference equations", Math. of Comp., 26(1972), pp. 859-868.

[15] Wilkinson, J.H., "Error analysis of direct methods of matrix inversion", J. ACM, 8(1961), pp. 281-330.

[16] Wilkinson, J.H., discussion at Gatlinburg VI Symposium on Numerical Algebra, December, 1974.

A Recursive Analysis of Dissection Strategies[†]

DONALD J. ROSE, GREGORY F. WHITTEN

Applied Mathematics, Aiken Computation Laboratory
Center for Research in Computing Technology
Harvard University
Cambridge, Massachusetts 02138

1. INTRODUCTION

An idea used frequently in the top level design and analysis of computer algorithms is to partition a problem into smaller subproblems whose individual solutions can be combined to give the solution to the original problem. Furthermore, if the subproblems have a structure essentially identical to the original problem, this process can be described recursively, and the partitioning proceeds until the subproblems can be solved trivially. Aho, Hopcroft, and Ullman [AhoHU74] call such an algorithmic strategy "divide and conquer;" however, we prefer the label "structural recursion" and view such recursive algorithms as a special case of dividing and conquering or general partitioning. In this paper we will present a simple formalism for analyzing "structurally recursive" algorithms and then apply this analysis to study the strategy known as nested dissection [Geo73, BirG73, RosW74] for ordering sparse matrices whose graphs are $n \times m$ grids.

We begin in §2 with some notation describing "grid graphs" of various types. Then in §3 we present our structural recursion formalism and derive immediately the asymptotic complexity (order of magnitude) of several numerical algorithms including $n \times n$ nested dissection. We note that while many significant algorithms can be explained recursively at the top level (see[AhoHU74], Chaps. 2.6, 2.7, 6, 7) few numerical algorithmists have adopted this view perhaps due to some combination of devotion to FORTRAN (although this is not an essential restriction as we shall see) and obstinacy in maintaining a "bottom-up" viewpoint.

[†] This work was supported in part by the Office of Naval Research under contract N00014-75-C-0243.

In §4 we apply the formalism of §3 to give a detailed analysis of nested dissection strategies on $n \times m$ grids. Frequently we are able to obtain precisely the higher order terms in the complexity expansions, hence explaining the differences in several variants. Section 5 summarizes the design of an implemented FORTRAN package for solving the sparse symmetric linear equations of grids efficiently. This design attempts to handle the nonnumerical phase of ordering and fill-in representation with particular care before executing a specialized numerical phase of factorization and back solution. We tested this FORTRAN package by comparing nested dissection with a recently proposed iterative method for solving second order self-adjoint variable coefficient elliptic PDEs on rectangular regions. This particular iterative method reported in [Ban75a] is based on a splitting using a fast separable direct method (see also [BanR75, Ban75b]) and is theoretically (asymptotic complexity) faster than nested dissection. Our experiments confirm the theoretical advantage over nested dissection for rectangular grids of any size. However, the usefulness of nested dissection for higher order or non-self-adjoint PDEs on possibly irregular regions still needs to be examined because of the potentially more general nature of direct elimination.

2. GRID GRAPH NOTATION

We will now introduce <u>grid graphs</u> of various types. A <u>graph</u> $G = (V,E)$ will be a pair where V is a finite set called <u>vertices</u> and E is a set of (distinct) pairs from V called <u>edges</u>. We will assume some familiarity with the now standard graph theoretic model of elimination [Ros73] and recall that the graph of a $p \times p$ symmetric matrix $M = (m_{ij})$, $G(M) = (V,E)$, consists of p vertices $V = \{v_i\}_{i=1}^{p}$ and edges $\{v_i, v_j\} \in E$ if and only if $m_{ij} \neq 0$ and $i < j$. Here $G(M)$ is regarded as ordered (since V is ordered); when G is regarded as unordered, it represents the equivalence class PMP^T. Different orderings of V (i.e., different P) lead to different elimination costs.

Following the notation in [HofMR73], we define the 5 point discretization $n \times m$ grid graph, $G^5_{n,m} = (V,E)$ (or G^5_n if $n = m$), embedded in the plane as follows:

$$
\begin{aligned}
&V = \{(i,j) \mid 1 \le i \le n \text{ and } 1 \le j \le m\}, \\
(2.1)\qquad &E = \{\{(i,j), (i,j+1)\} \mid 1 \le i \le n \text{ and } 1 \le j \le m-1\} \\
&\qquad \cup \{\{(i,j), (i+1,j)\} \mid 1 \le i \le n-1 \text{ and } 1 \le j \le m\},
\end{aligned}
$$

The 9 point discretization $n \times m$ grid graph, $G^9_{n,m} = (V,E^9)$ (or G^9_n if $n = m$), contains $G^5_{n,m}$ as a subgraph since E^9 is defined as

$$(2.2) \qquad E^9 = E \cup \{\{(i,j),\ (i+1,j+1)\} \mid 1 \le i \le n-1 \text{ and } 1 \le j \le m-1\} \cup \{\{(i,j),\ (i-1,j+1)\} \mid 2 \le i \le n \text{ and } 1 \le j \le m-1\}.$$

Where the distinction between $G^5_{n,m}$ and $G^9_{n,m}$ is not needed, we will refer to $n \times m$ grid graphs as $G_{n,m}$. Also for convenience any translate of $G_{n,m}$ will be regarded as $G_{n,m}$; hence we can also denote an embedded $n \times m$ grid graph as

$$(2.3) \qquad G[(i,j);(k,\ell)]$$

where $k-i = n-1 \ge 0$ and $\ell-j = m-1 \ge 0$. For example the grid graphs $G^5_{3,3}$ and $G^9_{4,3}$ are shown in Fig. 1.

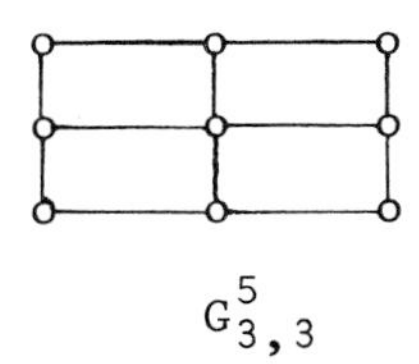
$G^5_{3,3}$

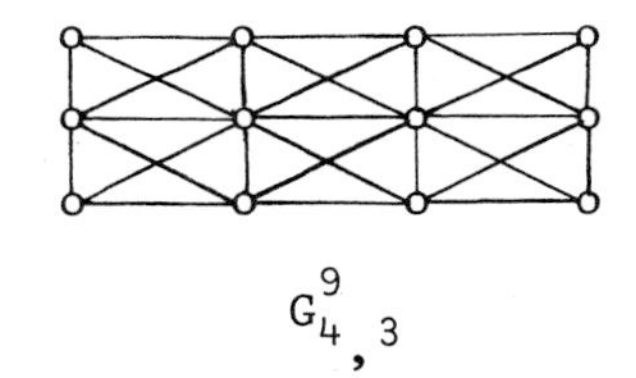
$G^9_{4,3}$

Figure 1. Grid Graphs

Given any graph $G_{n,m}$ as in (2.1) or (2.2) there will be four <u>boundary lines</u>; i.e., sets of vertices

$$(2.4) \qquad \begin{aligned} L_1 &= \{(1,j) \mid 1 \le j \le m\}, & L_2 &= \{(n,j) \mid 1 \le j \le m\}, \\ L_3 &= \{(i,1) \mid 1 \le i \le n\}, & L_4 &= \{(i,m) \mid 1 \le i \le n\}. \end{aligned}$$

Recall that a <u>clique</u> in a graph $G = (V,E)$ is a subset of vertices $W \subseteq V$ each distinct pair of which is an edge in E. For any $W \subseteq V$ let $\mathscr{C}(W) = \{\{u,v\} \mid u,v \in W \text{ and } u \ne v\}$. Hence $\mathscr{C}(W) \subseteq E$ iff W is a clique. We now define graphs which contain $G_{n,m}$ (that is, are supergraphs of $G_{n,m}$) since in these graphs some subset of the lines L_i of (2.4) will be cliques. Hence again letting $G_{n,m} = (V,E)$ we define

$$(2.5a) \qquad A_{n,m} = (V, E \cup A), \text{ where } A = \mathscr{C}(L_1 \cup L_3);$$

(2.5b) $B_{n,m} = (V, E \cup B)$, where $B = \mathscr{C}(L_1 \cup L_2 \cup L_3)$;

(2.5c) $C_{n,m} = (V, E \cup C)$, where $C = \mathscr{C}(L_2 \cup L_3 \cup L_4)$; and

(2.5d) $D_{n,m} = (V, E \cup D)$, where $D = \mathscr{C}(L_1 \cup L_2 \cup L_3 \cup L_4)$.

Graphs $A_{n,m}$, $B_{n,m}$, $C_{n,m}$, $D_{n,m}$ are shown in Fig. 2. Essentially they contain $G_{n,m}$ (either 5 or 9 point) but have certain boundary lines "cliqued" together, such sets of vertices will be called <u>boundary cliques</u>. Again any translate or reflection (about horizontal or vertical boundary lines) of these graphs will be considered the same graph. See Fig. 2. In each graph we will call the <u>interior</u> <u>(vertices)</u> those vertices not in any boundary clique. All vertices of $G_{n,m}$ are interior vertices.

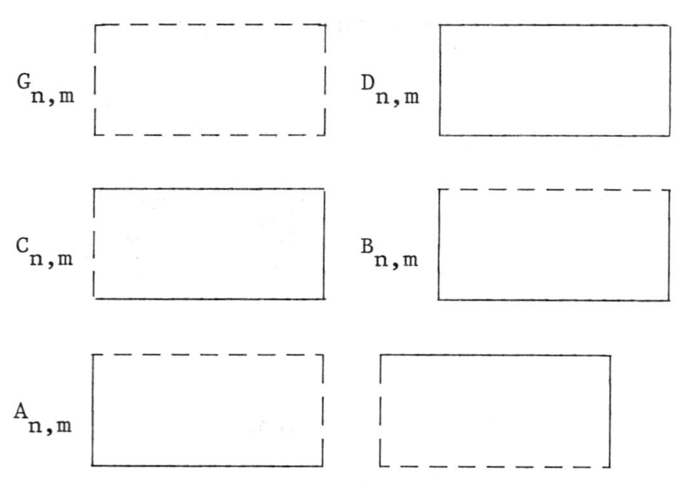

Figure 2.
Graphs $A_{n,m}$, $B_{n,m}$, $C_{n,m}$, and $D_{n,m}$. Vertices on solid lines are cliques in their respective graphs.

3. STRUCTURAL RECURSION FORMALISM

To motivate the results in this section and begin our analysis of nested dissection consider the grid graph $G_n = (V,E)$. We wish to determine the elimination complexity in arithmetic and storage of computing the LDL^t decomposition of some matrix M whose graph is G_n. In choosing a specific M we will equivalently be choosing an elimination ordering, say $\alpha:\{1, 2, \ldots, n\} \leftrightarrow V$, for the vertices of G_n.

To start the analysis for G_n we will first examine the grid graph D_n, where $n = 2^\ell+1$. Notice that the vertex induced subgraph of the interior vertices of D_n is simply the grid graph G_{n-2}. Regarding the boundary clique of D_n as ordered last we can start ordering the following interior vertices backwards; we order

(3.0) $S_\ell = H_\ell \cup V_\ell$, where $H_\ell = \{(i,2^{\ell-1}+1) \mid 1 < i < n\}$ and $V_\ell = \{(2^{\ell-1}+1,j) \mid 1 < j < n\}$.

S_ℓ is called a separator. At this point we have four copies of the grid graph G_{m-2}, where $m = 2^{\ell-1}+1$, with vertices unordered; these graphs being subgraphs of four copies of G_m with previously ordered boundary lines.

Applying the elimination theory in [Ros73] we note that since the subgrid G_{m-2} is connected, elimination of the vertices in G_{m-2} will cause the boundary lines of G_m to be a clique due to the fill-in caused by the elimination process.* Thus, we may regard the subgrids G_m as being subgrids D_m with previously ordered boundary lines. Hence, it follows that to finish ordering G_{n-2} we merely proceed recursively ordering backwards the four copies of G_{m-2} (D_{m-2} in the filled-in graph of G_n) beginning with the "separators", say $S_{\ell-1}$, as in (3.0). A detailed algorithm is given in Sec. 5; see also [RosW75]. Furthermore, if we define $D(n-1)$ as the complexity of eliminating the interior vertices of D_n with ordering α it follows that

(3.1) $D(2m) = 4D(m) + K(m)$

where $m = 2^\ell = n-1$ and $K(m)$ is the complexity of eliminating the separator set S_ℓ as in (3.0).

Our goal in this section is to analyze recursion equations like (3.1) and more complicated equations which arise from removing the restrictions on n. We begin by studying properties of equations of the form

(3.2) $T(ax) = cT(x) + K(x)$

---

*A very minor modification of this argument is necessary for the 5-point grid graph due to the connectivity of the "corner" vertices.

where, in application, constants a and c and variable x will usually be integers, although it is useful to regard x as a real variable. The following two properties are easy to verify.

Proposition 1 (nonuniqueness). Let $T(x)$ be any function satisfying (3.2). Then $T(x) + \alpha x^p$ where $p = \log_a c$ satisfies (3.2) for any α. (In application α will be chosen to satisfy initial conditions. x^p will be called the homogeneous solution since it solves (3.2) when $K(x) \equiv 0$.)

Proposition 2 (superposition). Let $T_i(x)$ satisfy (3.2) when $K(x) = K_i(x)$, $1 \leq i \leq n$. Then $T(x) = \sum \alpha_i T_i(x)$ satisfies (3.2) when $K(x) = \sum \alpha_i K_i(x)$.

Of particular interest to us will be the solution to (3.2) when $K(x)$ has the form

$$K(x) = x^p (\log x)^j, \quad j \text{ integer.}$$

In this case we have

Theorem 1. Let $K(x) = x^p(\log_a x)^j$, j integer in (3.2). Then $T(x)$ satisfies (3.2) where

(i) if $p \neq \log_a c$,

$$T(x) = x^p \sum_{k=0}^{j} \alpha_k (\log_a x)^k \quad \text{with} \quad \alpha_j = \frac{1}{(a^p - c)} \quad \text{and} \tag{3.3}$$

$$\alpha_\ell = \frac{-a^p}{(a^p - c)} \sum_{k=\ell+1}^{j} \binom{k}{\ell} \alpha_k, \quad \ell = j-1, \ldots, 0;$$

(ii) if $p = \log_a c$,

$$T(x) = x^p \sum_{k=1}^{j+1} \alpha_k (\log_a x)^k \quad \text{with} \quad \alpha_{j+1} = \frac{1}{c(j+1)} \quad \text{and} \tag{3.4}$$

$$\alpha_\ell = \frac{-1}{\ell} \sum_{k=\ell+1}^{j+1} \binom{k}{\ell-1} \alpha_k, \quad \ell = j, \ldots, 1.$$

Proof. In case (i) assume a solution of the form (3.3) and use the method of undetermined coefficients to obtain the triangular system determining the α_k. Case (ii) is similar.

Note that Proposition 2 and Theorem 1 imply that if

(3.5a) $K(x) = \sum_p \sum_j \beta_{pj} x^p (\log x)^j$ then

(3.5b) $T(x) = \sum_p \sum_j \gamma_{pj} x^p (\log x)^j$;

that is, formal sums of the form (3.5a) are closed under $\Gamma : K \to T$ given by (3.3) and (3.4).

Returning to the grid graph D_n, we now consider the function $D(n-1)$ (which measures the complexity of eliminating the interior of D_n) for general n. Using simple embedding arguments we can define lower and upper bound functions, say $L(n)$ and $U(n)$ respectively, such that

(3.6) $L(n) \leq D(n) \leq U(n)$

where $L(n)$ and $U(n)$ are determined by

(3.7a) $U(2n-1) = U(2n) = c\, U(n) + K_u(n)$ and

(3.7b) $L(2n+1) = L(2n) = c\, L(n) + K_\ell(n)$

with $L(2) = U(2) = D(2)$ and $c = 4$.
Note that if $K_u \equiv K_\ell \equiv K$ of (3.1) for all $n \geq 2$ we have

(3.8) $L(n^*) = D(n^*) = U(n^*)$

for $n^* = 2^\ell$ while, in general, (3.6) holds. We relate the functions U and L to functions T satisfying (3.2) (where $a = 2$ for convenience) in the following Propositions.

<u>Proposition 3</u>. Let $U(n)$ and $T^*(x)$, respectively satisfy

$U(2n-1) = U(2n) = c\, U(n) + K_u(n)$ and

$T^*(2x) = c\, T^*(x) + K^*(x)$

with $c \geq 0$ and
(i) $K^*(x) \geq K_u(x+1)$,
(ii) K^* and T^* nondecreasing, and
(iii) $T^*(1) \geq U(2)$.
Then $T^*(n-1) \geq \bar{U}(n)$ for all $n \geq 2$.

<u>Proof</u>. We use induction on n, condition (iii) insuring the case $n = 2$. Assuming $2 \leq n \leq m-2$ we consider $n = m$. Suppose $m = 2j$. Then

$$T^*(m-1) - U(m) = c(T^*(j-\tfrac{1}{2})-B(j))+(K^*(j-\tfrac{1}{2})-K_u(j))$$
$$\geq c(T^*(j-1)-B(j))+(K^*(j-1)-K_u(j))$$
(using (ii))
≥ 0 (using (i) and the induction hypothesis).

Similarly for $m = 2j-1$ we find

$$T^*(m-1)-U(m) = c(T^*(j-1)-B(j))+(K^*(j-1)-K_u(j)) \geq 0.$$

Proposition 4. Let $L(n)$ and $T_*(x)$ respectively satisfy

$$L(2n+1) = L(2n) = c\,L(n) + K_\ell(n) \quad \text{and}$$

$$T_*(2x) = c\,T_*(x) + K_*(x) \quad \text{with} \quad c \geq 0 \quad \text{and}$$

(iv) $K_*(x+1) \leq K_\ell(x)$,

(v) K_* and T_* nondecreasing, and

(vi) $T_*(4) \leq L(2)$.

Then $T_*(n+1) \leq L(n)$ for all $n \geq 2$.

Proof. The proof is similar to that for Proposition 3. Note that (vi) implies that $T_*(4) \leq L(3)$ and $T_*(3) \leq L(2)$ since $T_*(3) \leq T_*(4) \leq L(2) = L(3)$.

From (3.6) and Propositions 3 and 4 we obtain the string of inequalities

$$(3.9) \quad T_*(n) \leq T_*(n+1) \leq L(n) \leq D(n) \leq U(n) \leq T^*(n-1) \leq T^*(n)$$

for all $n \geq 2$. The significance of (3.9) is that, while equations (3.7a,b) are complicated, the equations for T_* and T^* are covered by Theorem 1 and have the simple solutions exhibited there. In view of Propositions 1, 3, 4 and Theorem 1 we may set $T^*(1) = U(2) = L(2) = T_*(4)$, to determine T^* and T_* satisfying (3.9). Furthermore, if K^* and K_* are of the form of $K(x)$ in Theorem 1 or (3.5) and $\frac{K^*}{K_*} \to 1$ as $x \to \infty$ while $\frac{x^p}{K^*}$ is bounded as $x \to \infty$, $p = \log_a c$, then $\frac{T^*}{T_*} \to 1$ as $x \to \infty$, and the leading term (or terms) in the asymptotic expansion for $D(n)$ can be determined precisely. However, if the homogeneous solution x^p grows faster than K^* , then, in general, the initial conditions on T^* and T_* will determine different constants for the leading terms of T^* and T_*, and we must be content with the bounds for $D(n)$ given by (3.9).

Recall that in Propositions 2 and 3 we have taken a of (3.2) as $a = 2$. A more general analysis is possible, necessitating minor changes in the functional definitions of U(n) and L(n) of (3.7a,b) and in (i)-(vi) of Propositions 3 and 4 to obtain results of the character of (3.9). In most applications, however, the $a = 2$ is natural. Note that in (3.7a,b) we have also taken $n = 2$ as the "initial" condition and have used this choice explicitly in Propositions 3 and 4. Again such a restriction is inessential and we can extend our results to cover initial conditions set at $n = n_o$ (say). Such a generalization is important since it will allow imposing the natural initial condition for a particular problem and will also enable us to analyze hybrid algorithms; for example, methods employing one (structurally recursive) algorithm while $n \geq n_o$ and then another algorithm for $n < n_o$.

We present a more comprehensive analysis of our formalism in [RosW75]. Some of the discussion above and the application of the formalism are illustrated in the following examples.

Example 1: Fast Matrix Multiplication

In [Str69] Strassen presented a method for multiplying 2 $n \times n$ matrices over an arbitrary ring using 7 multiplications and 18 additions. Hence if m is an even integer, $m = 2j$, two $m \times m$ matrices can be multiplied in 7 $j \times j$ matrix multiplications and 18 $j \times j$ matrix additions. For $n = 2^k$ we may apply the process recursively obtaining the equation

$$S_1(2n) = 7S_1(n) + 18n^2, \quad S_1(2) = 25,$$

where $S_1(n)$ is the total number of arithmetic operations to multiply two $n \times n$ matrices using this method. Applying Theorem 1 and Proposition 1 we obtain

$$S_1(n) = \alpha n^{\log_2 7} - 6n^2 \quad \text{and} \quad S_1(2) = 7\alpha - 24 = 25$$

giving $\alpha = 7$.

Since 7 is a relatively large coefficient we might consider using Strassen's method for $n > \ell$ and ordinary matrix multiplication for $n = \ell$ where $n = \ell 2^k$ and attempt to find an optimum ℓ. We then have

$$S_2(2n) = 7S_2(n) + 18n^2, \quad S_2(\ell) = 2\ell^3 - \ell^2.$$

We obtain

$$S_2(n) = \alpha(\ell) n^{\log_2 7} - 6n^2,$$

$$S_2(\ell) = \alpha(\ell)\ell^{\log_2 7} - 6\ell^2 = 2\ell^3 - \ell^2$$

so

$$\alpha(\ell) = \frac{2\ell^3+5\ell^2}{\ell^{\log_2 7}}.$$

The integer valued function α obtains its minimum at $\ell = 10$ with $\alpha(10) = 3.89569$ while for ℓ restricted to $\ell = 2^j$ α is minimized at $\ell = 8$ with $\alpha(8) = 3.91836$.

When m is an odd integer, $m = 2j-1$, we can embed 2 $m\times m$ matrices in 2 $(m+1)\times(m+1)$ matrices and multiply the two matrices using the block Strassen algorithm on $j\times j$ submatrices. This leads to an algorithm for fast matrix multiplication for general n with complexity equation

(3.10) $\quad S_3(2n-1) = S_3(2n) = 7S_3(n) + 18n^2,$

assuming that the zero operations due to embedding are counted (otherwise S_3 is an upper bound as in (3.6)-(3.9)). As in the analysis for S_2 we will combine (3.10) with the initial conditions

(3.11) $\quad S_3(\ell) = 2\ell^3-\ell^2 \quad \text{for} \quad n_o \le \ell \le 2(n_o-1);$

that is, we use ordinary multiplication for $1 \le \ell \le 2(n_o-1)$. A minor modification of Proposition 3 which assumes that

(iii) $\quad T^*(\ell-1) = S_3(\ell) \quad \text{for} \quad n_o \le \ell \le 2(n_o-1)$

gives $T^*(n-1) \ge S_3(n)$ for all $n \ge n_o$. Solving for T^* yields

$$T^*(n) = \alpha(n_o^*)n^{\log_2 7} - 6n^2 - \frac{36}{5}n - 3$$

where $\alpha(n_o^*)$ is the maximum of $\alpha(\ell-1)$ for $n_o \le \ell \le 2(n_o-1)$. Our calculations give $n_o = 19$ with $\alpha(n_o^*) = \alpha(18) = 4.61819$ indicating ordinary multiplication for $n < 37$.

Example 2: Nested Dissection

We now finish the analysis of the grid graph D_n which has been used to motivate this section. A more detailed analysis of nested dissection for the grid graph G_n is sketched in the next section, where we shall see that the recursion equation for D_n is the first of five recursion equations needed for the entire analysis. As far as the order of the asymptotic complexity is concerned, however, note that the analysis for D_n is sufficient since the elimination complexity for the interior of D_n certainly upper bounds the elimination complexity of G_{n-2}.

We must first determine $K(m)$ of (3.1) which is the complexity of eliminating the separator set as described above. $K(m)$ will be defined for even integers m consistent with (3.7a,b) and (3.8). To compute the multiplication and storage costs for eliminating the appropriate separator, say S, which has $4m-3$ vertices we count operations as in [Ros73] to obtain in the 9 point case*

$$(3.12a) \qquad \sum_S d_i^2 = 2\sum_{j=6m}^{7m-2} j^2 + \sum_{j=8m}^{10m-2} j^2 \qquad \text{and}$$

$$(3.12b) \qquad \sum_S d_i = 2\sum_{6m}^{7m-2} j + \sum_{8m}^{10m-2} j.$$

We then have $K_\mu(m) = \frac{1}{2}\sum_S d_i(d_i+3)$ and $K_\sigma(m) = \sum_S (d_i+1)$ as

the multiplication and space complexity respectively of eliminating S.

At this point we could solve for $D(m)$ (when $m = 2^\ell$) and $T_*(m)$ and $T^*(m)$ (for general m) using Propositions 3 and 4 and Theorem 1 with $D(2)=L(2)=U(2)=44$ = for multiplications or s or $D(2) = L(2) = U(2) = 8$ for storage. However the homogeneous solution to equations of the form (3.2) with $a = 2$ and $c = 4$ is $O(x^2)$. Since $K_\mu(m) = O(m^3)$ and $K_\sigma(m) = O(m^2)$ the initial condition does not affect the higher order term in either count. Indeed the discussion following (3.9) implies that if $K_1(x) \sim K_\mu(x)$ and $K_2(x) \sim K_\sigma(x)$ and $T_1(x)$ and $T_2(x)$ satisfy $T_i(2x) = 4T_i(x) + K_i(x)$, $i = 1, 2$, then $D_u(m) \sim T_1(m)$ and $D_\sigma(m) \sim T_2(m)$. Hence we compute

$$I_1 = 2\int_{6m}^{7m} x^2 dx + \int_{8m}^{10m} x^2 dx =$$

$$\tfrac{2}{3}(7^3-6^3)m^3 + \tfrac{1}{3}(10^3-8^3)m^3 = 247\tfrac{1}{3}m^3$$

and similarly $I_2 = 31m^2$ and take $K_1 = (123\frac{2}{3})x^3$ and $K_2(x) = 31x^2$. We obtain $D_\mu(m) \sim 30\frac{11}{12}m^3$ and $D_\sigma(m) \sim 7\frac{3}{4}m^2\log_2 m$.

Hence nested dissection requires $O(m^3)$ arithmetic operations and $O(m^2 \log m)$ space (for general m) for $m\times m$

---

*In the 5 point case vertices in S are not adjacent to the corner vertices.

grids. We remark that it is also possible to combine nested dissection with the fast LU decomposition of Bunch and Hopcroft [BunH74] (which applies Strassen's fast matrix multiplication) to obtain a $K_\mu = O(n^{\log_2 7})$ for eliminating S. This leads to a "fast" nested dissection algorithm requiring $O(n^{\log_2 7})$ arithmetic operations.

Example 3: Generalized Marching Algorithm

The generalized marching algorithm of Bank and Rose [BanR75,Ban75b] for solving elliptic boundary value problems on $m\times n$ finite difference grids can be viewed as another structurally recursive algorithm. In this case the arithmetic complexity equations take the form

$$T(2x) = 2T(x) + c_1 mx \tag{3.13a}$$

with the initial marching condition

$$T(k) = c_2 mk. \tag{3.13b}$$

Solving (3.13a) gives

$$T(x) = \left(\frac{c_1}{2}\right) mx\log_2 x + \alpha x, \tag{3.14a}$$

and by using the initial condition (3.13b) we solve for α giving

$$\alpha = c_2 m - \left(\frac{c_1}{2}\right) m\log_2 k. \tag{3.14b}$$

Thus, the arithmetic complexity is

$$T(n) = \left(\frac{c_1}{2}\right) mn\log_2\left(\frac{n}{k}\right) + c_2 mn. \tag{3.15}$$

Example 4: Minimal Storage Nested Dissection

For G_n the LDL^t factorization requires $O(n^2\log n)$ storage using a nested dissection ordering; however, Eisenstat, Schultz, and Sherman [EisSS75] have shown that these orderings lead to an elimination method that requires only $O(n^2)$ storage by saving only enough of the LDL^t factorization to continue with the elimination. The arithmetic complexity $T(x)$ of this method can be stated in terms of the complexity for ordinary nested dissection as

$$T(2x) = 4T(x) + G(2x).$$

Taking $G(n) = cn^3$ as the complexity for nested dissection on G_n

$$T(x) \sim \frac{1}{4} G(2x) = 2G(x).$$

Using "fast" nested dissection where $G(n) = cn^{\log_2 7}$

$$T(x) \sim \frac{1}{3} G(2x) = 2\tfrac{1}{3} G(x).$$

The asymptotic storage complexity for D_n satisfies

$$S(2x) = S(x) + cx^2 \quad \text{giving}$$
$$S(n) = \frac{1}{3} cn^2.$$

By examining additional recursion equations a more careful analysis of the storage complexity for G_n can be performed.

4. ANALYSIS OF NESTED DISSECTION

In this section we analyze the complexity of several nested dissection strategies for the grid graphs $G^9_{n,m}$, G^9_n, and G^5_n. To analyze the elimination complexity for $G^9_{n,m}$ we need to extend the formalism of §3 to higher dimensional recursion equations. (For a more complete discussion of the higher dimensional formalism see [RosW75].) Furthermore, when $n >> m$ we will present a hybrid nested dissection ordering that reduces the elimination complexity of $G_{n,m}$ from $O(n^3)$ to $O(nm^2)$ arithmetic operations and time, and from $O(n^2 \log_2 n)$ to $O(nm \log_2 m)$ storage. In the special case where $n = m$ we will show how to improve the complexity bounds of nested dissection by modifying the ordering algorithm. We also include brief analyses of diagonal nested dissection for G^5_n and nested dissection for the three dimensional grid graph $G_{n,n,n}$.

In §3 we presented the complexity analysis of D_n; the extension to $G_{n,m}$ is similar because the formalism from §3 extends to the higher dimensions. Consider the grid graph $G_{2n-1,2m-1}$ which has no boundary clique. If we order the separator V_s of $G_{2n-1,2m-1}$ last, as follows: (i) order $\{(i,m) \mid 0 < i < 2n\}$, (ii) order $\{(n,j) \mid 0 < j < m\}$, and (iii) order $\{(n,j) \mid m < j < 2m\}$; then there will be four copies of $G_{n-1,m-1}$ with vertices unordered. (We will use this ordering scheme for the sample algorithms.) Furthermore, each copy of $G_{n-1,m-1}$ can be considered as the interior of $A_{n,m}$ with its boundary clique already ordered (see Fig. 4.1a) because the elimination theory considers the original and filled-in graphs to be the same with respect to space and arithmetic complexity. Thus, the elimination complexity of $G_{2n-1,2m-1}$ can be stated in terms of the complexity for eliminating the interior of $A_{n,m}$

as in the following (recursion) equation

$$(4.1) \qquad G(2n,2m) = 4\,A(n,m) + k_G(n,m)$$

where $G(2n,2m)$ is the elimination complexity for $G_{2n-1,2m-1}$, $A(n,m)$ for the interior of $A_{n,m}$, and $k_G(n,m)$ for the separator V_s.

If we consider the grid graph $G_{p-1,q-1}$ where either p or q is not divisible by 2, (4.1) is not applicable; however, the higher dimensional analogs of the formalism in §3 allow us to find continuous upper and lower bounding functions T^*_G and T_{*G}. Furthermore, the formalism shows that k^*_G and k_{*G} associated with T^*_G and T_{*G}, respectively, are asymptotically equal implying that T^*_G, G, and T_{*G} are asymptotically equal; call this G'. Thus, we can use the integral approximations for k^*_G and k_{*G}, say k'_G, corresponding to the ordering of the separator to achieve the asymptotic bounds by taking

$$(4.2a) \qquad k_G(n,m) = 2\int_0^m \frac{(x+2n)^i}{i}\,dx + \int_0^{2n} \frac{x^i}{i}\,dx,$$

$i = 1$ for storage, 2 for multiplications.

This yields the (recursion) equation

$$(4.2b) \qquad G'(2n,2m) = 4\,A'(n,m) + k'_G(n,m)$$

where $A'(n,m)$ is the asymptotic complexity for eliminating the interior of $A_{n,m}$. To simplify the notation we will denote the asymptotic complexity $G'(n,m)$ as $G(n,m)$, dropping the primes where no distinction is necessary.

Similarly, to eliminate the interior of $A_{2n,2m}$ we order the separator V_s of $A_{2n,2m}$ which leaves four copies of $G_{n-1,m-1}$ with vertices unordered. The above subgrids are embedded in $A_{n,m}$, $B_{n+1,m}$, $C_{n,m+1}$, and $D_{n+1,m+1}$ whose boundary cliques are already ordered last (see Fig. 4.1b). Thus, to eliminate the interior of $A_{2n,2m}$ we obtain the asymptotic complexity from the recursion equation

$$(4.3a) \qquad k_A(n,m) = \int_0^m \frac{(x+2n+m)^i}{i}\,dx + \int_0^m \frac{(x+4n+m)^i}{i}\,dx + \int_0^{2n} \frac{(x+2n+2m)^i}{i}\,dx$$

(4.3b) $$A(2n,2m) = A(n,m) + B(n,m) + C(n,m) + D(n,m) + k_A(n,m)$$

where $k_A(n,m)$ is the asymptotic elimination complexity for V_s and $A(n,m)$, $B(n,m)$, $C(n,m)$, and $D(n,m)$ for the interiors of $A_{n,m}$, $B_{n+1,m}$, $C_{n,m+1}$, and $D_{n+1,m+1}$ respectively.

The asymptotic elimination complexity for the interiors of $B_{n+1,m}$, $C_{n,m+1}$, and $D_{n+1,m+1}$ may be found in a similar way giving their respective recursion equations (see Fig. 4.1cde) as:

(4.4a) $$k_B(n,m) = \int_0^m \frac{(x+2n+2m)^i}{i}\,dx + \int_0^m \frac{(x+4n+2m)^i}{i}\,dx + \int_0^{2n} \frac{(x+2n+4m)^i}{i}\,dx$$

(4.4b) $$B(2n,2m) = 2\,B(n,m) + 2\,D(n,m) + k_B(n,m)$$

(4.5a) $$k_C(n,m) = 2\int_0^m \frac{(x+4n+m)^i}{i}\,dx + \int_0^{2n} \frac{(x+4n+2m)^i}{i}\,dx$$

(4.5b) $$C(2n,2m) = 2\,C(n,m) + 2\,D(n,m) + k_C(n,m)$$

(4.6a) $$k_D(n,m) = 2\int_0^m \frac{(x+4n+2m)^i}{i}\,dx + \int_0^{2n} \frac{(x+4n+4m)^i}{i}\,dx$$

(4.6b) $$D(2n,2m) = 4\,D(n,m) + k_D(n,m)$$

where the k's are the asymptotic elimination complexities for the separator sets. By solving (4.2)-(4.6) we get the following asymptotic bounds for multiplications

(4.7) $$D(n,m) = \frac{19}{12}m^3 + 9nm^2 + 14n^2m + \frac{19}{3}n^3$$

(4.8) $$C(n,m) = \frac{11}{12}m^3 + \frac{17}{3}nm^2 + \frac{32}{3}n^2m + \frac{19}{3}n^3$$

(4.9) $$B(n,m) = \frac{19}{12}m^3 + \frac{49}{6}nm^2 + \frac{31}{3}n^2m + \frac{11}{3}n^3$$

(4.10) $$A(n,m) = \frac{11}{12}m^3 + \frac{215}{42}nm^2 + \frac{57}{7}n^2m + \frac{11}{3}n^3$$

(4.11) $$G(n,m) = \frac{1}{2}m^3 + \frac{59}{21}nm^2 + \frac{32}{7}n^2m + 2n^3$$

and for storage

(4.12) $$G(n,m) = A(n,m) = B(n,m) = C(n,m) = D(n,m) = \frac{5}{4}m^2\log_2 m + 2nm\log_2 nm + \frac{5}{2}n^2\log_2 n.$$

The analysis of the elimination complexity for G_n is clearly a subcase of that for $G_{n,m}$; hence, we can find continuous upper and lower bounding functions $T_G^*(n)$ and $T_{*G}(n)$ for the discrete bounding functions $U_G(n)$ and $L_G(n)$ which bound $G(n)$, the elimination complexity for G_{n-1}. Furthermore, the above functions

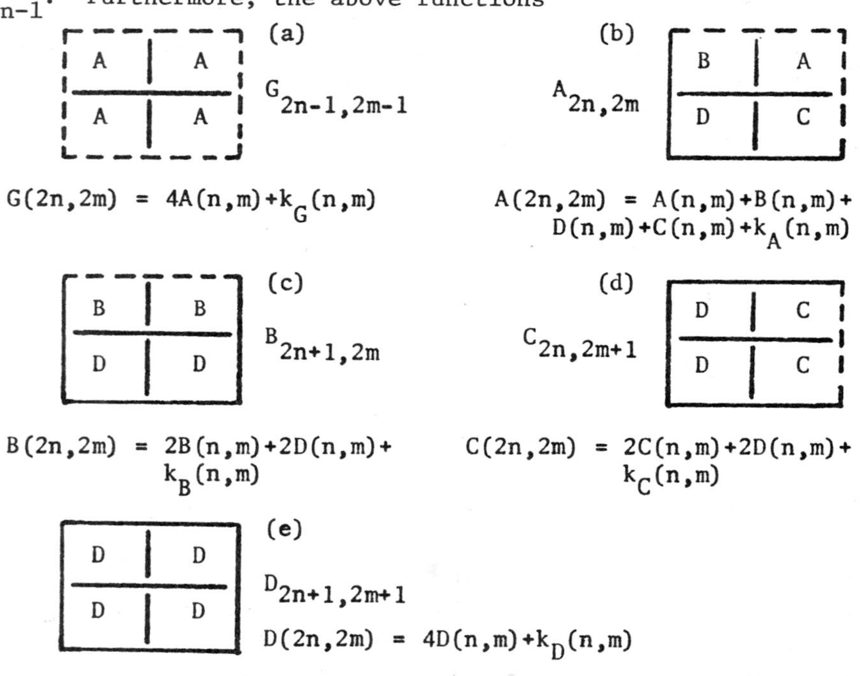

Figure 4.1. Grid graphs and their recursion equations.

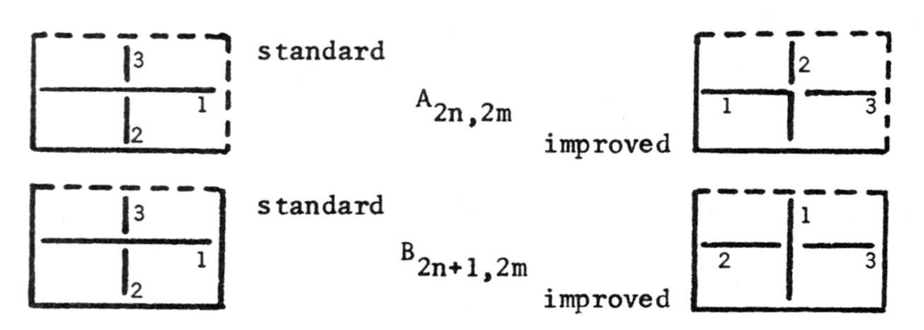

Figure 4.2. Order lines 1,2,3 of separator last to first.

are asymptotically equal. Thus, for the asymptotic multiplication complexity

(4.13) $G(n) = 9\,\frac{37}{42}\,n^3$ and for storage

(4.14) $G(n) = 7\,\frac{3}{4}\,n^2\log_2 n$.

For the special case where $n = 2^k$ we give the following table for the actual elimination complexity of G^9_{n-1}.

n-1	$G_\sigma(n)$	$G_\sigma(n)$	$\frac{G_\sigma(n)}{n \log n}$	$\frac{G_\mu(n)^{n-1}}{n^3}$
3	30	61	2.103	2.259
7	354	1,543	2.573	4.499
15	2,778	21,327	3.160	6.319
31	17,666	227,351	3.711	7.632
63	99,450	2,127,959	4.192	8.510
127	518,578	18,576,183	4.601	9.069
255	2,569,386	156,033,463	4.943	9.410
511	12,282,786	1,282,670,647	5.228	9.613
1,023	57,203,610	10,417,406,007	5.467	9.730

$$G_\sigma(n) = \frac{31}{4}\,n^2\log_2 n - 23n^2 + n\log_2 n + 45n - 8\log_2 n - 22$$
$$G_\mu(n) = \frac{415}{42}\,n^3 - 17n^2\log_2 n - \frac{163}{12}\,n^2 + \frac{67}{2}\,n\log_2 n - \frac{163}{6}\,n + 0\log_2 n + \frac{643}{21}$$

Table 4.1. Multiplication counts, $G_\mu(n)$, and storage counts, $G_\sigma(n)$, for G^9_{n-1}.

We can improve the asymptotic multiplication bounds by ordering the separators of $A_{2n,2m}$ and $B_{2n+1,2m}$ in a different order (see Fig. 4.2). Modifying the ordering of $A_{2n,2m}$ alone gives the complexity bounds reported in George [Geo73] of

(4.15) $G(n) = 9\,\frac{11}{21}\,n^3$.

If we also modify the ordering of $B_{2n+1,2m}$, we get a slightly better bound of

(4.16) $G(n) = 9\,\frac{43}{84}\,n^3$.

Also, see Duff, Erisman, and Reid [DufER75] for a discussion and empirical results on several nested dissection modifications.

Returning to the grid graph $G_{n,m}$ we see that when $n >> m$ the arithmetic complexity is $O(n^3)$. In order to reduce the complexity we introduce a hybrid nested dissection ordering algorithm that divides $G_{n,m}$ into $k \geq 2$ (integer) $G_{n/k,m}$ subgrids. We will first consider the special case where $n = ki-1$ and $m = j-1$. If the vertical lines $V_s = \{(pk,q) \mid 0 < p < i\}$ for $0 < q < j$ are ordered last, then the unordered vertices will induce k copies of $G_{i-1,j-1}$. In the filled-in graph of $G_{n,m}$ $k-2$ copies of $G_{i-1,j-1}$ will be subgrids of $S_{i+1,j-1}$ and the remaining two copies will be subgrids of $R_{i,j-1}$, where $R_{n,m} = (V, E \cup \mathscr{C}(L_1))$ and $S_{n,m} = (V, E \cup \mathscr{C}(L_1 \cup L_3))$ as in (2.4)-(2.5). This yields the equation

$$(4.17) \qquad G(i,j,k) = 2\,R(i,j) + (k-2)\,S(i,j) + k_G(i,j,k)$$

where $G(i,j,k)$, $R(i,j)$, and $S(i,j)$ are the complexities for eliminating the interiors of $G_{ki-1,j-1}$, $R_{i,j-1}$, and $S_{i+1,j-1}$ and $k_G(i,j,k)$ is the complexity for eliminating V_s. Similarly, we can find the complexity for eliminating the interiors of $R_{i,j-1}$ and $S_{i+1,j-1}$ by solving the equations

$$(4.18) \qquad R(2i,2j) = 2\,A(i,j) + 2\,B(i,j) + k_R(i,j) \quad \text{and}$$

$$(4.19) \qquad S(2i,2j) = 4\,B(i,j) + k_S(i,j).$$

As before the asymptotic complexity of $G(i,j,k)$ is the same for all the bounding functions; hence, we use integral approximations for $k_G(i,j,k)$, $k_R(i,j)$, and $k_S(i,j)$. For example,

$$(4.20) \qquad k_G(i,j,k) = \frac{(k-2)}{p}\int_0^j (x+j)^p\,dx + \frac{1}{p}\int_0^j x^p\,dx$$

where $p = 1$ for storage and $p = 2$ for multiplications when the $k-1$ vertical lines in V_s are eliminated last from left to right. Thus, the asymptotic solutions to the recursion equations (4.18)-(4.20) are for multiplications

$$(4.21) \qquad R(i,j) = \frac{11}{12}j^3 + \frac{32}{7}ij^2 + \frac{118}{21}i^2j + 2i^3$$

$$(4.22) \qquad S(i,j) = \frac{19}{12}j^3 + \frac{22}{3}ij^2 + \frac{20}{3}i^2j + 2i^3$$

$$(4.23) \qquad G(i,j,k) = 2ki^3 + \frac{(231k-294)j^3 + (616k-464)ij^2 + (560k-176)i^2j}{84}$$

and for storage

(4.24) $G(i,j,k) = k(\frac{5}{4} j^2 \log_2 j + \frac{5}{2} i^2 \log_2 i + 2ij \log_2 ij)$.

In our analysis k is an integer parameter that we have to choose. Empirical results show that for the special case where $j=i$, giving the grid graph $G_{ki-1,i-1}$, the best value of c in $G(\frac{i}{c},i,ck)$ (ck subgrids $G_{i/c-1,i-1}$) is approximately 1.75. However, when $c=2$ the results appear to be at most 5% worse. For example, the multiplication complexities of $G_{ki-1,i-1}$ for $c=1$ and $c=2$ are

$$G(i,i,k) = \frac{1575k-934}{84} i^3 \quad \text{and}$$

$$G(\tfrac{i}{2},i,2k) = \frac{700k-285}{42} i^3 \quad \text{which yields}$$

$$G(i,i,k) - G(\tfrac{i}{2},i,2k) = \frac{25k-52}{12} i^3 .$$

By formulating the appropriate recursion equations we can compute the asymptotic complexities for other dissection strategies. The grid graph G_n^5 may be ordered using a diagonal nested dissection scheme [BirG73]. This results in three recursion equations with the final solution for the elimination complexity as

$$G_\mu(n) = \frac{251}{48} n^3 \quad \text{multiplications and}$$

$$G_\sigma(n) = \frac{31}{8} n^2 \log_2 n \quad \text{storage.}$$

Similarly, by analyzing 9 recursion equations we compute the elimination complexities for the three dimensional grid graph $G_{n,n,n}$ using an ordering scheme analogous to that of (4.1)-(4.5) as follows

$$G_\mu(n) = \frac{60497}{17360} n^6 \quad \text{multiplications} \quad (\sim 3.485 n^6) \quad \text{and}$$

$$G_\sigma(n) = \frac{51}{4} n^4 \quad \text{storage} \quad (12.75\ n^4)$$

By a more careful ordering of the separators we can get slightly better bounds of

$$G_\mu(n) = \frac{536567}{156240} n^6 \quad \text{multiplications} \quad (\sim 3.434\ n^6) \quad \text{and}$$

$$G_\sigma(n) = \frac{761}{60} n^4 \quad \text{storage} \quad (\sim 12.683\ n^4)$$

(Complete details of the above analyses may be found in [RosW75] in addition to several other dissection strategies for $G_{n,m}^5$ and $G_{i,j,k}$.)

5. IMPLEMENTATION OF NESTED DISSECTION

In section 4 we describe the "top level" of the recursive nested dissection ordering algorithms for $G_{n,m}$. To compute the entire ordering and fill-in of the reordered linear system we continue the recursion until trivial subproblems are reached. In general, recursion is difficult to implement in a nonrecursive language such as FORTRAN, but the recursive process that we want to simulate can be viewed as a depth first traversal of the subgrids generated by the recursion. This type of recursion can be implemented with a single last-in first-out stack.

We will first discuss the ordering and fill-in computation for $G[(i,j);(k,\ell)]$ where $k-i \geq 0$ and $\ell-j \geq 0$ (see Equation 2.3). To implement the ordering algorithm we will use a stack SUBGRID, a top of stack pointer TOS, and two stack manipulating procedures POP and PUSH. Furthermore, W, S, E, and N will correspond to the i,j,k and ℓ's that identify the subgrid that is being processed by the algorithm. A counter CURRENT contains the new label for current vertex being ordered (vertices are labeled in reverse order, i.e. nm to 1). The basic ordering and fill-in algorithm follows:

```
Algorithm GRIDORDER (i,j,k,ℓ)
begin comment place G[(i,j);(k,ℓ)] onto the stack
     initialize:  TOS ← 0;  PUSH (i,j,k,ℓ);
     do while  TOS ≠ 0
     begin  comment  remove top subgrid from stack;
          POP  (W,S,E,N);
          comment  compute midpoints of the sides of the grid;
          MIDPOINT  (W,S,E,N,WE,SN);
          comment  reverse order and compute fill-in for
               {(p,q) | (p=WE and S ≤ q ≤ N) or
                        (q=SN and W ≤ p ≤ E)};
          ORDERFILL (W,S,E,N,WE,SN);
          comment  push the four subgrids onto the stack;
          PUSH(W,S,WE-1,SN-1);  PUSH (WE+1,S,E,SN-1);
          PUSH (WE+1,SN+1,E,N); PUSH (W,SN+1,WE-1,N);
     end;
end;

procedure  POP (W,S,E,N)
begin comment  remove top subgrid from stack;
     TOS ← TOS - 4;
     W ← SUBGRID(TOS);  S ← SUBGRID(TOS+1);
     E ← SUBGRID(TOS+2); N ← SUBGRID(TOS+3);
end;
```

```
procedure PUSH (W,S,E,N)
begin comment do not stack improper subgrids;
    if W ≤ E and S ≤ N then
    begin
        SUBGRID(TOS) ← W; SUBGRID(TOS+1) ← S;
        SUBGRID(TOS+2) ← E; SUBGRID(TOS+3) ← N;
        TOS ← TOS + 4;
    end;
end;
```

The overall space and time complexity of the above algorithm GRIDORDER if $n = k-i+1$ and $m = \ell-j+1$ is $O(nm)$ assuming that MIDPOINT is $O(1)$ and ORDERFILL is $O(1)$ per vertex ordered. The maximum size of the stack SUBGRID is less than $4(3\lceil \log_2(\min(n,m)+1) \rceil +1)$.

The procedure MIDPOINT simply computes the midpoints of (S,N) and (W,E) with a rounding heuristic. We need a two dimensional array GRID $(i-1:k+1, j-1:\ell+1)$ that has the values

GRID(p,q) = 0 if (p,q) has not yet been ordered or is not in the grid,
GRID(p,q) = ordered vertex number > 0, otherwise.

The following algorithm shows one of the many possible heuristics.

```
procedure MIDPOINT (W,S,E,N,WE,SN)
begin comment round toward highest labeled vertex;
    if GRID(W,S-1) ≥ GRID(W,N+1) then
        SN ← int((S+N)/2) else SN ← int((S+N+1)/2);
    if GRID(W-1,S) ≥ GRID(E+1,S) then
        WE ← int((W+E)/2) else WE ← int((W+E+1)/2);
end;
```

The complexity of MIDPOINT is clearly $O(1)$. Empirical results support the theoretical analysis that selective rounding is beneficial in reducing operations and storage. Similar results were observed in [DufER75].

The procedure ORDERFILL is the most complicated part of the ordering algorithm. In section 4 we present an ordering scheme for the vertices in the separator

$$V_s = \{(p,q) \mid (p = WE \text{ and } S \leq q \leq N) \text{ or } (q = SN \text{ and } W \leq p \leq E)\}$$

that orders the vertices first in the horizontal and then in the vertical. The complexity of the ordering algorithm is just $O(1)$ per vertex ordered.

The fill-in computation for the vertices in V_s parallels the complexity analysis for the n m grid in section 4. To compute the various k(n,m) terms it is necessary to know exactly where the fill-in occurs. We use this information to compute the fill-in in ORDERFILL. Thus, if we are ordering the subgrid G[(W,S);(E,N)] there will be a clique that contains the boundary lines of G[(W-1,S-1);(E+1,N+1)] that are in the original grid G[(1,1);(n,m)]. (For the five point grid graph the corner boundary vertices will be absent from the clique.) Furthermore, if the vertices on each of the boundary lines are ordered sequentially we can describe this boundary clique by at most four pairs of vertex labels. The algorithm ORDERFILL follows in a skeletal form below. CLEEK is an eight element array that describes the boundary clique. (The boundary lines must be ordered sequentially.) NC is the number of vertices in the boundary cliques.

```
procedure  ORDERFILL (W,S,E,N,WE,SN)
begin comment  order {(p,q) | q = SN and  W ≤ p ≤ W};
     CLIQUE  (W,S,E,N,CLEEK,NC);
     LASTVERTEX ← CURRENT;
     for  p ← W  step 1 until E do
     begin  GRID(p,SN) ← CURRENT;
          FILLIN (CURRENT,LASTVERTEX,CLEEK,NC);
          CURRENT ← CURRENT - 1;
     end;
     comment order {(p,q) | p = WE and SN+1 ≤ q ≤ N};
     CLIQUE  (W,SN+1,E,N,CLEEK,NC);
     LASTVERTEX ← CURRENT;
     for  q ← SN+1  step  1  until  N  do
     begin  GRID(WE,q) ← CURRENT;
          FILLIN (CURRENT,LASTVERTEX,CLEEK,NC);
          CURRENT ← CURRENT - 1;
     end;
     comment  order  (p,q) | p = WE and S ≤ q ≤ SN-1 ;
     CLIQUE (W,S,E,SN-1,CLEEK,NC);
     LASTVERTEX ← CURRENT;
     for  q ← S  step  1  until  SN-1  do
     begin  GRID(WE,q) ← CURRENT;
          FILLIN (CURRENT,LASTVERTEX,CLEEK,NC);
          CURRENT ← CURRENT - 1;
     end;
end;
```

The procedures CLIQUE and FILLIN are not described in full detail because they are dependent upon the data structures used to store the fill-in of the matrix. CLIQUE determines the boundary lines of $G[(W-1,S-1);(E+1,N+1)]$ that are in the original grid $G_{n,m}$. If the boundary lines in the boundary clique are ordered sequentially, then CLIQUE requires O(1) space and time; otherwise, it can be implemented so that it requires O(1) space and time per vertex ordered with respect to the call to ORDERFILL. The procedure FILLIN to describe the fill-in for vertex (row) CURRENT of the graph just needs to save a pointer to the information computed by the previous call to CLIQUE because the filled-in row is completely defined by the pair of vertices CURRENT and LASTVERTEX and a maximum of four pairs of vertices contained in the array CLEEK. It is possible to compute extra fill-in because we are assuming that the vertices CURRENT through LASTVERTEX are a clique which may not be the case if the separator that is being ordered is the entire subgrid that is currently being processed.

To extend the ordering algorithm to the case where the grid graph $G_{n,m}$ is considered to be composed of k subgrids of approximate size $G_{n/k,m}$, we will call GRIDORDER k times as illustrated in the following algorithm.

```
Algorithm  DISSECTION (n,m,k)
begin  comment  compute ordering and fill-in for G(n,m)  which
                has been divided into  k  subgrids;
    if  k > n+1  then  k ← n+1;
    CURRENT ← n * m;  EAST ← 0;
    REAST ← 0.0; RSTEP ← float (n+1)/k;

    for  kth ← 1  step  1  until  k  do
    begin  comment  order  kth  subgrid;
        WEST ← EAST ; REAST ← REAST + RSTEP;
        EAST ← int(REAST+0.5);
        comment  order vertical line EAST if in grid;
        if  EAST ≤ n  then
        begin comment  line EAST is cliqued to line WEST,
                  CLIQUE is used in a nonstandard fashion
                  for the above;
            LASTVERTEX ← CURRENT;
            CLIQUE (WEST+1,1,n,m,CLEEK,NC);
            for  q ← 1  step  1  until  m  do
            begin  GRID(EAST,q) ← CURRENT;
                  FILLIN (CURRENT,LASTVERTEX,CLEEK,NC);
                  CURRENT ← CURRENT - 1;
            end;
        end;
        comment  use nested dissection on subgrid;
```

```
        GRIDORDER (WEST+1,1,EAST-1,m);
    end;
end;
```

The space and time complexity of DISSECTION is just O(nm) because GRIDORDER's complexity is O(1) per vertex ordered.

The FORTRAN IV implementation of the ordering and fill-in algorithm is similar to the algorithms described above. The procedure CLIQUE is the most complex part of the algorithm because it has to resolve on which boundary line the corner boundary vertices are ordered. The numeric phase of the nested dissection implementation is a specialized sparse LDL^t decomposition algorithm similar to [Gus72,She75]. The major differences are in the types of data structures required to store the fill-in structure of the LDL^t decomposition. Our data structures describe the fill-in with O(nm) storage and reduce the list processing overhead to $O(nm \log_2(\min(n,m)))$ time for the LDL^t decomposition and to O(nm) time for the back solution of the linear system (see [RosW75]).

ACKNOWLEDGEMENTS

Part of this research was conducted while the authors were summer visitors at Argonne National Laboratory. James R. Bunch contributed to our understanding of the Bunch-Hopcroft and Strassen algorithms, and Randolph E. Bank provided his codes for our numerical experiments. We are grateful for the use of MIT MATHLAB MACSYMA system for symbolic computation.

REFERENCES

[AhoHU74] A.V. Aho, J.E. Hopcroft, and J.D. Ullman. The Design and Analysis of Computer Algorithms. Addison-Wesley Publishing Co., Reading, Massachusetts(1974).

[Ban75a] R.E. Bank. Marching Algorithms for Elliptic Boundary Value Problems. Ph.D. Thesis, Harvard University (1975).

[Ban75b] R.E. Bank. Marching algorithms for elliptic boundary value problems II: the non-constant coefficient case. SIAM J. Numer. Anal., submitted (1975).

[BanR75] R.E. Bank and D.J. Rose. Marching algorithms for elliptic boundary value problems I: the constant coefficient case. SIAM J. Numer. Anal., submitted (1975).

[BirG73] G. Birkhoff and J.A. George. Elimination by nested dissection. Complexity of Sequential and Parallel Numerical Algorithms. Academic Press, N.Y. (1973).

[BunH74] J. Bunch and J.E. Hopcroft. Triangular factorization and inversion by fast matrix multiplication. Math. Comp. 28:125(1974) p. 231-236.

[DufER75] I.S. Duff, A.M. Erisman, and J.K. Reid. On George's nested dissection method. SIAM J. Numer. Anal., to appear.

[EisSS75] S.C. Eisenstat, M.H. Schultz, and A.H. Sherman. Application of an element model for Gaussian elimination. This volume.

[Geo73] J.A. George. Nested dissection of a regular finite element mesh. SIAM J. Numer. Anal., 10(1973) p. 345-363.

[Gus72] F.G. Gustavson. Some basic techniques for solving sparse systems of linear equations. Sparse Matrices and Their Applications. D.J. Rose and R.A. Willoughby, eds. Plenum Press, N.Y. (1972) p. 41-52.

[HofMR73] A.J. Hoffman, M.S. Martin, and D.J. Rose. Complexity bounds for regular finite difference and finite element grids. SIAM J. Numer. Anal. 10(1973) p. 364-369.

[MACSYMA] MATHLAB group. MACSYMA Reference Manual. Project MAC, M.I.T., Cambridge, Massachusetts (1974).

[Ros73] D.J. Rose. A graph-theoretic study of the numerical solution of sparse positive definite systems of linear equations. Graph Theory and Computing, R. Read, ed. Academic Press, N.Y. (1973) p. 183-217.

[RosW74] D.J. Rose and G.F. Whitten. Automatic nested dissection. Proc. ACM Conference (1974) p. 82-88.

[RosW75] D.J. Rose and G.F. Whitten. Analysis and implementation of dissection strategies. Center for Research in Computing Technology Technical Report, to appear (1975).

[She75] A.H. Sherman. On the efficient solution of sparse systems of linear and nonlinear equations. Ph.D. dissertation, Yale University Research Report #46, (1975).

[Str69] V. Strassen. Gaussian elimination is not optimal. Numerische Mathematik 13(1969) p. 354-356.

Applications of an Element Model for Gaussian Elimination[1]

S. C. EISENSTAT,[2] M. H. SCHULTZ,[2]
and A. H. SHERMAN[3]

1. INTRODUCTION

Consider the system of linear equations

(1.1) $A x = b$

where A is an N×N sparse symmetric positive definite matrix such as those that arise in finite difference and finite element approximations to elliptic boundary value problems in two and three dimensions. A classic method for solving such systems is Gaussian elimination. We use the k*th* equation to eliminate the k*th* variable from the remaining N-k equations for k = 1,2,...,N-1 and then back-solve the resulting upper triangular system for the unknown vector x. Equivalently, we form the U^TDU decomposition of A and successively solve the triangular systems

(1.2) $U^T z = b, D y = z, U x = y.$

Unfortunately, as the elimination proceeds, coefficients that were zero in the original system of equations become nonzero (or fill-in), increasing the work and storage required. The purpose of this paper is to introduce a new graph-theoretic model of such fill-in; to establish lower bounds for the work and storage associated with Gaussian elimination; and to present a minimal storage sparse elimination algorithm that significantly reduces the storage required.

In section 2, we review the graph-theoretic elimination model of Parter and Rose and introduce a new element model

[1] This work was supported in part by NSF Grant GJ-43157 and ONR Grant N00014-67-A-0097-0016.

[2] Department of Computer Science, Yale University, New Haven, Connecticut 06520.

[3] Department of Computer Science, University of Illinois, Urbana, Illinois 61801.

of the elimination process. In section 3, we use this model to give simple proofs of inherent lower bounds for the work and storage associated with Gaussian elimination, generalizing similar results of George and Hoffman, Martin, and Rose. Last, in section 4, we show how the element model, combined with the rather unusual idea of recomputing rather than saving the factorization, leads to a minimal storage sparse elimination algorithm that requires significantly less storage than regular sparse elimination (e.g. $O(n^2)$ vs. $O(n^2 \log n)$ for the five- or nine-point operator on an $n \times n$ mesh).

2. AN ELEMENT MODEL FOR GAUSSIAN ELIMINATION

In this section, we shall review the graph-theoretic elimination model of Gaussian elimination suggested by Parter [6] and extensively developed by Rose [7] and introduce a new element model, cf. Eisenstat [1], Sherman [9].

Given an irreducible $N \times N$ symmetric positive definite matrix $A = (a_{ij})$, we can represent the zero-nonzero structure of A by a graph $G(A) = (V,E)$ as follows: The vertex v_i of $G(A)$ corresponds to the *ith* row/column of A; the edge (v_i,v_j) is an edge of $G(A)$ if and only if $a_{ij} \neq 0$. We shall model Gaussian elimination on the matrix A as a sequence $G^{(k)}$ of such graphs.

Let $G^{(1)} = G(A)$. Then the graph $G^{(k+1)}$ is derived from $G^{(k)}$ as follows: Corresponding to using the *kth* equation to eliminate the *kth* variable from the remaining $N-k$ equations, we add any edges necessary to make all vertices adjacent to the *kth* vertex v_k pairwise adjacent and then delete v_k and all edges incident to it. Thus the graph $G^{(k)}$ represents the nonzero structure of the lower $N-k+1 \times N-k+1$ submatrix of A just before the *kth* variable is eliminated. From this observation, the following operation and storage counts are immediate:

Theorem (Rose [7]): Let d_k denote the degree of vertex v_k in the elimination graph $G^{(k)}$, i.e. at the time it was eliminated. Then the number of multiplies required to form the U^TDU decomposition of A is given by

$$W = \sum_{k=1}^{N} \frac{1}{2} d_k(d_k+3)$$

and the number of off-diagonal nonzero entries in U is given by

$$S = \sum_{k=1}^{N} d_k.$$

The element model emulates Gaussian elimination as a sequence of transformations on the graph G(A) and the collection E of maximal cliques of vertices in G(A) which we shall refer to as *elements*. Initially, $\tilde{G}^{(1)} = G(A)$, $E^{(1)} = E$, and all vertices are marked uneliminated. Then $\tilde{G}^{(k+1)}$ and $E^{(k+1)}$ are derived from $\tilde{G}^{(k)}$ and $E^{(k)}$ as follows: Corresponding to using the *kth* equation to eliminate the *kth* variable from the remaining N-k equations, we mark the *kth* vertex v_k as eliminated and add any edges necessary to make all vertices adjacent to v_k pairwise adjacent (no vertices or edges are deleted); all the elements of $E^{(k)}$ containing v_k are merged into a new element and then deleted from $E^{(k)}$. Note that the subgraph of $\tilde{G}^{(k)}$ induced by the uneliminated vertices is just $G^{(k)}$ so that the number of uneliminated vertices adjacent to v_k at the time it is eliminated is d_k as above. Also, elements in $E^{(k)}$ are cliques in $\tilde{G}^{(k)}$ (though not necessarily maximal cliques). A vertex will be said to be an *exterior vertex* in $\tilde{G}^{(k)}$ if it belongs to more than one element; otherwise it belongs to exactly one element and is said to be an *interior vertex*.

As an example, consider the elimination process for the nine-point operator on a 3×3 grid:

```
1-5-2
| | |
7-8-9
| | |
3-6-4
```

Figure 2.1.

Initially, the elements correspond to the individual mesh squares or elements, since these form the only maximal cliques in G(A). Hence the name "element model" (cf. George [4]). When we eliminate vertex v_1, there is no change in the element graph (except to mark v_1 eliminated) since it is an interior vertex. The same is true when we eliminate v_2, v_3, and v_4. When we eliminate v_5, however, the two elements containing v_5 are merged, as also happens when we eliminate v_6. Now eliminating v_7 causes the two remaining elements to be merged into the final element, which consists of all the vertices. Eliminating vertices v_8 and v_9 has no further effect since they are interior vertices.

We note the following properties of the element model for future use. Throughout the elimination process, the number of elements that contain a given vertex never increases. Thus an eliminated vertex is necessarily an interior vertex, and, correspondingly, an exterior vertex is uneliminated. (However, an interior vertex need not have been eliminated.) Initially, the largest element contains at most d_{max} vertices, where d_{max} is the maximum degree of any vertex in G(A) and no vertex belongs to more than $2^{d_{max}}$ elements. (See Eisenstat [1] for an example where a vertex actually belongs to $2^{\frac{1}{2}(d_{max}-1)}$ elements.) At the end of the elimination, the only element consists of all the vertices.

3. LOWER BOUNDS FOR GAUSSIAN ELIMINATION

In recent years, there has been a great deal of interest in computational complexity, particularly in analyzing the number of fundamental operations inherent in a computation. Along these lines, George [4] and Hoffman, Martin, and Rose [5] have used the elimination graph model to prove lower bounds for the work and storage associated with Gaussian elimination on symmetric positive definite matrices whose graphs are certain regular planar grids. In this section, we shall describe how the element model can be used to give simpler proofs of these results that extend easily to some irregular grids in two and three dimensions. For further details, see Eisenstat [1].

First we shall define the model of computation in which these lower bounds are valid. Given an irreducible N×N symmetric positive definite matrix A, we seek to solve the system of linear equations (1.1) using Gaussian elimination. The work associated with the elimination process will be taken as the number of multiplies to factor A; the storage as the number of nonzero entries in the matrix U.

The work and storage required for Gaussian elimination is directly related to the order in which the variables are eliminated. Thus, instead of solving the original system (1.1), we might prefer to solve the permuted system

$$(3.1) \qquad P A P^T y = P b, \; P^T y = x$$

for some given permutation matrix P so as to reduce the associated work and storage. Indeed, a great deal of research has been done toward discovering good ordering strategies (e.g. minimum degree [7] and nested dissection [4]). Note that the permutation P does not change the structure of the graph G(A), other than relabelling the vertices and thus changing the order of elimination. Since the lower bounds we shall derive will be independent of such considerations, they will be valid for all possible orderings.

As in section 2, let G(A) = (V,E) denote the graph associated with A, and let d_{max} denote the maximum degree of any vertex in G(A). We shall now make the following additional assumption about the matrix A, or equivalently the graph G(A):

"Isoperimetric" Inequality: There exist constants $K > 0$ and $0 < \alpha, \beta \le 1$ such that, given any subset S of V with $|S| \le \beta N$, we have

$$|\partial S| \ge K |S|^{\alpha}$$

where

$$\partial S = \{v \in S : \exists w \in V{-}S \text{ such that } v \text{ and } w \text{ are adjacent in } G(A)\}$$

is the *boundary* of S.

This assumption is closely related to the classical isoperimetric inequality, which relates the area A and perimeter P of a plane figure

$$P^2 \ge \frac{\pi A}{4},$$

and the isovolumetric inequality, which relates the volume V and surface area S of a three-dimensional region

$$S^3 \ge 36\pi V^2.$$

Indeed, proofs of this property for particular grids follow

the classical proofs. As an example, for the five- or nine-point operator on an n×n grid, we have

$$|\partial S| \geq |S|^{1/2} \quad \text{if } |S| \leq \frac{n^2}{2}$$

while for the seven- or twenty-seven-point operator on an n×n×n grid we have

$$|\partial S| \geq |S|^{2/3} \quad \text{if } |S| \leq \frac{n^3}{6}$$

cf. Eisenstat [1].

Lemma (Eisenstat [1]): Assume that the graph G(A) satisfies an isoperimetric inequality. Then there exists a $K(\beta N/2^{d_{max}})^{\alpha}$ clique in some elimination graph $G^{(k)}$.

Proof: At the start of the elimination process, the maximum size of any element is d_{max}, the maximum degree of any vertex in G(A). At the end of the elimination process, there is exactly one element of size N, namely the set of all vertices. Therefore, at some point during the elimination, the first element of size > βN was created. This element was created by merging all the elements containing some vertex v_k. As we saw in section 2, v_k could belong to at most $2^{d_{max}}$ elements at this point and thus at least one such say e*, must contain more than $\beta N/2^{d_{max}}$ vertices. Consider the boundary vertices of e*. There are at least $K(\beta N/2^{d_{max}})^{\alpha}$ such vertices by the isoperimetric inequality, all are uneliminated vertices at this stage, and they are all pairwise adjacent. Thus they form a clique in $G^{(k)}$.

QED

Corollary: The bandwidth of the matrix A is at least

$$K(\beta/2^{d_{max}})^{\alpha}\, N^{\alpha}.$$

Corollary: The total work required to factor the matrix A using band elimination is at least

$$\frac{1}{2}\, K^2(\beta/2^{d_{max}})^{2\alpha}\, N^{(2\alpha+1)};$$

the number of nonzeroes in the band of A is at least

$$K(\beta/2^{d_{max}})^{\alpha}\, N^{(\alpha+1)}.$$

The proof follows directly from the standard operation and

storage counts for band elimination in terms of the number of equations and the bandwidth.

Corollary: The total work required to factor the matrix A is at least

$$\frac{1}{6} K^3 (\beta/2^{d_{max}})^{3\alpha} N^{3\alpha};$$

the number of nonzeroes in U is at least

$$\frac{1}{2} K^2 (\beta/2^{d_{max}})^{2\alpha} N^{2\alpha}.$$

The proof follows from the Lemma as in Hoffman, Martin, and Rose [5].

We shall now examine the consequences of these simple results. For planar grids with $\sim n^2$ vertices and bounded degree, such as the five- or nine-point operator on an $n \times n$ grid:

(1) the bandwidth is at least $O(n)$
(2) the work and storage are at least $O(n^4)$ and $O(n^3)$ respectively for band elimination
(3) the work and storage are at least $O(n^3)$ and $O(n^2)$ respectively for sparse elimination (a more careful analysis gives $O(n^2 \log n)$ for the storage; see Eisenstat [1]).

For three-dimensional grids with $\sim n^3$ vertices and bounded degree, such as the seven- or twenty-seven-point operator on an $n \times n \times n$ grid:

(4) the bandwidth is at least $O(n^2)$
(5) the work and storage are at least $O(n^7)$ and $O(n^6)$ respectively for band elimination
(6) the work and storage are at least $O(n^6)$ and $O(n^4)$ respectively for sparse elimination.

4. MINIMAL STORAGE SPARSE ELIMINATION

One of the major disadvantages of Gaussian elimination for solving sparse systems of linear equations is the amount of storage required. For example, to solve the nine-point finite difference operator on an $n \times n$ grid requires at least $O(n^2 \log n)$ storage whereas an iterative method would require only $O(n^2)$ storage. In this section, we present a variation of sparse Gaussian elimination that trades a significant reduction in storage for a modest increase in work. For ease of exposition, we shall restrict attention to the nine-point operator on an $n \times n$ grid with $n = 2^t - 1$, but the results are valid for more general finite difference and finite element grids, cf. Eisenstat, Schultz, and Sherman [3], Sherman [9]. For similar results on band elimination, see Eisenstat, Schultz, and Sherman [2], Sherman [9].

Two basic concepts are used in achieving this reduction: First, rather than save the entire factorization, we shall throw most of the nonzero entries of U away and recompute them as needed during the back-solution; second, we use an element merge tree to specify a divide-and-conquer elimination ordering and to keep track of those entries of U that are being saved during each step of the calculation.

Suppose we were to perform Gaussian elimination in such a way that the last 2n+1 variables to be eliminated correspond to the vertices on a dividing cross as shown in Figure 4.1. At the end of the elimination process, we would have generated all the coefficients in the upper triangular system of equations that remains, and we have merely to solve this system for the vector of unknowns x. Yet suppose we had saved only those coefficients in the last 2n+1 rows. Then we could only solve for the last 2n+1 variables, i.e. the values of the unknowns on the dividing cross. This is enough, however, to split our original $n \times n$ problem into four smaller $\sim n/2 \times \sim n/2$ problems of the same form, which we can now solve in the same fashion.

Figure 4.1.

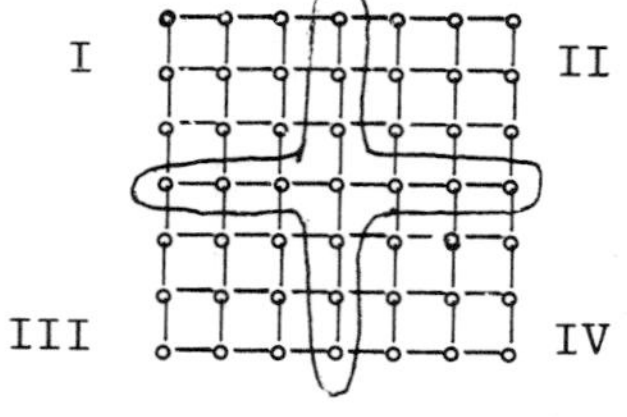

Of course, we will have to do more work than we would have had we saved the entire factorization. But how much more? The nested dissection ordering of George [4] orders the variables on the dividing cross last and requires approximately Cn^3 multiplies for some fixed constant C. Thus the cost of the whole procedure is just Cn^3 plus the cost of solving four $\sim n/2 \times \sim n/2$ problems. Letting $\theta(n)$ denote this cost, we have that

$$\theta(n) \simeq Cn^3 + 4\theta(n/2)$$

and that $\theta(n) = Cn^3$ for n sufficiently small. (After all, when the amount of storage for the factorization gets small enough, we may as well save all of it.) The solution to this recurrence relation is

$$\theta(n) \approx 2Cn^3.$$

Thus we are doing twice as much work but, as we shall see, the savings in storage will be much more significant.

The *element merge tree* is based on the element model introduced in section 2. Recall that initially there were a number of elements, or maximal cliques of vertices, and that as the elimination progressed these elements were merged into larger and larger elements until only a single element containing the entire set of vertices remained. Given an elimination ordering, we construct the corresponding element merge tree as follows: The nodes in the tree represent elements that were created during the elimination process; the root of the tree is the final element consisting of the entire set of vertices; the node (i.e. element) e_1 is a son of another node e_2 if and only if e_1 was merged into e_2 when some vertex v_k was eliminated. The merge tree for the 3×3 nine-point operator of Figure 4.2a is given in Figure 4.2b.

Figure 4.2a.

```
1—5—2
|  |  |
7—8—9
|  |  |
3—6—4
```

Figure 4.2b.

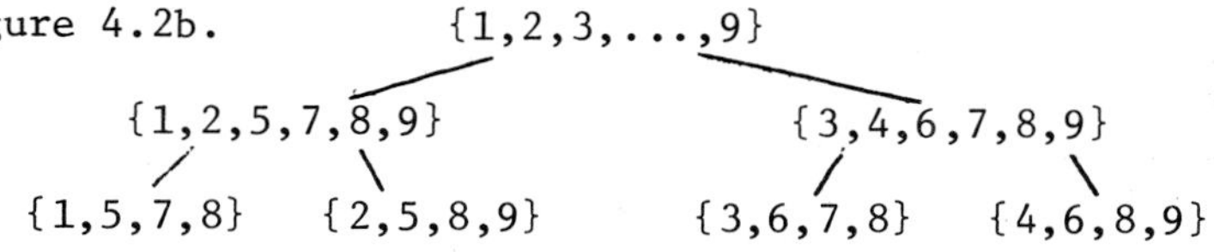

```
                    {1,2,3,...,9}
                  /               \
        {1,2,5,7,8,9}             {3,4,6,7,8,9}
          /       \                 /        \
   {1,5,7,8}   {2,5,8,9}     {3,6,7,8}   {4,6,8,9}
```

We will now specify the ordering of vertices and the corresponding element tree which we will use to make the storage required for our procedure $O(n^2)$ as opposed to the $O(n^2 \log n)$ for the factorization. Beginning with the entire grid, we break it into four equal-size elements using a dividing cross as in Figure 4.1. (Note that the vertices on the dividing cross belong to more than one element since they are exterior vertices.) The last vertices to be eliminated will be the center vertex followed by the other vertices on the dividing cross. We then order the interior vertices in element I, followed by those of element II, element III, and element IV. The vertices within each of these are eliminated in an analogous fashion: The last vertices to be eliminated are the center vertex followed by the other vertices on a dividing cross; we then order the vertices in each of the four subelements, and so on.

The ordering that results can be shown to be equivalent to the nested dissection ordering in terms of the work and storage required (cf. Eisenstat, Schultz, and Sherman [3], Rose and Whitten [8], Sherman [9]), and the element merge trees are identical, although the actual orderings are completely different. Thus the divide-and conquer order produces an $O(n^3)$ elimination scheme. We now show that the elimination can be carried out in such a way that the coefficients in the last 2n+1 rows can be computed using only $O(n^2)$ storage.

Consider again the element merge tree. The nodes at the bottom or zeroth level correspond to elements in the original graph G(A), all of which contain exactly $4 = (2^0+1)^2$ vertices. At the first level, the elements were formed by merging four bottom-level elements and thus each contains exactly $9 = (2^1+1)^2$ vertices. In general, at the k*th* level the elements were formed by merging four elements on the (k-1)*th* level and, by induction, contain precisely $(2^k+1)^2$ vertices.

We shall say that a particular element in the element tree is active at a particular point in the elimination if it has not yet been merged into another element. Then the only entries in the triangular factor U that we shall be saving at that point are those corresponding to pairs of exterior vertices in active elements. Note that the final element is active once it is created and is never deactivated, so that the coefficients in the last 2n+1 rows of U will be available to solve for the unknowns on the dividing cross at the end of the elimination.

Since there are at most four active elements at any level in the element merge tree (other than the zeroth) at any stage of the elimination by virtue of the recursive definition of the ordering, we can easily bound the total storage required: At the k*th* level from the bottom, there are at most $\ell = 4 \cdot 2^k$ exterior vertices per element, which require at most $\ell(\ell+1)/2$ nonzero entries of U to be saved; thus the total storage is at most

$$\sum_{k=0}^{t-2} 4 \cdot \frac{1}{2} (4 \cdot 2^k) (4 \cdot 2^k + 1) \simeq \frac{32}{3} n^2.$$

A more careful analysis using two-way dissection show that the total storage can be reduced to $9n^2$, cf. Eisenstat, Schultz, and Sherman [3], Sherman [9].

Note that this surprising result does not really violate the results of section 3; the storage for U still is $O(n^2 \log n)$ — we just aren't saving all the entries. Another point worth noting is that the algorithm can be implemented to run in time $O(n^{\log_2 7})$ if the elimination of variables on dividing crosses is done in blocks using Strassen's algorithm for matrix multiplication and inversion [10]. Again, this does not violate the results of section 3 since we are doing block rather than point elimination.

REFERENCES

[1] S. C. Eisenstat. Complexity bounds for Gaussian elimination.

[2] S. C. Eisenstat, M. H. Schultz, and A. H. Sherman. Minimal storage band elimination.

[3] S. C. Eisenstat, M. H. Schultz, and A. H. Sherman. Minimal storage sparse elimination.

[4] J. A. George. Nested dissection of a regular finite element mesh. *SIAM Journal on Numerical Analysis* 10: 345-363, 1973.

[5] A. J. Hoffman, M. S. Martin, and D. J. Rose. Complexity bounds for regular finite difference and finite element grids. *SIAM Journal on Numerical Analysis* 10:364-369, 1973.

[6] S. V. Parter. The use of linear graphs in Gaussian elimination. *SIAM Review* 3:119-130, 1961.

[7] D. J. Rose. A graph-theoretic study of the numerical solution of sparse positive definite systems of linear equations. In R. Read, editor, *Graph Theory and Computing*, 183-217. Academic Press, 1972.

[8] D. J. Rose and G. F. Whitten. Automatic nested dissection. *Proceedings of the ACM National Conference*, 82-88, 1974.

[9] A. H. Sherman. On the efficient solution of sparse

systems of linear and nonlinear equations. PhD dissertation, Yale University, 1975.
[10] V. Strassen. Gaussian elimination is not optimal. *Numerische Mathematik* 13:354-356, 1969.

AN OPTIMIZATION PROBLEM ARISING FROM TEARING METHODS

Alberto Sangiovanni-Vincentelli*
Dept. of Electrical Engineering and
Computer Sciences
University of California, Berkeley, Ca 94720

1. INTRODUCTION

When a linear system of algebraic equations $Ax = b$, $A \in \mathbb{R}^{n^2}$ and sparse, $x \in \mathbb{R}^n$, $b \in \mathbb{R}^n$, has to be solved, Gaussian Elimination (or one of its modification like LU decomposition) can be conveniently used. However, due to the particular structure of A, sometimes it is possible to save computation time and/or storage requirements implementing tearing methods [1]. Tearing consists mainly of two parts: at first, the solution of the system $A^*x = b$ is computed, where A^* has been obtained from A zeroing some elements, then this solution is modified to take into account the real structure of the original system.

It has to be noted that this method may be necessary, even though not convenient, when it is not possible to process the original system $Ax = b$ due to its dimension w.r.t. the dimension of the available computer. In order to optimize the performances of tearing we have to minimize its overall computational complexity. As above mentioned, the two fundamental steps are: (i) computation of the solution x^* of $A^*x = b$, (ii) computation of the solution $\bar{x}$ of $Ax = b$ given x^*.

The complexity of (i) depends on the structure of A^*, the complexity of (ii) depends on the rank of $A-A^*$. Therefore, we have to face a two criteria decision making

---

*On leave from Istituto di Electrotecnica ed Electronica, Politecnico di Milano, Milano, Italy.

problem. Due to the difficulty of these problems, a general strategy to obtain "good" solutions consists in changing a set of criteria into a set of constraints. In our particular case, we fix a "convenient" structure of A* and try to minimize the rank of A–A*. A choice frequently done is to require that A* is in Inferior Triangular Form (ITF) [2] (Fig. 1a). It is immediate to note that

min rank(A–A*) (or more precisely symbolic rank(A–A*))[14]
subject to

A* is in ITF

is equivalent to the problem of finding a nonsymmetric permutation of A, PAQ, such that PAQ is in an optimum Bordered Triangular Form (BTF) [2], i.e., in BTF with minimum bord [3] (Fig. 1b).

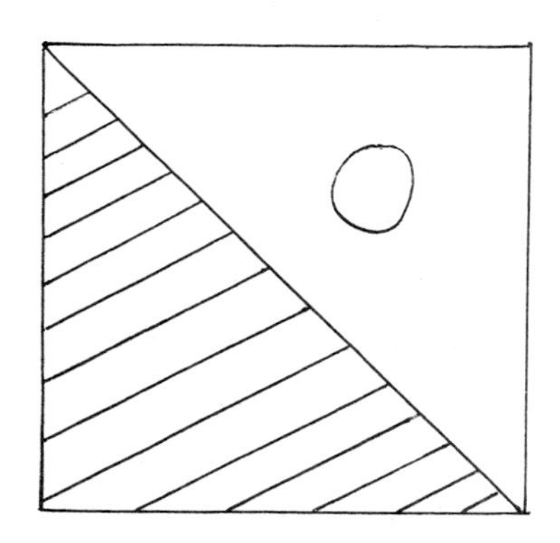

Figure 1a.

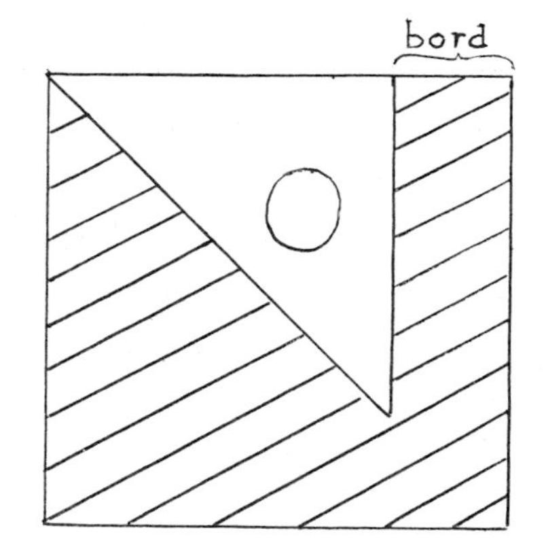

Figure 1b.

This problem has been deeply investigated when a particular permutation strategy is used, i.e., when $Q = P^T$. In this case, a graph theoretic interpretation can be given and some graph algorithms have been proposed [3,4,5]. The determination of an optimal nonsymmetric permutation has been proposed as an open question in [3]. In this paper, a graph theoretic interpretation of the problem based on a bipartite graph associated to A is given and a heuristic algorithm is proposed. In particular, the paper is organized as follows: in Section 2 some preliminary definitions and remarks are given; in Section 3 a graph theoretic interpretation of the problem is given; in Section 4 an algorithm based on bipartite bipartite graphs is proposed, its correctness proved and its complexity discussed; in Section 5 some concluding remarks are introduced.

2. PRELIMINARY DEFINITIONS AND REMARKS

A graph theoretical background is presented in this Section. All the undefined terms are to be understood according to Harary [6]. Let $G = (X,U)$ $(G = (X,E))$ be a *(di)graph* with a set of *vertices* or *nodes* X and a set of (directed) *edges* or *arcs* $U = \{\{x_i,x_j\} | x_i,x_j \in X\}$ $(E = \{(x_i,x_j) | x_i,x_j \in X\})$. A simple (directed) *path* $\mu(x_i,x_j)$ $x_i \neq x_j$, of length ℓ is an ordered sequence of distinct vertices. $\mu(x_i,x_j) = x_i \langle x_i = p_o,p_1,\ldots,p_\ell = x_j \rangle \{p_k,p_{k+1}\} \in U((p_k,p_{k+1}) \in E)$ $k = 0,\ldots,\ell-1$.
A simple (directed) cycle η of length ℓ is an ordered sequence of vertices $\eta = \langle p_o = \bar{x},p_1,\ldots,p_\ell = \bar{x} \rangle$ such that the ordered sequence $\langle p_o,p_1,\ldots,p_{\ell-1} \rangle$ is a simple path and $\{p_{\ell-1},p_\ell\} \in U((p_{\ell-1},p_\ell) \in E)$. The *section* (di)*graph* defined on a subset $Y \subset X$ is $G(Y) \triangleq (Y,U(Y))$ $(= (Y,E(U))$ where $U(Y) = \{\{x_i,x_j\} \in U | x_i,x_j \in Y\}$ $(E(Y) = \{(x_i,x_j) \in E | x_i,x_j \in Y)$.

Given a digraph $G = (X,E)$, the *reversion of an arc* (x_i,x_j) is performed by replacing it with an edge (x_j,x_i).

Let G be a directed graph. G is said to be strongly connected if for each pair of vertices $x_i,x_j \in X$, there exist a simple path $\mu_1(x_i,x_j)$ and a simple path $\mu_2(x_j,x_i)$. It has to be noted that the trivial graph constituted by one node only is considered to be strongly connected. Let $\pi = \{X_1,\ldots,X_q\}$ be a partition of the nodes X. If the section graphs $G_i = (X_i,E_i) = G(X_i)$, $i = 1,\ldots,q$, are strongly connected and if no G_i is a proper subgraph of a strongly connected subgraph of G, then the G_i's are called the strongly connected components of G.

A bipartite (di)graph $B = (S,T,U)$ $(B = (S,T,E))$ is a (di) graph $B = (X,U)$ $(B = (X,E))$ such that $S \cup T = X$, $S \cap T = \emptyset$ and the section (di)graphs B(S) and B(T) are both vertex graphs. Given a bipartite (di)graph $B = (S,T,U)$ $(B, = (S,T,E))$ a matching I is a subset of U(E) such that no two edges in I are incident to the same node. A node $x \in X$ is said covered if there is an edge in I that incides in it. A complete matching is a matching such that all the nodes of the bipartite (di)graph are covered. A maximum cardinality matching is a matching containing a maximum number of edges. Given a bipartite graph and a matching I an alternating simple path $\lambda(x_i,x_j)$ is a simple path such that if the edges $\{p_k,p_{k+1}\}$ with k even are in I, the edges with k odd are not in I or vice versa. An alternating simple cycle ρ is an alternating simple path $\lambda(x_i,x_j)$ in which $x_i = x_j$. Given a bipartite graph B and a matching I, B is ρ-acyclic w.r.t. I if it has no alternating cycles w.r.t. I. Given a bipartite graph $B = (S,T,U)$ a dumb bell is a couple of nodes $d = [s,t]$ such that $s \in S$, $t \in T$ and $\{s,t\} \in U$. Given a bipartite graph B and a complete matching I, the fundamental dumb bell set D(I) w.r.t. I is the set of dumb bells individuated by the complete matching I. Given a bipartite graph $B = (S,T,U)$ a set of dumb bells D and a dumb bell $d_i \in D$, $(d_i = [s_i,t_i])$ the set of dumb bells

$$\text{S-adj}(d_i) = \{d \in D,\ d = [s,t] \mid \{s,t_i\} \in U\}$$

is called the S-adjacent to d_i dumb bell set; the set of dumb bells

$$\text{T-adj}(d_i) = \{d \in D,\ d = [s,t] \mid \{s_i,t\} \in U\}$$

is called the T-adjacent to d_i dumb bell set; the S-degree of d_i, $Sd(d_i)$, is the cardinality of the set S-adj(d_i), the T-degree of d_i, $Td(d_i)$ is the cardinality of the set T-adj(d_i)

A set of dumb bells D individuates the following node sets: $S(D) = \{s \in S \mid \exists d \in D,\ d = [\bar{s},t]\}$ $T(D) = \{\bar{t} \in t \mid \exists d \in D,\ d = [s,\bar{t}]\}$, $X(D) = S(D) \cup T(D)$. Let B be a bipartite graph with a complete matching I, then a set of dumb bells F is an essential dumb bell set of B, if the section graph B(Y), where Y = X-X(F) admits a complete matching I and is ρ-acyclic w.r.t. I. An essential dumb bell set of minimum cardinality MF is called minimum dumb bell essential set of B.

3. THE OPTIMUM BORDERED TRIANGULAR FORM AND BIPARTITE GRAPHS

It has to be noted that the "structural information" of a matrix $A \in \mathbb{R}^{n^2}$ can be efficiently coded with a graph. In particular, given a matrix $A = |a_{ij}| \in \mathbb{R}^{n^2}$, a bipartite graph B[A] can be associated to A as follows: B[A] = (S,T,U), with $|S| = |T| = n$, and $\{s_i,t_j\} \in U$ iff (if and only if) $a_{ij} \neq 0$; $i, j = 1,\ldots,n$. Bipartite graphs are particularly suited when a "non symmetric permutation strategy" has to be chosen. In fact, a non symmetric permutation on A, PAQ, has as effect the reordering of the rows and of the columns of A. By definition of B[A] it follows that B[PAQ] is isomorphic to B[A]. Then the structural properties of B[A] are maintained. The optimum tearing problem has a graph theoretic interpretation based on these theorems, which we state without proof due to lack of space.

Theorem 3.1. A matrix $A \in \mathbb{R}^{n^2}$ can be put in inferior triangular form iff B[A] admits a complete matching I and is ρ-acyclic w.r.t. I. ⌑

Theorem 3.2. The determination of an optimum nonsymmetric permutation on $A \in \mathbb{R}^{n^2}$, PAQ, such that PAQ is in an optimum BTF is equivalent to the determination of a minimum dumb bell essential set MF in B[A]. ⌑

It has to be noted that the problem of determining an MF in B[A] is a NP complete problem [7]. In fact, it can be trivially observed that if this problem could be solved in a polynomial bounded time, then the minimum feedback vertex set problem, which is proved to be NP-complete [7], could be solved in a polynomial bounded time. In order to solve NP complete problems, branch and bound techniques are implemented, or suboptimal algorithms are devised. In this paper the second way is followed, due to the simpler and often more efficient performances of this kind of algorithms.

It has to be noted that the request of the existence of a complete matching in B[A] (Y) increases heavily the difficulty of devising an algorithm. On the other hand, this request cannot be removed, being related with the constraint of non zero elements on the main diagonal of A_{11}, which assures the "solvability" of the torm system.

4. THE DUMB BELL ESSENTIAL SET ALGORITHM

As remarked in the previous section, an algorithm for the selection of a minimum dumb bell essential set is not computationally feasible due to the enormous amount of computation time and storage needed. Then, we are looking for a heuristic procedures able to find an essential set F of as low as possible cardinality. In order to devise a heuristic procedure to achieve our goal, recall that if F is an essential dumb bell set, then B[A] (Y), where Y = X-X(F), admits a complete matching and is ρ-acyclic. Now let us consider the following theorem proved in [8].

Theorem 4.1. Let B be a bipartite graph with a complete matching I. B is ρ-acyclic w.r.t. I iff it is possible to number the dumb bell $d \in D(I)$ so that

$$d_j \in S\text{-adj}(d_i) \Rightarrow j > i \qquad (4.1)$$

¤

Then, if we order the dumb bells in B[A](Y) according to Theorem 4.1 and we order the dumb bells in F as the last ones in B[A], the dumb bell d_1 in B[A] has in general small S-degree (often the minimum S degree w.r.t. the other dumb bells) and large T-degree (often the maximum T-degree w.r.t. the other dumb bells) [9]. The same result obviously holds for the graph $B[A](X-X(d_1))$ w.r.t. the dumb bells d_2. Moreover, the dumb-bells in F have large both the S-degree and T-degree.

We use these remarks in order to devise a sequential choice procedure. In fact, the large complexity of covering problems as well as other difficult combinatorial problems arises from the necessity of using back track procedures. Then in order to lower the computational complexity we have to renounce to the back track procedures and devise sub-optimal sequential choices.

In this case, if we name $\bar{F}$ the set of dumb bells not in F, we try to assign dumb bells in $\bar{F}$ with small S-degree and large T-degree, and dumb bells in F with large S-degree and large T-degree.

Selected the choice rule, we have to face the complete matching problem, i.e., we have to assure that the choice done at a certain step will not affect the possibility of finding a complete matching in the remaining steps.

The basic idea is: (1) to begin with a complete matching in B[A]. (2) to consider as candidate elements the dumb bells corresponding to edges which are in I or which can be inserted in a complete matching. (3) if an element correspondent to an edge not in I has been chosen according to the heuristic rules adopted, to modify I in order to insert the new edge.

The set of elements which can be inserted at a certain step of the selection procedure is identified by the following Proposition which can be considered a consequence of a theorem in [10 pg. 123]

Proposition 4.1. Given a bipartite graph B = (S,T,U) and a complete bipartite matching I_1, any other possible bipartite complete matching in B, I_h, can be obtained from I_1 as follows: individuate one or more disjoint simple alternating cycles (e.g., m) ρ^j w.r.t. I_1. Let I_1' be the set of edges in I_1 but not incident in the nodes of the alternating cycles, and $\bar{I}$ the set of edges incident in the vertices of the alternating cycles but not in I_1. Then $I_h = I_1' \cup \bar{I}$. ¤

Suppose now to have a complete matching I in B and direct the edges in B as follows:

$$\{s_i,t_j\} \in U \begin{cases} (t_j,s_i) \in E \text{ iff } \{s_i,t_j\} \in I \\ (s_i,t_j) \text{ otherwise} \end{cases} \tag{4.2}$$

Let $\bar{B}_I$ = (S,T,E) be the obtained directed bipartite graph.

Proposition 4.2. There is a one-to-one correspondence between simple directed cycles in $\bar{B}_I$ and simple alternating cycles in B w.r.t. I. ¤

Now it is possible to prove the fundamental theorem.

<u>Theorem 4.1</u>. Let B = (S,T,U) a bipartite graph and I a complete matching in it. Let $\{s_i, t_j\}$ be an edge not in I. There exists a complete matching I' in B such that $\{s_i, t_j\}$ $\in$ I' iff (s_i, t_j) belongs to a strongly connected component of $\bar{B}_I = (S, T, E)$.

Proof. If part: by definition of strongly connected components there exists a simple path $\mu(t_j, s_i)$ in $\bar{B}_I$. Then $\langle s_i, \mu(t_j, s_i) \rangle$ is a simple cycle in $\bar{B}_I$. By Propositions 4.1 and 4.2, it is possible to obtain a complete matching I' such that $\{s_i, t_j\} \in I'$.

Only if part: If there exists a complete matching I' such that $\{s_i, t_j\} \in I'$, then by Proposition 4.1, there exists an alternating cycle containing $\{s_i, t_j\}$. By Proposition 4.2, there exists in $\bar{B}_I$ a simple directed cycle such that $\exists k$, $p_k = s_i$, $p_{k+1} = t_j$. By definition of strongly connected component, there exists a strongly connected component of $\bar{B}_I$ such that (s_i, t_j) is present in it. ⌑

The algorithm we propose is based on finding the strongly connected components of a directed graph. It has to be noted that there is a partial ordering of the strongly connected components of a digraph $O^* = \langle G_1, \ldots, G_q \rangle$ defined as follows:

$$\forall i = 1, \ldots, q-1, \ \forall j = i+1, \ldots, q, \ \forall x \in X_i, \ \forall y \in X_j, \ (x,y) \notin E.$$

Since we are dealing with bipartite directed graph obtained via (4.2), we need the definition of modified trivial strongly connected component. Let $\bar{B}_t$ be the set of the trivial strongly connected components of $\bar{B}_I$, then the set of the modified trivial connected components is obtained from $\bar{B}_t$ as follows:

$$\bar{B}_{tm} = \{\bar{B}(s_h \cup t_j) \mid s_h \in \bar{B}_t, \ t_j \in \bar{B}_t, \ (t_j, s_h) \in E\}$$

Now, Algorithm DES (Dumb-bell Essential Set) is described. In this Algorithm we use a stack α in which the strongly connected components are stored, and three arrays of dimension n, τ, F and $\bar{F}$. The elements in the essential set F are put in the array F and the elements not in the essential set are put in the array $\bar{F}$. It is assumed that a complete matching in B[A] is given.

ALGORITHM DES

STEP 0. Initialization. Set $\bar{B}^* = \bar{B}_I$. α is empty as well as τ, F and $\bar{F}$.

STEP 1. Find the strongly connected components of $\bar{B}^*$ and put them on the top of the stack α according to the partial ordering O*.

STEP 2. If α is void STOP. Otherwise pickup the first element on the top of the stack α, $B_{\alpha 1}$, delete it from α and set $\bar{B}^* = B_{\alpha 1}$.

STEP 3. Selection rule. a. Pickup the first element t_1^* in the array τ such that it belongs to $\bar{B}^*$. If none of the elements in τ belongs to $\bar{B}^*$ or if τ is void, go to STEP 3b. Choose a dumb bell $d^* = [s^*,t_1^*]$ such that its S-degree is maximum and put it the array F. If some ties occur pick up any of them. Go to STEP 4. b. Find the set of dumb bells in B* with minimum S-degree and between them the one with maximum T-degree. If more than one have the same "maximum" T-degree, we proceed as follows: let $d_1^o,\ldots,d_k^o,\ldots,d_m^o$ be the selected dumb bells. For each d_k^o, consider the dumb bells $d_{k,i}$ which have as S node s_k^o and compute the T-degree of each of them. Add the obtained numbers and associate the sum to $d_k^o \cdot d^* = [s^*,t^*]$ is the dumb bell with the maximum associated number. Put the T nodes adjacent to s* in τ and d* in $\bar{F}$.

STEP 4. If d* is a trivial strongly connected component go to STEP 2.

STEP 5. If the edge corresponding to d* is directed from S to T, find a path μ(t*,s*) and accomplish the reversion of the arcs in the cycle $\rho = \langle s^*, \mu(t^*,s^*) \rangle$ in $\bar{B}^*$ and $\bar{B}_I$. Otherwise continue

STEP 6. Set $\bar{B}^* = \bar{B}^*(X^*-X(d^*))$. Go to STEP 1. ⌑

Theorem 4.3. The set of dumb bells F is a dumb bell essential set of B[A].

Proof. According to the definition of essential set, we prove that (i) $F \cup \bar{F}$ individuate a complete matching B[A] (ii) B[A](X($\bar{F}$)) is ρ-acyclic

(i) By definition of strongly connected component and by STEP 4 and 6, STEP 3 is executed n times. Then $|F \cup \bar{F}| = n$. By STEP 2, 4, 5, 6, by Propositions 4.1, 4.2 and by Theorem 4.2, the set $F \cup \bar{F}$ individuates a complete matching in B[A].

(ii) According to STEP 3a and 3b, if we number the dumb bells put in $\bar{F}$ according to the order in which they are chosen, Theorem 4.1 holds and the Theorem is proved.

The complexity of an implementation of Algorithm DES is discussed now.

It is immediate to observe that STEPS 1, 3 and 5 are dominant in complexity, so we concentrate our complexity analysis on these steps. STEP 1 can be implemented with Tarjan algorithm [11] or Gustavson algorithm [12]. Both of them are O(|X|,|E|) if |X| is the number of nodes of the considered digraph and E is the number of its edge. The data structure used in [11] can be applied in DES as well, while the data structure used in [12] has to be modified with the addition of new arrays. STEP 1 is executed in the worst possible case n-1 times on graphs of decreasing size. For this reason the complexity of STEP 1 is estimated to be $O(n^2, n\ell)$ where ℓ is the number of nonzero elements in A.

STEP 3 consists mainly in the selection of the maximum and/or of the minimum in a list of at most n elements and is executed n times. Therefore its global complexity is $O(n^2)$.

STEP 5 consists mainly in finding a directed path between two vertices. If a depth first search is strategy is used on a directed graph stored as in [11], the complexity of STEPS is $O(|X|, |E|)$ in the worst case this STEP is executed n times. Therefore the overall complexity is $O(n^2, n\ell)$.

It is immediate to conclude that an implementation of algorithm DES has complexity not greater than $O(n^2, n\ell)$.

Remark 1. If we assume that at the beginning a complete matching in B is not provided, Hopcroft and Karp algorithm [13] or Gustavson algorithm [12] can be used.

Remark 3. In order to improve the cardinality of F, we can use a local minimum strategy applying the following Refinement Algorithm.

R Algorithm

STEP 0. Let $F = \{d_1^*, \ldots, d_k^*\}, i = 0$.

STEP 1. $i = i+1$. If $i = k+1$ STOP

STEP 2. If $\bar{B}_I(X - X(F) \cup X(d_i^*))$ is acyclic, then $F = F - d_i^*$, and go to STEP 1. Otherwise go directly to STEP 1.

The obtained F is a minimal dumb bell essential set in the sense that no proper subset of F is a dumb bell essential set.

5. CONCLUDING REMARKS

In this paper, an optimization problem arising from the application of tearing methods has been investigated: the determination of a nonsymmetric permutation on a matrix $A \in \mathbb{R}^{n^2}$, PAQ, such that PAQ is in an optimum Bordered Triangular Form. A graph theoretic interpretation based on bipartite graph has been proposed and, as the problem is NP-complete, a heuristic algorithm has been described. The algorithm has been proved to be correct and its complexity has been estimated to be $O(n^2, n\ell)$ where ℓ is the number of nonzero elements in A.

Further work is needed to determine how far it is the solution given by the heuristic algorithm from the optimum one. Moreover, it is important to devise new heuristic procedures such that more flexibility is allowed introducing the possibility of limited backtracking. The parallelism between the nonsymmetric permutation problem and the symmetric permutation one suggests that some optimal reduction rules can be successfully implemented. These problems will be discussed in a future paper.

REFERENCES

1. G. Kron, Special Issue on G. Kron's Work in J. Franklin Inst., vol. 286, Dec. 1968.
2. R. P. Tewarson, Sparse Matrices, New York: Academic Press, 1973.
3. L. K. Cheung and E. S. Kuh, "The Bordered Triangular Matrix and Minimum Essential Set of a Digraph," IEEE Trans. on Circuits and Systems, vol. CAS-21, pp. 633-639 Sept. 1974.
4. A. Kevorkian and J. Snoek, "Decomposition in Large Scale Systems: Theory and Applications in Solving Large Sets

of Nonlinear Simultaneous Equations," in D. M. Himmemblau Ed., Decomposition of Large Scale Problems. North Holland Publ. Comp., 1973.

5. L. K. Cheung and E. S. Kuh, "Feedback Vertex and Edge Sets of a Digraph and Applications to Sparse Matrices," Memo ERL-M419, University of California, Berkeley, February 1974.

6. F. Harary, Graph Theory, Reading, Mass., Addison-Wesley, 1969.

7. R. M. Karp, Reducibility Among Combinatorial Problems in R. E. Miller and J. W. Thatcher Ed., Complexity of Computer Computations, New York: Plenum, 1972.

8. A. Sangiovanni-Vincentelli, "Bipartite Graphs and Optimal Tearing of a Sparse Matrix," Proc. 1975 Midwest Symposium, Montreal, August 1975.

9. K. Bhat and B. Kinariwala, "Optimum and Suboptimal Methods for Permuting a Matrix into a Bordered Triangular Form," unpublished manuscript.

10. C. Berge, Graphs and Hypergraphs, North-Holland, 1973.

11. R. Tarjan, Depth First Search and Linear-Graph Algorithms, SIAM J. Comp., vol. 1, no. 2, pp. 146-160, 1972.

12. F. Gustavson, "Finding the Block Lower Triangular Form of a Sparse Matrix," in J. Bunch and D. Rose (eds.) Sparse Matrix Computations, Academic Press, 1975.

13. J. E. Hopcroft and R. M. Karp, An $n^{5/2}$ Algorithm for Maximum Matchings in Bipartite Graphs, SIAM J. Compt., vol. 2, 1972.

14. D. Konig, Theorie der Endlichen und Unendlichen Graphen, Chelsea, New York, 1950

II

Eigenvalue Problems

A BIBLIOGRAPHICAL TOUR OF THE LARGE, SPARSE GENERALIZED EIGENVALUE PROBLEM

G. W. Stewart*

To A. S. Householder with thanks for [ho75]

Abstract

This paper surveys the literature on algorithms for solving the generalized eigenvalue problem $Ax = \lambda Bx$, where A and B are real symmetric matrices, B is positive definite, and A and B are large and sparse.

1. Introduction. This paper concerns the problem of computing the nontrivial solutions of the equation

$$Ax = \lambda Bx, \tag{1}$$

where A and B are real symmetric matrices of order n and B is positive definite. Such problems arise in many engineering and scientific applications (e.g. see [lan66, bat73a], where the eigenvalues λ usually represent some kind of vibration. The matrices A and B can be very large, but when they are large they are usually quite sparse. Accordingly, a good deal of effort has been directed toward developing algorithms that take advantage of this sparsity.

Because of limitations of space, this paper is little more than a commentary on the bibliography to follow. The reader is assumed to be familiar with the general literature in numerical linear algebra. For background see [ho64, wi65, st73], and for the Russian point of view see [fa63]. For a comprehensive bibliography of numerical algebra see [ho75]. The bibliography in this paper is by no means complete. One reason for this is that relevant articles may be found scattered throughout the scientific and engineering literature, and the task of tracking them all down is impossibly large. But more important is that there are

* Department of Computer Science, University of Maryland.
This work was supported in part by the Office of Naval Research under Contract No. N00014-67-A-0218-0018.

really very few basic ideas in the subject. It may well be that a definitive method is to be found in a back issue of some obscure journal, but it is my experience that most "new" methods in old journals are variants of the inverse power method, coordinate relaxation, or the method of steepest descent.

2. Background. The algebraic theory of the generalized eigenvalue problem (1) is well understood (see, e.g., [lan66]). The problem has n real eigenvalues $\lambda_1, \lambda_2, \ldots, \lambda_n$, corresponding to a set of linearly independent eigenvectors $x_1, x_2, \ldots, x_n$. The eigenvectors may be chosen to be B-orthonormal; that is $X^T B X = I$, where $X = (x_1, x_2, \ldots, x_n)$. The smallest eigenvalue is the minimum of the Rayleigh quotient

$$\rho(x) = x^T A x / x^T B x, \tag{2}$$

a fact which is the basis for some of the algorithms to be discussed later.

The perturbation theory for this problem is also fairly well understood, but the results are scattered throughout the literature [fal64, fal65, st72, st75a, cra75]. When B is well conditioned with respect to inversion [wi65], the problem (1) behaves very much like a symmetric eigenvalue problem and presents few theoretical or numerical difficulties. When B is ill conditioned, there usually appear a number of very large eigenvalues which vary sharply with small changes in A and B. In addition to these so-called infinite eigenvalues there will be smaller eigenvalues which are insensitive to perturbations in A and B. The separation of these eigenvalues from the large ones is a difficult numerical problem.

A second pathology occurs when A and B have a common null vector x. In this case it is seen from (1) that any number λ is an eigenvalue. When A and B are near matrices with a common null vector, the problem has a very ill-conditioned eigenvalue which need not be large. This phenomenon is discussed more fully in [st72] and [st75a].

One source of this pathology is the widely used Rayleigh Ritz method. Described in matrix terms, one starts with a set of vectors $Y = (y_1, y_2, \ldots, y_k)$ and forms the matrices

$\tilde{A} = Y^TAY$ and $\tilde{B} = Y^TBY$. If Z is the matrix of eigenvectors of the eigenvalue problem $\tilde{A}z = \mu\tilde{B}z$, then the columns of YZ may provide good approximate eigenvectors. For example, if the column space of Y is the space spanned by the eigenvectors $x_1, x_2, \ldots, x_k$, then the columns of YZ will be eigenvectors corresponding to $\lambda_1, \lambda_2, \ldots, \lambda_k$. This makes Rayleigh-Ritz refinement an especially good technique for accelerating methods that hunt for a subspace corresponding to a set of eigenvalues. However, when the columns of Y are nearly dependent, the matrices $\tilde{A}$ and $\tilde{B}$ will have a common approximate null vector and hence an ill conditioned eigenvalue. Since many generalized matrix eigenvalue problems originate from the Rayleigh-Ritz method applied to continuous operators, it is not surprising that the pathology mentioned above occurs with some frequency. I know of no good study of just what can be computed in this case.

3. <u>The Dense Generalized Eigenvalue Problem</u>. Because many of the numerical techniques for the dense problem can be generalized to the sparse problem and because the Rayleigh-Ritz method requires the solution of a dense problem, we shall begin our survey of numerical methods with a brief review of techniques for solving the dense problem. Further information may be found in [ku64, wi65, pet70, wi71] and in [pet74], which unfortunately exists only as a manuscript.

An important method for solving (1) is to reduce it to an ordinary eigenvalue problem, which can then be solved by standard techniques [wi65, wi71, sm74]. This can be accomplished in several ways. First, from (1) is follows that

$$B^{-1}Ax = \lambda x, \tag{3}$$

so that the eigenvalues and eigenvectors of $B^{-1}A$ are the same as those of (1). However, this technique destroys the symmetry of the problem. Both symmetry and work can be saved by computing the Cholesky factorization of B., i.e. by finding a lower triangular L such that $B = LL^T$ [ma65]. Then with the transformation

$$y = L^Tx \tag{4}$$

equation (1) becomes

$$L^{-1}AL^{-T}y = \lambda y, \tag{5}$$

so that y and λ are an eigenvector and eigenvalue of the symmetric matrix $L^{-1}AL^{-T}$. The eigenvectors of the original problem can be recovered from (4). Programs for accomplishing this reduction may be found in [ma68] and [sm74].

When B is well conditioned, the reduction (5) is an excellent way of solving the generalized eigenvalue problem. When B is ill conditioned, however, either $B^{-1}A$ or $L^{-1}AL^{-T}$ will be large, and the small, well-conditioned eigenvalues of the problem will be computed inaccurately. This problem can be mitigated by pivoting in the Cholesky reduction of B, in which case the matrix $L^{-1}AL^{-T}$ will have a graded structure that often permits the successful calculation of the smaller eigenvalues. Better yet, one can diagonalize B by an orthogonal similarity, $Q^TBQ = D$, with the diagonal elements in ascending or descending order, and then form the graded matrix $D^{-1/2}Q^TAQD^{-1/2}$. However, these methods are not foolproof [pet74].

Another approach is to attempt to remove the singular part of B before performing an inversion. Algorithms of this kind have been described in [pet70], [fi72], and [pet74]. The problem with this approach is that it cannot handle the case where B is ill conditioned enough to effect the computations but not so ill conditioned that its singular part can be ignored.

The QZ algorithm [mo73] works directly with A and B without an inversion and is not affected by ill-conditioning in B. But it destroys symmetry in the process.

When only a few eigenvalues are required, they can sometimes be found by the method of Sturm sequences, of which more in the next section.

An eigenvector corresponding to the approximate eigenvalue λ may be calculated by the inverse power method [wi65, pet70]. Here, starting with some suitable vector x_0, one iterates according to the formula

$$(A-\lambda B)x_{i+1} = Bx_i. \tag{6}$$

If λ is close to an eigenvalue, only one or two iterations are required to produce a good approximate eigenvector. Advantage may be taken of the symmetry of $A-\lambda B$ by using special programs designed for symmetric indefinite systems [par70, bu71, aa71]. Although the system (6) is quite ill conditioned, no special precautions are needed [wi65]. However, some care must be taken to obtain linearly independent eigenvectors corresponding to tight clusters of eigenvalues (cf. [pet71] where programs are given for the ordinary eigenvalue problem).

4. The Dense Banded Problem. In this section we shall be concerned with banded matrices whose bandwidths are small enough to allow the computation of many LU decompositions. It should be noted that many problems from structural mechanics are naturally banded, and with appropriate massaging their bandwidths can be reduced to the point where the methods of this section can be applied. For a survey of bandwidth reduction techniques see [cut73].

When only a few eigenvalues are required, the technique of choice is the method of Sturm sequences [gi54, ba67, pet69, gu72, bat73b, gu74]. It is based on the following observation. Let A_k and B_k denote the leading principal submatrices of order k of A and B, and let $p_k(\lambda) = \det(A_k - \lambda B_k)$. Then the number of sign agreements in consecutive terms of the Sturm sequence $p_0(\lambda), p_1(\lambda), \ldots, p_n(\lambda)$ is equal to the number of eigenvalues greater than or equal to λ. By evaluating the Sturm sequence for different values of λ, one can locate any eigenvalue in an interval that can be made as small as rounding error will permit.

The usual method of evaluating the Sturm sequence is a variant of Gaussian elimination with pivoting (programs may be found in [ma67]). Unfortunately, the pivoting destroys the symmetry of the problem and increases the storage requirements. It has been observed in practice that elimination without pivoting usually works, and in [pet74] some reasons for this phenomenon are adduced. But special precautions must be taken to make the technique without pivoting safe.

The process of finding an eigenvalue can be speeded up in several ways. In the first place the information accumulated in finding one eigenvalue can be used to help locate the others [pet69, ba67]. Second, since the eigenvalues are the zeros of $p_n(\lambda)$, which must be computed as part of the Sturm sequence, it is natural to use a zero finding scheme to accelerate the location of the eigenvalue. An elegant combination of bisection and the secant method is described in [pet69]. Finally, once a moderately accurate approximation has been found, one can shift to the inverse power method [bat73b, gu74].

When all of the eigenvalues of the problem are required, it may be more economical to reduce the problem to a banded, ordinary eigenvalue problem. This can be accomplished by the algorithm in [cra73], which transforms $L^{-1}AL^{-T}$ as it is formed so that its band structure is preserved. The resulting banded eigenvalue problem can be further reduced as in [rut63, sc63].

Eigenvectors for selected eigenvalues can be found using the inverse power method. Because A and B are banded, the calculations can be reduced considerably [ma67].

5. <u>The Problem when an LU Factorization is Possible</u>. In this section we shall consider problems where it is possible to factor A, B, or $A-\lambda B$ into the product LU of a lower and an upper triangular matrix, or equivalently to solve linear systems involving A, B, or $A-\lambda B$. As in Section 3, the approach is to reduce the generalized eigenvalue problem to an ordinary eigenvalue problem. However, this cannot be done explicitly, since the matrices in (3) and (5) will not in general be sparse. Thus we are left with two questions: how does one calculate the LU decomposition of a large sparse matrix, and how does one solve the resulting eigenvalue problem?

The decomposition of large sparse matrices has been the main topic of three symposia on sparse matrices [wil68, re71, ro72], and their proceedings give extensive bibliographies on the subject. Here we only summarize some of the more important points. First the matrix in question must be represented in some way that dispenses with the zero elements yet allows new elements to be inserted as they are generated by the elimination process that produces the decomposition [kn68, cur71, cut73, rh74]. During the elimination process, pivoting must be performed for two reasons: to preserve sparsity and to insure numerical stability. Since these goals are often at odds, some compromise must be made [cur71, er74, du74]. The elimination process gives complete information on how new nonzero elements fill in, and this information can be used to advantage when the LU decomposition of a different matrix having the same sparsity structure must be computed [er72,gus72]. For example time may be saved in decomposing $A = \lambda B$ for different values of λ. However, this approach precludes pivoting for stability. Whether some of the techniques for symmetric indefinite systems mentioned in Section 3 can be adapted to the large, sparse, banded problems that are typically encountered here is an open question. FORTRAN programs for decomposing sparse matrices are given in [re72, rh73].

The solution of the ordinary eigenvalue problems to which the generalized eigenvalue problem is reduced cannot be accomplished directly, since (say) the matrix $L^{-1}AL^{-T}$ of (5) will not be sparse. However, the product $y = L^{-1}AL^{-T}x$ can be formed for any vector x by solving $L^Tu = x$, computing $v = Au$, and solving $Ly = v$. Thus the problem becomes one of finding the eigenvectors of a matrix C when one can only form the product Cx.

Surveys of the large eigenvalue problem may be found in [ru72a] and [st74], and some of the techniques discussed in [fa63] and [wi65] are also applicable. Most of these methods find only a few of the eigenvalues and corresponding eigenvectors; however, in most applications that is all that is required. The two classes of methods that are of greatest interest here are methods of simultaneous iteration and Lanczos methods.

Methods of simultaneous iteration [bau57, la59, le65, dau68, rut69, st69, cl70, rut70, cl71, je71, bat72, je73, je75,st75b] are variants of the power method in which one works with several vectors simultaneously. Specifically, starting with an $n \times r$ matrix Q_0, one generates a sequence Q_0, Q_1, Q_2 according to the formula $Q_{i+1} = CQ_iS_i$. Here S_i is chosen to keep the columns of Q_{i+1} independent (e.g. when C is symmetric, S_i is usually taken to be the upper triangular matrix that makes the columns of Q_{i+1} orthonormal). Under rather general hypotheses the column space of Q_i will contain an approximation to the eigenvector of C corresponding to the j-th largest eigenvalue ($j \leq r$). This approximation is accurate to terms of order $|\lambda_{r+1}/\lambda_j|^i$ (here $|\lambda_1| \geq |\lambda_2| \geq \ldots \geq |\lambda_n|$), and it can be recovered by a Rayleigh-Ritz refinement [rut69, st69, je75, st75a]. In the symmetric case, convergence to the dominant eigenvector can be considerably speeded up by using Chebyshev polynomials [rut69]. Programs based on simultaneous iteration are tricky to construct, and the reader is advised to consult the literature for details (especially [rut70]).

The Lanczos algorithm [lanc50] for symmetric matrices requires only the ability to form the product Cx in order to construct a sequence of tridiagonal matrices $T_1, T_2, \ldots, T_n$

such that T_n is similar to C. Although early work on the algorithm regarded it as an exact method for solving the eigenvalue problem [fa63, wi65], it is now realized that at a very early stage the matrices T_i contain accurate approximations to the largest and smallest eigenvalues of C, thus making it a powerful iterative method [pa72]. Recently block variants of the method have been proposed [cu74, un75]. The reader is warned that in its original form the method is numerically unstable. But methods for controlling these instabilities have been developed [pa70, ka74], and I suspect that Lanczos algorithms will become the preferred way of computing the extreme eigenvalues of symmetric matrices.

In the dense eigenvalue problem the inverse power method is ordinarily used to compute an eigenvector corresponding to an accurate approximation to an eigenvalue. In the large, sparse problem, on the other hand, the inverse power method is more frequently used to make visible otherwise inaccessible eigenvalues. Specifically, the eigenvalues of the matrix $(B^{-1}A-\lambda I)^{-1} = (A-\lambda B)^{-1}B$ are $(\lambda_i-\lambda)^{-1}$ $(i=1,2,\ldots,n)$ so that the eigenvectors corresponding to eigenvalues near λ are the dominant ones of $(A-\lambda B)^{-1}B$. This means that the inverse power method combined with simultaneous iteration can be used to investigate the eigenvalues lying in an interval centered about λ. The matrix $(A-\lambda B)^{-1}B$ is not symmetric; however its eigenvalues are real, so that many of the techniques in [rut70] can be used [mc75]. An interesting variant [jen72] computes vectors one at a time, shifting occasionally to improve convergence.

For the Lanczos algorithm one requires symmetry. In the inverse power method this can be obtained by working with the matrix $L^T(A-\lambda B)^{-1}L$, but at the cost of computing two decompositions, one of B and one of $A - \lambda B$.

6. <u>The Problem when No LU Factorization is Possible.</u> When the matrices A and B are so large that it is no longer possible to compute an LU factorization, it is not possible to construct a linear operator whose eigenvectors are simply related to those of the original problem. The approach usually taken in this case is to construct a functional $\varphi(u)$ that attains a minimum at one or more of the eigenvectors, and then to minimize φ numerically. This is usually done by successive searches; given u_k, a direction p_k is chosen and α_k is determined so that φ is smaller at $u_{k+1} = u_k + \alpha_k p_k$ than at u_k.

A widely used class of methods results from trying to minimize the Rayleigh quotient (2) by cyclically adjusting one coordinate at a time, which amounts to taking for the directions p_k the unit vectors $e_1, e_2, \ldots, e_n, e_1, e_2, \ldots$. The method is fairly old [pe40, co48, cr51, fa63], although a paper by Nesbet seems to have sparked the current interest in the subject [ne65, ka66, sh70, be70, ni72, fa173, sh73, ru74, sc74a, sc74b]. It is easy to calculate α_k so that u_{k+1} is minimized along p_k, and in this form the algorithm is sometimes called coordinate relaxation [fa63]. However, this is asymptotically equivalent to adjusting cyclically in i the i-th coordinate of u_k so that u_{k+1} satisfies the equation $a_i^T u_{k+1} = \rho(u_k) b_i^T u_{k+1}$, where a_i^T and b_i^T are the i-th rows of A and B. In the early stages coordinate relaxation reduces the Rayleigh quotient somewhat faster, but it requires more work [ka65, sh73].

Coordinate relaxation has been analyzed in a number of places [fa63, ka65, ru74, sc74b]. Since the Rayleigh quotients $\rho(u_1), \rho(u_2), \ldots$ form a decreasing sequence that is bounded below, they must approach a limit, which can be shown to be an eigenvalue. If $\rho(u_0)$ is less than the second smallest eigenvalue λ_2, then the $\rho(u_k)$ must approach the smallest eigenvalue λ_1, and the u_k will converge in direction to the corresponding eigenvector x_1. Even when $\rho(u_0) > \lambda_2$, experience indicates that convergence to x_1 will usually be obtained, although an example in [sc74b] suggests that in some cases the iteration can appear to be converging to x_2. The method can be overrelaxed in the usual way, and much of the theory of relaxation methods for linear systems [var62, yo71] goes through for the method of coordinate relaxation [ka65, ke65, ru74, sc74b].

Since the larger eigenvalues are not minima of the Rayleigh quotient, a deflation procedure must be employed in order to use the method of coordinate relaxation to compute the larger eigenvalues and their eigenvectors [fa173, sh73, sc74a]. Methods of orthogonalization or projection do not seem to work very well, since they destroy the sparsity of the correcting vectors e_i. The usual deflation is to work with the matrix $A_\ell = A + \sigma_1 B x_1 x_1^T B + \ldots + \sigma_\ell B x_\ell x_\ell^T B$, which shifts the eigenvalues $\lambda_1, \lambda_2, \ldots, \lambda_\ell$ to $\lambda_1+\sigma_1, \ldots, \lambda_\ell+\sigma_\ell$. With appropriate choice of the σ_i these new eigenvalues can be made larger than $\lambda_{\ell+1}$. Of course one actually uses the previously computed approximations to $x_1, \ldots, x_\ell$ in forming A_ℓ. Deflation processes have been extensively analyzed for the ordinary eigenvalue problem (see, e.g., [wi65]), and it is known that an accurate eigenvector is not always necessary. The main objection to deflation by shifting is the ad hoc nature of the shifts σ_i.

Various forms of gradient methods to minimize the Rayleigh quotient have been proposed. One step methods, such as the method of steepest descent and its variants (e.g. [he51]) do not seem to decrease the Rayleigh quotient enough to justify the expense of updating the entire vector. However, the method of conjugate gradients [he52,f164] has been applied to the generalized eigenvalue problem with considerable success [br66, fo68, ge71, fr72a] (the following summary is based on [ru72a]). In this method the first search direction is taken to be the negative of the gradient of $\rho(x_0)$: $p_0 = -g_0 \equiv -2(Ax_0-\rho(x_0)Bx_0)/x_0Bx_0$. At the k-th step, α_k is chosen to minimize $\rho(u_k+\alpha p_k)$, and p_{k+1} is chosen in the form $p_{k+1} = -g_{k+1} + \beta_k p_k$ in such a way that it would be conjugate to the previous directions if the Rayleigh quotient were simply a quadratic form. There are several ways of choosing β_k: $\beta_k = g_{k+1}^T g_{k+1}/g_k^T-g_k$ [f164], $\beta_k = g_{k+1}(g_{k+1}-g_k)/g_k^T g_k$ [po69], or β_k can be chosen so that p_{k+1} is conjugate to p_k with respect to the second derivative of ρ at u_{k+1} [da67, ge71]. Since $\rho(x)$ is homogenous in x, its second derivative is indefinite at the minimum, and it is natural to constrain the iteration by fixing a single variable [br66, fo68]; however, experience indicates that this is unnecessary or even deleterious [ge71], although without some form of constraint the usual convergence theory [co72, kam72] does not apply.

Limited experiments reported in [ru74] indicate that the conjugate gradient method may be better than the method of coordinate relaxation. Moreover, a natural projection strategy allows the method to get at the larger eigenvalues [ge71]. For these reasons, conjugate gradient methods should be regarded as an important area for further research.

There are a number of methods [al61, wa63, bl67, mc72, ro73] that attempt to minimize a functional, such as $\varphi(x) = \|Ax-\rho(x)Bx\|_2^2$, that is minimized by any eigenvector. This permits one to get directly at the interior eigenvalues of the problem; however, the functional φ above is more difficult to minimize than the Rayleigh quotient (in fact it is the square of the norm of the gradient of ρ), and I suspect the same is true of the other functionals. An interesting approach to calculating interior eigenvalues is to solve the equations of the inverse power method by the method of conjugate gradients [ru72b]. In this connection the conjugate gradient algorithm in [pa75], which can be applied to indefinite systems, may be useful.

References

[aa71] Aasen, J.O., On the reduction of a symmetric matrix to tridiagonal form, BIT 11 (1971), 233-242.

[al61] Altman, N., An iterative method for eigenvalues and eigenvectors of matrices, Bull. Acad. Polon. Sci. Ser. Sci. Math. Astronom. Phys. 9 (1961), 639-644, 727-732, 751-755.

[ba67] Barth, W., Martin, R.S., and Wilkinson, J.H., Calculation of the eigenvalues of a symmetric tridiagonal matrix by the method of bisection, Numer. Math. 9 (1967), 386-393, also in [wi71] II/5.

[bat72] Bathe, K.-J., and Wilson, E.L., Large eigenvalue problems in dynamic analysis, ASCE J., Eng. Mech. Div. 98 (1972), 1471-1485.

[bat73a] Bathe, K.-J., and Wilson, E.L., Solution methods for eigenvalue problems in structural mechanics, Int. J. Numer. Meth. Eng. 6 (1973), 213-226.

[bat73b] Bathe, K.-J., and Wilson, E.L., Eigensolution of large structural systems with small bandwidth, ASCE J., Eng. Mech. Div. 99 (1973), 467-480.

[bau57] Bauer, F.L., Das Verfahren der Treppeniteration und verwandte Verfahren zur Lösung algebraischer Eigenwertprobleme, Z. Angew. Math. Phys. 8 (1957), 214-235.

[be70] Bender, C.F., and Skavitt, I., An iterative procedure for the calculation of the lowest real eigenvalue and eigenvector of a nonsymmetric matrix, J. Comput. Phys. 6 (1970), 146-149.

[bl67] Blum, E.K., The computation of eigenvalues and eigenvectors of a completely continuous self-adjoint operator, J. Comput. Syst. Sci. 1 (1967), 362-370.

[br66] Bradbury, W.W., and Fletcher, R., New iterative methods for solution of the eigenproblem, Numer. Math. 9 (1966), 259-267.

[bu71] Bunch, J.R., and Parlett, B.N., Direct methods for solving symmetric indefinite systems of linear equations SIAM J. Numer. Anal. 8 (1971), 639-655.

[cl70] Clint, M., and Jennings. A., The evaluation of eigenvalues and eigenvectors of real symmetric matrices by simultaneous iteration, Comput. J. 13 (1970), 68-80.

[cl71] Clint, M., and Jennings, A., A simultaneous iteration method for the unsymmetric eigenvalue problem, J. Inst. Math. Appl. 8 (1971), 111-121.
[co72] Cohen, A. F., Rate of convergence of several conjugate gradient algorithms, SIAM J. Numer. Anal. 9 (1972) 248-259.
[coo48] Cooper, J.L.B., The solution of natural frequency equations by relaxation methods, Quart. Appl. Math. 6 (1948), 179-183.
[cr51] Crandall, S.H., On a relaxation method for eigenvalue problems, J. Math. Physics 30 (1951), 140-145.
[cra73] Crawford, C.R., Reduction of a band-symmetric generalized eigenvalue problem, Comm. ACM 16 (1973), 41-44.
[cra75] Crawford, C.R., A stable generalized eigenvalue problem, Erindale College, University of Toronto, Manuscript (1975), to appear SIAM J. Numer. Anal.
[cu74] Cullum, J., and Donath, W.E., A block generalization of the symmetric S-step Lanczos algorithm, IBM Thomas J. Watson Research Center, RC 4845 (#21570) (1974).
[cur71] Curtis, A.R., and Reid, J.K., The solution of large sparse unsymmetric systems of equations, J. Inst. Math. Appl. 8 (1971), 344-353.
[cut73] Cuthill, E., Several strategies for reducing the bandwidth of matrices, in [ros72], 157-166.

[da67] Daniel, J.W., Convergence of the conjugate gradient method with computationally convenient modifications, Numer. Math. 10 (1967), 125-131.
[dau68] Daugavet, V.A., Variant of the stepped exponential method of finding some of the first characteristic values of a symmetric matrix, USSR Comput. Math. Math. Phys. 8 (1968), 212-223.
[du74] Duff, I.S., and Reid, J.K., A comparison of sparsity ordering for obtaining a pivotal sequence in Gaussian elimination, J. Inst. Math. Appl. 4 (1974), 281-291.

[er72] Erisman, A.M., Sparse matrix approach to the frequency domain analysis of linear passive electrical networks, in [ros72], 31-40.
[er74] Erisman, A.M., and Reid, J.K., Monitoring stability of the triangular factorization of a sparse matrix, Numer. Math. 22 (1974), 183-186.

[fa63] Faddeev, D.K., and Faddeeva, V.N., Computational Methods of Linear Algebra, W.H. Freeman and Co., San Francisco (1963).

[fal64] Falk, S., Einschliessungssätze für die Eigenwerte normaler Matrizenpaare, Z. Angew. Math. Mech. 44 (1964), 41-55.

[fal65] Falk, S., Einschliessungssätze für die Eigenvektoren normaler Matrizenpaare, Z. Angew. Math. Mech. 45 (1965), 47-56.

[fal73] Falk, S., Berechnung von Eigenwerten und Eigenvektoren normaler Matrizenpaare durch Ritz-Iteration, Z. Angew. Math. Mech. 53 (1973), 73-91.

[fe74] Feler, M.G., Calculation of eigenvectors of large matrices, J. Comput. Phys. 14 (1974), 341-349.

[fl64] Fletcher, R., and Reeves, C.M., Function minimization by conjugate gradients, Comput. J. 2 (1964), 149-153.

[fi72] Fix, G., and Heiberger, R., An algorithm for the ill-conditioned eigenvalue problem, SIAM J. Numer. Anal. 9 (1972), 78-88.

[fo68] Fox, R.L., and Kapoor, M.P., A minimization method for the solution of the eigenproblem arising in structural dynamics, Proc. 2nd Conf. Matrix Meth. Struct. Mech., Wright-Patterson AFB, Ohio (1968).

[fr72a] Fried, I., Optimal gradient minimization scheme for finite element eigenproblems, J. Sound Vib. 20 (1972), 333-342.

[ge71] Geradin, M., The computational efficiency of a new minimization algorithm for eigenvalue analysis, J. Sound Vib. 19 (1971), 319-331.

[gi54] Givens, J.W., Numerical computation of the characteristic values of a real symmetric matrix, Oak Ridge National Laboratory, ORNL-1574 (1954).

[gu72] Gupta, K.K., Solution of eigenvalue problems by Sturm sequence method, Int. J. Numer. Meth. Eng. 4 (1972), 379-404.

[gu74] Gupta, K.K., Eigenproblem solution by a combined Sturm sequence and inverse iteration technique, Int. J. Numer. Meth. Eng. 8 (1974), 877-911.

[gus72] Gustavson, F.G., Some basic techniques for solving sparse systems of linear equations, in [ros72], 41-52.

[he51] Hestenes, M.R., and Karush, W., Solutions of $Ax = \lambda Bx$, J. Res. Nat. Bur. Standards 47 (1951), 471-478.

[he52] Hestenes, M.R., and Stiefel, E., Methods of conjugate gradients for solving linear systems, J. Res. Nat. Bur. Standards 49 (1952), 409-436.

[ho64] Householder, A.S., The Theory of Matrices in Numerical Analysis, Blaisdell, New York (1964).

[ho75] Householder, A.S., KWIC Index for Numerical Algebra, Oak Ridge National Laboratory ORNL-4778 (1975).

[je71] Jennings, A., and Orr, D.R.L., Application of the simultaneous iteration method to undamped vibration problems, Int. J. Numer. Meth. Eng. 3 (1971), 13-24.

[je73] Jennings, A., Mass condensation and simultaneous iteration for vibration problems, Int. J. Numer. Meth. Eng. 6 (1973), 543-552.

[je75] Jennings, A., and Stewart, W.J., Simultaneous iteration for partial eigensolution of real matrices, J. Inst. Math. Appl. 15 (1975), 351-362.

[jen72] Jensen, Paul S., The solution of large symmetric eigenproblems by sectioning, SIAM J. Numer. Anal. 9 (1972), 534-545.

[ka66] Kahan, W., Relaxation methods for an eigenproblem, Stanford University, Dept. Scomp. Sci., TEch. Rep. CS-44 (1966).

[ka74] Kahan, W., and Parlett, B.N., An analysis of Lanczos algorithms for symmetric matrices, Electronics Research Laboratory, University of California, Berkeley, Memorandum No. ERL-M467 (1974).

[kam72] Kammerer, W.J., and Nashed, M.Z., On the convergence of the conjugate gradient method for singular linear operator equations, SIAM J. Numer. Anal. 9 (1972), 165-181.

[ke65] Keller, H.B., On the solution of singular and semidefinite linear systems by iteration, J. SIAM Ser. B 2 (1965), 281-290.

[kn68] Knuth, D.E., The Art of Computer Programming, V1, Fundamental Algorithms, Addison-Wesley, Reading, Mass. (1968).

[ku64] Kublanovskaja, V.N., and Faddeeva, V.N., Computational methods for the solution of a generalized eigenvalue problem, Amer. Math. Soc. Transl., Ser. 2, 40 (1964), 271-290.

[la59] Laasonen, P., A Ritz method for simultaneous determination of several eigenvalues and eigenvectors of a big matrix, Ann. Acad. Sci. Fenn., Ser. A.I. 265 (1959).

[lan66] Lancaster, P., Lambda-matrices and Vibrating Systems, Pergamon, Oxford (1966).

[lanc50] Lanczos, C., An iteration method for the solution of the eigenvalue problem of linear differential and integral operators, J. Res. Nat. Bur. Standards 45 (1950), 255-281.

[le65] Levin, A.A., On a method for the solution of a partial eigenvalue problem, USSR Comput. Math. and Math. Phys. 5 (1965), 206-212.

[ma65] Martin, R.S., Peters, G., and Wilkinson, J.H., Symmetric decomposition of a positive definite matrix, Numer. Math. 7 (1965), 362-383; also in [wi71] I/4.

[ma67] Martin, R.S., and Wilkinson, J.H., Solution of symmetric and unsymmetric band equations and the calculation of eigenvectors of bank matrices, Numer. Math. 9 (1967), 279-301; also in [wi71], I/6.

[ma68] Martin, R.S., and Wilkinson, J.H., Reduction of the symmetric eigenproblem $Ax = \lambda Bx$ and related problems to standard form, Numer. Math. 11 (1968), 99-110; also in [wi71], II/10.

[mc72] McCormick, S.F., A general approach to one-step iterative methods with application to eigenvalue problems, J. Comput. Sys. Sci. 6 (1972), 354-372.

[mc75] McCormick, S.F., and Noe, T., Simultaneous iteration for the matrix eigenvalue problem, Colorado State University, Department of Mathematics, Manuscript (1975).

[mo73] Moler, C.B., and Stewart, G.W., An algorithm for generalized matrix eigenvalue problems, SIAM J. Numer. Anal. 10 (1973), 241-256.

[ne65] Nesbet, R.K., Algorithm for diagonalization of large matrices, J. Chem. Phys. 43 (1965), 311-312.

[ni72] Nisbet, R.M., Acceleration of the convergence of Nesbet's algorithm for eigenvalues and eigenvectors of large matrices, J. Comput. Phys. 10 (1972), 614-619.

[pa70] Paige, C.C., Practical use of the symmetric Lanczos process with reorthogonalization, BIT 10 (1970), 183-195.

[pa72] Paige, C.C., Computational variants of the Lanczos method for the eigenproblem, J. Inst. Math. Appl. 10 (1972), 373-381.

[pa75] Paige, C.C., and Saunders, M.A., Solution of sparse indefinite systems of linear equations, SIAM J. Numer. Anal. 12 (1975), 617-629.

[par70] Parlett, B.N., and Reid, J.K., On the solution of a system of linear equations whose matrix is symmetric but not definite, BIT 10 (1970), 386-397.

[pe40] Pellew, A., and Southwell, R.U., Relaxation methods applied to engineering problems VI. The natural frequencies of systems having restricted freedom, Proc. Roy. Soc. (A) 175 (1940), 262-290.

[pet69] Peters, G., and Wilkinson, J.H., Eigenvalues of $Ax = \lambda Bx$ with band symmetric A and B, Comp. J. 12 (1969), 398-404.

[pet70] Peters, G., and Wilkinson, J.H., $Ax = \lambda Bx$ and the generalized eigenproblem, SIAM J. Numer. Anal. 7 (1970), 479-492.

[pet71] Peters, G., and Wilkinson, J.H., The calculation of specified eigenvectors by inverse iteration, in [wi71] II/18.

[pet74] Peters, G., and Wilkinson, J.H., Some algorithms for the solution of the generalized symmetric eigenvalue problem $Au = \lambda Bu$, Manuscript (1974).

[po69] Polyak, B.T., The conjugate gradient method in extremal problems, USSR Comp. Math. Math. Phys. 7 (1969), 94-112.

[re71] Reid, J.K. (ed.), Large Sparse Sets of Linear Equations, Academic Press, New York (1971).

[re72] Reid, J.K., FORTRAN subroutines for the solution of sparse systems of nonlinear equations, United Kingdom Atomic Energy Authority, Atomic Energy Research Establishment, AERE-R 7293 (1972).

[rh73] Rheinboldt, W.C., and Mesztenyi, C.K., Programs for the solution of large sparse matrix problems based on the arc-graph structure, University of Maryland, Computer Science Technical Report TR-262 (1973).

[rh74] Rheinboldt, W.C., and Mesztenyi, C.K., Arc graphs and their possible application to sparse matrix problems, BIT 14 (1974), 227-239.

[ro73] Rodrigue, G., A gradient method for the matrix eigenvalue problem $Ax = \lambda Bx$, Numer. Math. 22 (1973), 1-16.

[ros72] Rose, D.J., and Willoughby, R.A. (eds.), Sparse Matrices and Their Applications, Plenum Press, New York (1972).

[ru72a] Ruhe, A., Iterative eigenvalue algorithms for large symmetric matrices, University of Umeå, Department of Information Processing, Report UMINF 31.72 (1972).

[ru72b] Ruhe, A., and Wiberg, Torbjörn, The method of conjugate gradients used in inverse iteration, BIT 12 (1972) 543-554.

[ru74] Ruhe, A., SOR methods for the eigenvalue problem with large sparse matrices, Math. Comp. 28 (1974), 695-710.

[rut63] Rutishauser, H., On Jacobi rotation patterns, Proc. AMS Symposium in Applied Mathematics 15 (1963), 219-239.
[rut69] Rutishauser, H., Computational aspects of F. L. Bauer's simultaneous iteration method, Numer. Math. 13 (1969), 4-13.
[rut70] Rutishauser, H., Simultaneous iteration method for symmetric matrices, Numer. Math. 16 (1970), 205-223; also in [wi71] II/9.

[sc63] Schwarz, H.R., Tridiagonalization of a symmetric band matrix, Numer. Math. 12 (1963), 231-241; also in [wi71] II/18.
[sc74a] Schwarz, H.R., The eigenvalue problem $(A-\lambda B)x=0$ for symmetric matrices of high order, Comput. Meth. Appl. Mech. Eng. 3 (1974), 11-28.
[sc74b] Schwarz, H.R., The method of coordinate overrelaxation for $(A-\lambda B)x=0$, Numer. Math. 23 (1974), 135-151.
[sh70] Shavitt, I., Modification of Nesbet's algorithm for the iterative evaluation of eigenvalues and eigenvectors of large matrices, J. Comput. Phys. 6 (1970), 124-130.
[sh73] Shavitt, I., Bender, C.F., Pipano, A., Hosteny, R.P., The iterative calculation of several of the lowest or highest eigenvalue and corresponding eigenvectors of very large symmetric matrices, J. Comput. Phys. 11 (1973), 90-108.
[sm74] Smith, B.T., Boyle, J.M., Garbow, B.S., Ikebe, Y., Klema, V.C., and Moler, C.B., Matrix Eigensystem Routines --EISPACK Guide, Lecture Notes in Computer Science, V. 6, Springer, New York (1974).
[st69] Stewart, G.W., Accelerating the orthogonal iteration for the eigenvalues of a Hermitian matrix, Numer. Math. 13 (1969), 362-376.
[st72] Stewart, G.W., On the sensitivity of the eigenvalue problem $Ax = \lambda Bx$, SIAM J. Numer. Anal. 4 (1972), 669-686.
[st73] Stewart, G.W., Introduction to Matrix Computations, Academic Press, New York (1973).
[st74] Stewart, G.W., The numerical treatment of large eigenvalue problems, in Proc. IFIP Congress 74, North Holland Publishing Co. (1974), 666-672.
[st75a] Stewart, G.W., Gerschgorin theory for the generalized eigenvalue problem $Ax = \lambda Bx$, Math. Comp. 29 (1975), 600-606.
[st75b] Stewart, G.W., Simultaneous iteration for computing invariant subspaces of non-Hermitian matrices, to appear Numer. Math.

[un75] Underwood, R., An iterative block Lanczos method for the solution of large sparse symmetric eigenproblems, Stanford University, Department of Computer Science, CS 496 (1975).

[va71] Vandergraft, J.S., Generalized Rayleigh methods with applications to finding eigenvalues of large matrices, Lin. Alg. Appl. 4 (1971), 353-368.

[var62] Varga, R.S., Matrix Iterative Analysis, Prentice-Hall, Englewood Cliffs, New Jersey (1962).

[wa63] Wang, Jin-ru, A gradient method for finding the eigenvalues and eigenvectors of a self-adjoint operator, Acta Math. Sinica 13 (1963), 23-28; Chinese Math. 4 (1963), 24-30.

[wi65] Wilkinson, J.H., The Algebraic Eigenvalue Problem, Clarendon, Oxford (1965).

[wi71] Wilkinson, J.H., and Reinsch, C. (eds.), Handbook for Automatic Computation, v. II. Linear Algebra, Springer-Verlag, New York (1971).

[wil68] Willoughby, R.A. (ed.), Sparse Matrix Proceedings, IBM Research RA1, Yorktown Heights, New York (1968).

[yo71] Young, D.M., Iterative Solution of Large Linear Systems, Academic Press, New York (1971).

HOW FAR SHOULD YOU GO WITH THE LANCZOS PROCESS?[†]

W. KAHAN and B.N. PARLETT
Department of Mathematics
and
Computer Science Division
Department of Electrical Engineering
and Computer Sciences
University of California at Berkeley
Berkeley, California 94720

Abstract

The Lanczos algorithm can be used to approximate both the largest and smallest eigenvalues of a symmetric matrix whose order is so large that similarity transformations are not feasible. The algorithm builds up a tridiagonal matrix row by row and the key question is when to stop. An analysis leads to a stopping criterion which is inspired by a useful error bound on the computed eigenvalues.

1. INTRODUCTION

The Lanczos algorithm came back into prominence about five years ago [7,8] as the most promising way to compute a few of the eigenvectors of very large symmetric matrices. To be specific we think of computing the p smallest (or largest) eigenvalues of the $n \times n$ symmetric matrix A together with the associated eigenvectors. Typical values are $p = 3$, $n = 1000$.

The algorithm must be provided with a starting vector q_1 and then it builds up, one column per step, two auxiliary matrices. After j steps it will have produced $j \times j$ symmetric tridiagonal matrix T_j and an $n \times (j+1)$ matrix $Q_{j+1} = (q_1, q_2, \ldots, q_{j+1})$. Let $\theta_1, \ldots, \theta_p$ denote the p extreme eigenvalues of T_j.

What we want is that the $\{\theta_i\}$ should be good approximations to the wanted eigenvalues of A ($\alpha_1, \ldots, \alpha_p$, say) and we ask the following questions. Will the θ_i inevitably improve as j increases? How much or rather how little work is needed to compute an a posteriori bound on the errors in the θ_i?

We still do not know how best to use the Lanczos process. The surprising fact is that the θ_i are sometimes correct to 3 or 4 decimal figures even when j is as small as $\sqrt{n}$. What makes the method interesting is that such good fortune cannot be guaranteed. It depends on q_1, on the spread of the α_i, and on the precision of the arithmetic operations.

[†]The authors are pleased to acknowledge partial support from Office of Naval Research Contract N00014-69-A-0200-1017.

It is a pleasure to acknowledge the excellent pioneering work of Paige in his doctoral thesis [8] and the studies by Golub [2] and others [1] on the block version of the method.

Sparsity of A plays a simple but crucial role here. It permits the computation of Av, for any n-vector v, in only wn basic operations, where w is the average number of nonzero elements per row. Of more importance sparsity discourages the use of explicit similarity transformations which invariably destroy sparsity even when they preserve bandwidth. The only knowledge of A which the Lanczos algorithm demands is how A acts on selected vectors, an attractive property for sparse problems.

To get round some of the difficulties posed by the gaps in our understanding it has been proposed that the Lanczos algorithm be used iteratively as follows. Start with $q_1^{(1)}$ and run the algorithm for $j = j^{(1)}$ steps, compute the p best eigenvector approximations available from $T_j^{(1)}$ and $Q_j^{(1)}$ and take a weighted combination of them as a new starting vector $q_1^{(2)}$, run Lanczos for $j^{(2)}$ steps, compute a new $q_1^{(3)}$, and so on. The process can be continued until the $\{\theta_i\}^1$ computed at the end of each run converge to the desired number of figures.

Is this a good idea? How should the $j^{(i)}$ be chosen? We do not have definitive answers to the questions raised here but we do present a computable a posteriori error bound, an estimate for loss of orthogonality among the columns of Q_j, a tentative stopping criterion, and a useful way of analyzing the algorithm.

Standard Householder matrix conventions will be followed except that M* denotes the conjugate transpose of M and 1 denotes the identity matrix.

2. ERROR BOUNDS FOR THE EXACT LANCZOS ALGORITHM

When executed in exact arithmetic the matrix $Q_j \equiv (q_1,\ldots,q_j)$, which is generated column by column, is orthonormal:

$$Q_j^* Q_j = 1 \,, \tag{1}$$

and is part of an orthogonal matrix Q which reduces A to tridiagonal form. The algorithm, or rather its j-th step, is completely specified by the single matrix relation

$$(2) \qquad A \; Q_j = Q \begin{bmatrix} T_j \\ \beta_j \blacksquare \\ 0 \end{bmatrix} = Q_j \; T_j + \begin{bmatrix} 0 & \begin{matrix} x \\ x \\ \vdots \\ x \\ x \end{matrix} \end{bmatrix}$$

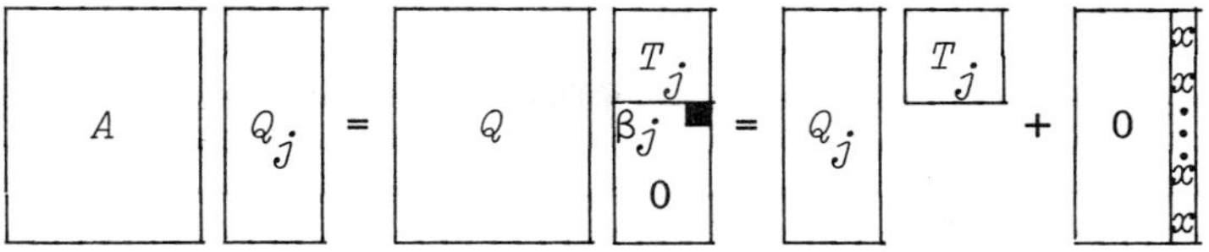

$$= Q_j T_j + r_j e_j^* \; ,$$

where $e_j^* = (0,\ldots,0,1)$ has j elements and $\|r_j\|^2 \equiv r_j^* r_j = \beta_j^2$. The residual vector

$$r_j \equiv \beta_j q_{j+1} = (AQ_j - Q_j T_j) e_j$$

is always uniquely determined by A, Q_j, and T_j. An essential characteristic of the Lanczos process is that the residual matrix $AQ_j - Q_j T_j$ has all its substance concentrated in its final column. An atractive feature of the process, not obvious from (2), is that only the two most recent columns of Q_k need be kept in fast memory at the k-th step, so q_{k-2} can be put out to secondary storage. Sometimes the early q vectors are discarded but this makes it very difficult to estimate the accuracy of our approximations. Equation (2) is a nice compact way of remembering the algorithm.

Suppose now that the Lanczos process is halted at the end of the j-th step. The approximations to the desired eigenvectors and values of A are made as follows. The p extreme eigenvalues of T_j are computed along with their normalized eigenvectors. We suppress the dependence on j and write

$$T_j c_i = c_i \theta_i \; , \quad i = 1,\ldots,p \; , \text{ i.e. } T_j C = C\Theta$$

where $$C = (c_1,\ldots,c_p) \; , \quad \Theta = diag(\theta_1,\ldots,\theta_p) \; .$$

Then we compute the Ritz vectors $v_i = Q_j c_i$, $i = 1,\ldots,p$. When Q_j is orthonormal these Ritz vectors are the best approximate eigenvectors, in the sense of residuals, that can be made from linear combinations of the columns of Q_j and the θ_i are the best approximations to the corresponding eigenvalues. There is a considerable body of knowledge concerning the accuracy of the exact Ritz approximations. The Kaniel-Paige theory [3] gives a priori error bounds on θ_i and v_i which tell how accuracy increases with j. Moreover there are refined a posteriori error bounds which can be computed when the algorithm is terminated. However even with exact calculations there is no inexpensive, feasible criterion which can tell us the best j at which to stop.

Let us look at some simple a posteriori bounds. They employ the spectral matrix norm $\|B\| \equiv \max\|Bv\|/\|v\|$, $v \neq 0$.

THEOREM 1 (Kahan, 1967). Let H be any p×p symmetric matrix with eigenvalues θ_i and let S be n×p and orthonormal. Then there are p eigenvalues $\alpha_{i'}$ of A such that, for $i = 1,\ldots,p$,

$$|\theta_i - \alpha_{i'}| \leq \|AS - SH\| .$$

We do not know which of A's eigenvalues are the $\alpha_{i'}$.

The proof is based on the Weyl/Wielandt monotonicity theorem and is given in [3].

We are not interested in all of T_j's eigenvalues, only p of them. In fact we have

$$T_j C = C\Theta .$$

Now apply Theorem 1 with $S = Q_j C$ to obtain the following result.

COROLLARY. Let $AQ_j - Q_j T_j = r_j e_j^*$ and $T_j C = C\Theta$ where Q_j and C are orthonormal and $\|r_j\| = \beta_j$. Then there are p eigenvalues $\alpha_{i'}$ of A such that for $i = 1,\ldots,p$,

$$|\theta_i - \alpha_{i'}| \leq \beta_j \|e_j^* C\| .$$

Thus when the last elements of some normalized eigenvectors of T_j are small good accuracy is obtained for their eigenvalues <u>even when β_j is not small</u>. Unfortunately we have no easy way of guaranteeing which eigenvalues of A are being approximated. This is an intrinsic limitation of the Lanczos method and will not be discussed in detail here.

The best way we know of testing which of A's eigenvalues $\{\alpha_k\}$ are being approximated by the θ_i $(i = 1,\ldots,p)$ is to perform a triangular factorization of $A - \theta_i$ into LDL^T and count the number ν_i of negative elements of D. By Sylvester's Inertia theorem there are ν_i of the α's less than θ_i.

3. PRACTICAL ERROR BOUNDS AND TERMINATION CRITERIA

From now on we let Q_j, T_j, etc. stand for the quantities stored in the computer. Because of roundoff error Q_j will not be orthonormal and the residual matrix will not have all its substance in the last column. Fortunately Theorem 1 can be generalized.

> THEOREM 2 (Kahan, 1967). Let H be any p×p matrix with eigenvalues θ_i, let S be any n×p matrix of full rank p; then there are p eigenvalues $\alpha_{i'}$ of A such that, for i = 1,...,p,
>
> $$|\theta_i - \alpha_{i'}| \leq \sqrt{2}\|AS-SH\| \cdot \|(S^*S)^{-1/2}\| .$$

Note that $\|(S^*S)^{-1/2}\|$ is the reciprocal of S's smallest singular value $\sigma_1(S) \equiv \sqrt{\lambda_1(S^*S)}$. A proof may be found in [3]. The factor $\sqrt{2}$ is believed to be superfluous.

In order to use Theorem 2 we observe that Q_j and T_j actually satisfy

$$AQ_j - Q_jT_j = F_j + r_je_j^* \tag{3}$$

where F_j accounts for roundoff error.

> COROLLARY. Let $T_jC = C\Theta$; then there are p eigenvalues $\alpha_{i'}$ of A such that, for i = 1,...,p,
>
> $$|\theta_i - \alpha_{i'}| \leq \sqrt{2}(\|F_j\| \cdot \|C\| + \beta_j\|e_j^*C\|)/\sigma_1(Q_jC) .$$

PROOF. Apply Theorem 2 with $S = Q_jC$. □

In order to use the corollary an upper bound is needed on $\|F_j\|$ and a lower bound on the singular values of Q_jC. With the best of current techniques C will be very close to orthonormal. In the sense of quadratic forms,

$$0 \leq C^*C \lesssim 1 \tag{4}$$

and hence, by the Cauchy interlace inequalities

$$\sigma_1(Q_j) \lesssim \sigma_1(Q_jC) .$$

Error analyses in [5] and [8] show that $\|F_j\|$ is small, like roundoff in $\|A\|$, and so the term $\|F_j\| \cdot \|C\|$ is completely dominated by $\beta_j\|e_j^*C\|$ for all realistic values of j. Thus in practice the bound in Corollary 1 is degraded by the factor $1/\sigma_1$.

Corollary 2 assures us that it is worthwhile to continue the Lanczos algorithm even after orthogonality among the q_i has been lost provided that they are still linearly independent. Experience suggests that good approximation to internal eigenvalues α_i can be obtained by continuing the algorithm indefinitely. However after $\sigma_1(Q_j) = 0$, i.e. when linear independence is lost, the user is faced with the identification problem, namely to say which of the θ_i do not approximate any of the α's and so are spurious. This problem can often be solved in particular applications but it is very troublesome to devise a procedure which will make this identification for the general case.

In order to escape this difficulty we would like to stop immediately $\sigma_1(Q_j) < 1/2$. However it is out of the question to compute this number at each step, or even update it. Instead we develop a computable lower bound on $\sigma_1(Q_j)$ and stop a Lanczos run as soon as it vanishes. If the approximations are not satisfactory we can form a new starting vector from the Ritz vectors and start another run.

Since C is not computed at each step there is no simple way of bounding $\sigma_1(Q_jC)$ rather than $\sigma_1(Q_j)$.

4. MONITORING LOSS OF ORTHOGONALITY

Because of finite precision arithmetic Q_j will not be orthonormal. Let us write

$$\|1 - Q_j^*Q_j\| \leq \kappa_j$$

and find some specific expressions for κ_j. Note that

$$\|Q_j\|^2 = \|Q_j^*Q_j\| = \|1 - (1-Q_j^*Q_j)\| \leq 1+\kappa_j ,$$

and, in the sense of quadratic forms

$$1-\kappa_j \leq Q_j^*Q_j \leq 1+\kappa_j .$$

Hence, while $\kappa_j < 1$,

$$\sqrt{1-\kappa_j} \leq \sqrt{\lambda_1(Q_j^*Q_j)} \equiv \sigma_1(Q_j) \leq \sigma_1(Q_jC) .$$

Our problem is thus reduced to finding a computable bound κ_j. This can be accomplished in a variety of ways. Here is one of the simplest. Observe that

$$1 - Q_{j+1}^*Q_{j+1} = \begin{pmatrix} 1-Q_j^*Q_j & -Q_j^*q_{j+1} \\ -q_{j+1}^*Q_j & 1-\|q_{j+1}\|^2 \end{pmatrix} .$$

Let

$$\|Q_j^*q_{j+1}\| \leq \zeta_j .$$

Then, by considering the quadratic forms we see that

$$\|1-Q_{j+1}^*Q_{j+1}\| \leq \left\| \begin{pmatrix} \kappa_j & \zeta_j \\ \zeta_j & \kappa_1 \end{pmatrix} \right\| .$$

Here we have used the fact that, by definition,

$$\|1-Q_1^*Q_1\| = |1 - \|q_1\|^2| \leq \kappa_1$$

but, actually, κ_1 is the bound on the error in normalizing <u>any</u> vector to the given precision, including q_{j+1}, and is taken as known.

Any computable bound ζ_j will yield a corresponding definition of κ_j, namely

$$\kappa_{j+1} \equiv \left\| \begin{pmatrix} \kappa_j & \zeta_j \\ \zeta_j & \kappa_1 \end{pmatrix} \right\|$$

$$= \frac{1}{2}\{\kappa_j + \kappa_1 + \sqrt{(\kappa_j-\kappa_1)^2+4\zeta_j^2}\} .$$

Since κ_1 is known we focus our attention on ζ_j and $Q_j^*q_{j+1}$.

5. AN EXPRESSION FOR $Q_j^*r_j$

We write

$$T_j = \begin{bmatrix} \alpha_1 & \beta_1 & & & \bigcirc \\ \beta_1 & \alpha_2 & \beta_2 & & \\ & & \cdot & \cdot & \cdot \\ & \bigcirc & & \beta_{j-1} & \alpha_j \end{bmatrix}$$

and let $F_j = (f_1,\ldots,f_j)$. Our bound ζ_j comes from a useful expression for $Q_j^*r_j$.

> LEMMA 1. Let Q_j and T_j satisfy (H): $AQ_j - Q_jT_j = F_j + r_je_j^*$. Then
>
> $$Q_j^*r_j = [(1-Q_j^*Q_j)T_j - (1-e_je_j^*)T_j(1-Q_j^*C_j)]e_j + F_j^*q_j + (q_j^*Aq_j-\alpha_j)e_j - Q_j^*f_j .$$

PROOF.

$$Q_j^*r_j = Q_j^*(AQ_j-Q_jT_j-F_j)e_j , \quad \text{using (H)},$$
$$= [(AQ_j^*)Q_j - Q_j^*Q_jT_j]e_j - Q_j^*f_j , \quad \text{using } A^*=A,$$

$$= [T_jQ_j^*Q_j + F_j^*Q_j + e_jr_j^*Q_j - Q_j^*Q_jT_j]e_j - Q_j^*f_1 \ , \quad \text{(H) again,}$$
$$= [-T_j(1-Q_j^*Q_j) + (1-Q_j^*Q_j)T_j]e_j + F_j^*q_j + e_jr_j^*Q_je_j$$
$$- Q_j^*f_j \ .$$

From the first line above

$$e_j^*Q_j^*r_j = q_j^*Aq_j - e_j^*Q_j^*Q_jT_je_j$$
$$= q_j^*Aq_j - \alpha_j + e_j^*(1-Q_j^*Q_j)T_je_j \ .$$

On transposing and rearranging the desired expression is obtained. □

In practice

$$q_{j+1} = r_j/\beta_j + g_j$$

where g_j accounts for roundoff error in the division by β_j and is always insignificant. In any case

$$Q_j^*q_{j+1} = Q_j^*r_j/\beta_j + Q_j^*g_j \ .$$

We observe that $\|Q_j^*q_{j+1}\|$ will not be small, like roundoff in 1 whenever, because of past errors, $\|Q_j^*r_j\| \doteq \|Q_j^*\|\|r_j\|$. In exact arithmetic $Q_j^*r_j = 0$ and this property fails in practice to the extent that there is cancellation in forming r_j; the resulting large relative error in the small vector r_j becomes significant when r_j is divided by β_j to produce q_{j+1}. As Paige points out in [7] this happening must be seen in perspective. We <u>want</u> cancellation to occur in the formation of r_j because cancellation is the harbinger of convergence of the θ_1 to an α.

6. A COMPUTABLE BOUND ON $Q_j^*q_{j+1}$

The expression for $Q_j^*r_j$ given in Lemma 1 is best split into two parts

$$b_j \equiv [(1-Q_j^*Q_j)T_j - (1-e_je_j^*)T_j(1-Q_j^*Q_j)]e_j \ ,$$
$$d_j \equiv F_j^*q_j + (q_j^*Aq_j - \alpha_j)e_j - Q_j^*f_j \ .$$

It turns out that our bounds on $\|b_j\|$ overwhelm those on $\|d_j\|$ as soon as $j > 4$. To bound $\|d_j\|$ a detailed error analysis of the Lanczos algorithm is needed. There is little incentive to present this because the resulting bound makes an insignificant contribution to ζ_j. We shall simply quote the results in [5] which are quite similar to those in [7].

The results are stated in terms of

$$\|A\|_E \equiv \left(\sum_\mu \sum_\nu |a_{\mu\nu}|^2\right)^{1/2} = [trace(A^*A)]^{1/2}$$

which arises in the general bound on the error in computing Av. However $\|A\|_E$ is just a convenient bound on the more realistic but less accessible quantity $\|\tilde{A}\|$ where $\tilde{A} = (|a_{ik}|)$. For large sparse matrices $\|\tilde{A}\|$ is usually much smaller than $\|A\|_E$. The following estimates are crude but adequate for our purposes: for $i \le j$,

$$\|f_i\| + \beta_i \|g_i\| < \kappa_1 \|A\|_E ,$$
$$|q_i^* A q_i - \alpha_i| < \kappa_1 \|A\|_E ,$$
$$\|Q_i\|^2 < 1 + \kappa_i ,$$
$$\|F_j\|_E < \sqrt{j}\, max\, \|f_i\| , \quad i \le j ,$$
$$\kappa_1 < 2(n+6)\varepsilon ,$$

where ε is the precision of the arithmetic facilities. Finally we quote

LEMMA 2. $\|d_j\| \le (\sqrt{j} + 3 + \kappa_j)\|A\|_E$.

Turning to $\|b_j\|$ we find

LEMMA 3. If $\|Q_i^* q_{i+1}\| \le \zeta_i$, $i < j$, then

$$\|b_j\|^2 \le \{\|(T_{j-1} - \alpha_j)\|\zeta_{j-1} + \beta_{j-1}(\zeta_{j-2} + 2\kappa_1)\}^2 + \{(3j+1)\kappa_1 \|A\|_E\}^2$$

PROOF. Partition the terms in b_j and observe

$$(1 - Q_j^* Q_j) T_j e_j = \begin{bmatrix} \{-Q_{j-1}^* q_j\} \\ 1 - \|q_j\|^2 \end{bmatrix} \alpha_j + \begin{bmatrix} -Q_{j-2}^* q_{j-1} \\ 1 - \|q_{j-1}\|^2 \\ -q_j^* q_{j-1} \end{bmatrix} \beta_{j-1} ,$$

$$T_j(1 - Q_j^* Q_j) e_j = \begin{bmatrix} -T_{j-1} Q_{j-1}^* q_j + e_{j-1}\beta_{j-1}(1 - \|q_j\|^2) \\ \alpha_j(1 - \|q_j\|^2) - \beta_{j-1} q_{j-1}^* q_j \end{bmatrix} .$$

The factor $(1-e_j e_j^*)$ simply annihilates the bottom element. Recall that $|1-\|q_i\|^2| < \kappa_1$, by definition of κ_1. Hence the top part of b_j is

$$(T_{j-1}-\alpha_j)Q_{j-1}^* q_j - \beta_{j-1}\begin{pmatrix} Q_{j-2}^* q_{j-1} \\ 0 \end{pmatrix} + \beta_{j-1} e_{j-1}(\|q_j\|^2 - \|q_{j-1}\|^2)$$

and the bottom element is

$$(1-\|q_j\|^2)\alpha_j - \beta_{j-1} q_j^* q_{j-1} \ .$$

By definition of the basic Lanczos steps

$$q_{j+1}\beta_j = Aq_j - \alpha_j q_j - \beta_{j-1} q_{j-1} - f_j + \beta_j g_j$$

then, subtracting and adding α_j,

$$|q_j^* q_{j+1}\beta_j| \leq |q_j^* A q_j - \alpha_j| + |1-\|q_j\|^2|\cdot|\alpha_j| + |\beta_{j-1} q_j^* q_{j-1}| + |q_j^*(f_j+\beta_j g_j)| \ .$$

Now suppose that $|q_i^* q_{i+1}\beta_i| \leq \psi_i$, $i \leq j$, then the error bounds quoted above Lemma 2 yield

$$\begin{aligned} \psi_j &\leq 3\kappa_1\|A\| + \psi_{j-1} \ , \\ &\leq 3j\kappa_1\|A\| \ , \text{ since } \psi_1 \leq 3\kappa_1\|A\| \ . \end{aligned}$$

Straightforward application of the triangle inequality gives the desired bound. □

Adding these results together we define

$$\begin{aligned} \omega_j &\equiv \{[\|(T_{j-1}-\alpha_j)\|\zeta_{j-1} + \beta_{j-1}(\zeta_{j-2}+2\kappa_1)]^2 \\ &\quad + (3j+1)^2\kappa_1^2\|A\|_E^2\}^{1/2} + (\sqrt{j}+3+\kappa_j)\|A\|_E \ , \\ \zeta_j &\equiv \omega_j/\beta_j + \sqrt{1+\kappa_j}\ \varepsilon \ . \end{aligned}$$

Then, by induction,

LEMMA 4. If $\zeta_{-1} = \zeta_0 = 0$, $\|Q_i^* q_{i+1}\| \leq \zeta_i$, $i < j$ then

$$\|Q_j^* q_{j+1}\| \leq \zeta_j \ .$$

PROOF. Apply Lemmas 2 and 3 and the error bounds to

$$Q_j^* q_{j+1} = Q_j^* r_j/\beta_j + Q_j^* g_j \ . \quad \square$$

Let us summarize the procedure. During the j-th step of a Lanczos run we compute

$$\alpha_j, \; r_j, \; \beta_j, \; q_{j+1} \; ,$$

then

$$\omega_j, \; \zeta_j, \text{ and } \kappa_{j+1} \; .$$

While $\kappa_j < 1$ the algorithm proceeds to step (j+1).

The extra work in updating κ_j is very slight and is dominated by the computation of $\|(T_{j-1}-\alpha_j)\|$. This may be bounded as follows.

$$\begin{aligned} \|(T_{j-1}-\alpha_j)\| &\le \max_{i<j}(\beta_{i-1} + |\alpha_i-\alpha_j| + \beta_i) \; , \\ &\le \bar{\beta} + \max_{k<j}|\alpha_k-\alpha_j| \; , \\ &\le \bar{\beta} + \max(|\underline{\alpha}-\alpha_j|, |\bar{\alpha}-\alpha_j|) \; , \end{aligned}$$

where

$$\underline{\alpha} = \min \alpha_i, \quad \bar{\alpha} = \max \alpha_i, \quad \bar{\beta} = \max_{i \le k}(\beta_{i-1}+\beta_i) \; .$$

Note that the third inequality involves no searching over j elements provided that $\underline{\alpha}$, $\bar{\alpha}$, and $\bar{\beta}$ are updated at each step; $\underline{\alpha} = \min(\underline{\alpha},\alpha_j)$, $\bar{\alpha} = \max(\bar{\alpha},\alpha_j)$, $\bar{\beta} = \max(\bar{\beta},\beta_{j-2}+\beta_{j-1})$. The second inequality costs (j-1) comparisons and is tighter. The first inequality is $\|(T_{j-1}-\alpha_j)\|_\infty$, except for the presence of β_{j-1}, and this can never exceed $\|(T_{j-1}-\alpha_j)\|$ by more than a factor of $\sqrt{2}$.

A more complicated procedure for monitoring loss of orthogonality is described in [5]. $1-Q_j^*Q_j$ is majorized by a $j\times j$ matrix W_j, called the scoreboard. The triangular factorization of $1-W_j$ is updated at each step and the Lanczos run continues until a nonpositive diagonal element appears. The cost is approximately $j^2/2$ arithmetic operations and, more seriously, j^2 storage locations.

The quantity κ_j grows exponentially and its use will certainly terminate Lanczos runs prematurely. This is not necessarily inefficient for the following reason. The Lanczos algorithm yields monotonically improving approximations. It thus can be used for calculations to low accuracy as well as high. It is important not to give the user unwanted figures when the extra cost is significant. If Lanczos is run until $|q_1^*q_{j+1}| > 0.1$ (say) then θ_1 may already have converged to too many figures.

In practice $\|A\|_E$ may not be readily available and the

estimate $\|(T_{j-1}-\alpha_j)\| + |\alpha_j|$ is used in its place. Sometimes the bound on $\|d_j\|$ is omitted for simplicity.

7. BEHAVIOR OF THE SPECTRUM OF T_j

In Figure 1 we show the lowest five eigenvalues of T_j, for j = 10,20,...,60, in a Lanczos run on a matrix A with eigenvalues $\alpha_i = i$, i = 1,2,...,253. Also shown is $\sigma_i(Q_j)$ and $\sqrt{1-\kappa_j}$. The starting vector q_1 was chosen to be rich in the first four eigenvectors.

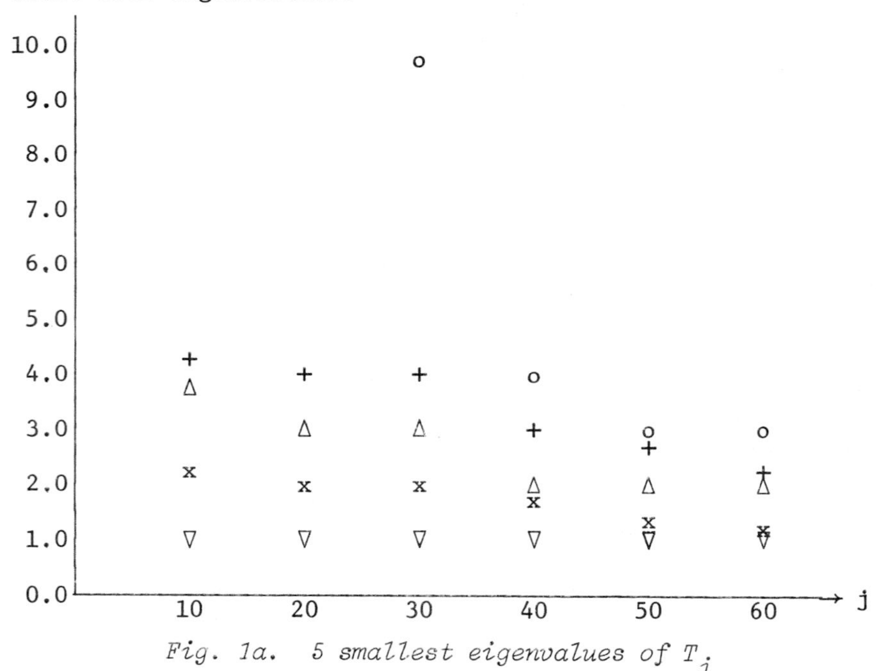

Fig. 1a. 5 smallest eigenvalues of T_j

Three phases may be distinguished in the run. Note that p = 4.

Early Phase: θ_i is not an accurate approximation to α_i for i = 1,...,p.

Middle Phase: One or more of the θ_i have converged to working accuracy and there is still a one-one correspondence between θ_i and α_i for i = 1,...,p.

Late Phase: The one-one correspondence has been lost. Among the θ_i occur multiple approximations to various α's as well as spurious values close to no α.

If q_1 had been chosen to be rich in eigenvectors 1, 3, and 4 the picture would have been more complicated. If p is

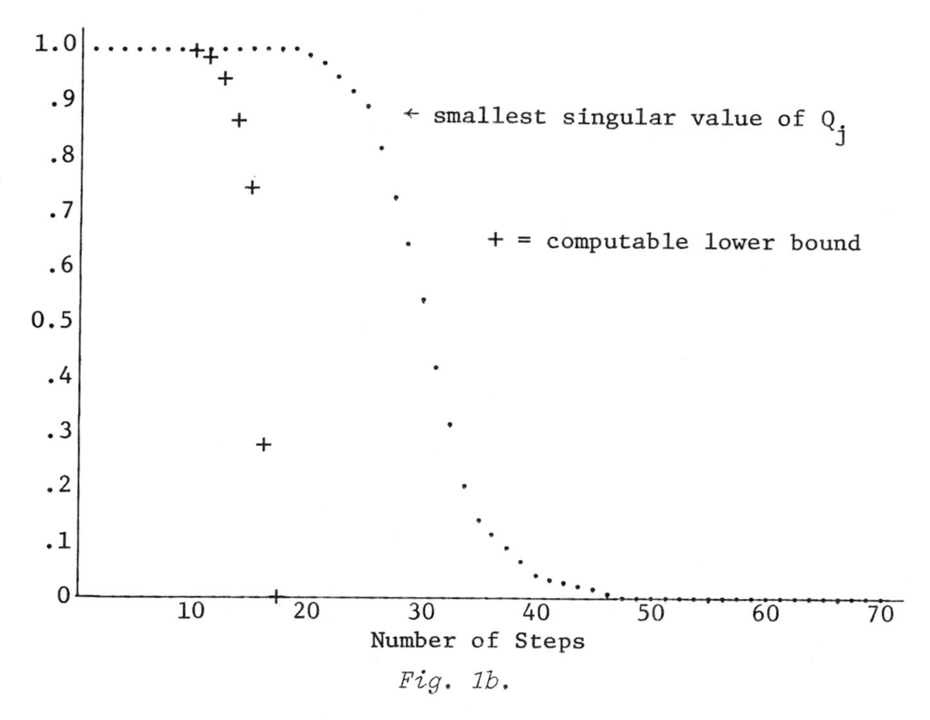

Fig. 1b.

too large then there may be multiple approximations to α_1 before any θ has converged to α_p.

Corollary 2 in Section 2 implies that the late phase cannot begin until linear independence is lost to working accuracy. Paige's error analyses [7] show that Phase 1 ends before orthogonality is lost to working accuracy.

It is usual for an outer eigenvalue to converge before an inner one but this cannot be taken for granted. The outcome depends on q_1.

We recommend that Lanczos algorithms be tailored to the problem in hand. The urge to write a universal Lanczos program should be resisted, at least until the process is better understood.

ACKNOWLEDGEMENT.

The authors wish to thank Mr. W. Steele for expert programming assistance.

REFERENCES

[1] Jane Cullum, W.E. Donath, A Block Lanczos Generalization of the Symmetric S-STEP Lanczos Algorithm, Technical Report RC 4845 (X 21570), Mathematics Department, IBM Research Center, Yorktown Heights, New York.

[2] G.H. Golub, R. Underwood and J.H. Wilkinson, The Lanczos Algorithm for the Symmetric Ax = λBx Problem, Report CS-72-270, Stanford University, 1972.

[3] W. Kahan, Inclusion Theorems for Clusters of Eigenvalues of Hermitian Matrices, Computer Science Report, University of Toronto, 1967.

[4] W. Kahan and B. Parlett, An Analysis of Lanczos Algorithms for Symmetric Matrices, Memorandum ERL-M467, Electronics Research Laboratory, University of California, Berkeley, 1974.

[5] S. Kaniel, Estimates for Some Computational Techniques in Linear Algebra, Mathematics of Computation 20 (1966), 369-378.

[6] C. Lanczos, An Iteration Method for the Solution of the Eigenvalue Problem of Linear Differential and Integral Operators, Journal of Research of the National Bureau of Standards 45 (1950), 255-282.

[7] C.C. Paige, Practical Use of the Symmetric Lanczos Process with Re-Orthogonalization, BIT 10 (1970), 183-195.

[8] C.C. Paige, The Computation of Eigenvalues and Eigenvectors of Very Large Sparse Matrices, Ph.D. Thesis, London University, Institute for Computer Science, Technical Note.

[9] C.C. Paige, Computational Variants of the Lanczos Method for the Eigenproblem, Journal of IMA 10 (1972), 373-381.

[10] J.H. Wilkinson, The Algebraic Eigenvalue Problem, O.U.P., 1965.

III

Optimization, Least Squares, and Linear Programming

OPTIMIZATION FOR SPARSE SYSTEMS

T. L. Magnanti

Sloan School of Management, M.I.T.

Sparse matrices with special structure arise in almost every application of large scale optimization. In linear programming, these problems usually are solved by pivoting procedures, most notably the simplex method, refined and modified in various ways to exploit structure. More recently iterative relaxation methods and dual ascent algorithms have been proposed for certain applications. In surveying several algorithms from each of these categories, this paper demonstrates the potential for investigating and applying sparse system techniques in optimization.

I. Introduction

Nearly every medium to large scale optimization model met in practice is sparse in the sense of:

Micro structure - Each decision variable appears only in a few of the constraints, less than 0.5% in many linear programming applications, or

Macro structure - The model as a whole has special structure; it is, for example, a network model or a model composed of submodels linked by time, by a few common constraints, or by a few common variables.

The literature studying such optimization problems is vast and expanding rapidly; several survey articles (Dantzig 68, Geoffrion 70a, Lasdon 76, Tomlin 72, and White 73) as well as several texts, conference proceedings, and collections (Cottle and Krarup 74, Himmelblau 73, Lasdon 70, Orchard-Hays 68, Reid 71, Rose and Willoughby 72, Willoughby 69 and Wismer 71) summarize much of the theory, applications and computer software development for optimizing sparse systems. Our purpose here is to bring together and discuss three recently studied areas - basis factorization in the simplex method, minimum cost network flow, and constructive duality theory - which have not been covered extensively in any of these sources.

Much of the latest research in optimization reflects a merging of numerical analysis, computer science, and mathematical programming, as, for example, in the evolution of conjugate direction methods for both unconstrained optimization and systems of equations. In the context of linear programming, LU and Cholesky factorization techniques for solving systems of linear equations now are being applied to the simplex method, both to exploit special structure and to ensure numerical stability.

The recent resurgence in research for minimum cost flow problems illustrates the importance of using efficient data management and manipulation techniques from computer science. The algorithms being designed for these problems are substantially the same as those used in the past, using the property that every basis is triangular, but further exploit the special structure of network data to develop efficient list processing techniques for streamlining calculations. Implementations based upon these ideas have been successful, reducing running time by an order of magnitude over previous implementations.

Constructive duality often involves clever formulation, and subsequent solution, of a dual problem formed to exploit problem structure. Since the dual problem is concave, but usually nondifferentiable, much of the research in this area is tied with nondifferentiable optimization using relaxation techniques, subgradient optimization, and extensions of standard nonlinear programming procedures designed for nondifferentiable functions.

In reviewing these topics, we consider only certain aspects of optimization for sparse systems. Those contributions previously covered in Dantzig's [68] and Geoffrion's [70a] comprehensive and insightful surveys and in Lasdon's [70] text only are included in this discussion if they bear directly on these later developments. Also, we only briefly mention applications of large scale optimization and we do not indicate much about how these topics specialize for and relate to control theory, integer programming, nonlinear programming or stochastic optimization. For discussions of some of these subjects, see the highly regarded collection, Geoffrion 72b, and the papers by Geoffrion 74a, Polak 73, and Wets 74.

II. Basis Factorization

When applied to the linear program

$$\text{minimize } \{cx : Ax = b, x \geq 0\},$$

the simplex method solves a sequence of linear systems

$$\pi B = \gamma, \quad By = \alpha, \quad \text{and} \quad Bw = b \tag{1}$$

where B is a nonsingular submatrix of the m by n matrix A and w and γ are column and row subvectors of x and c corresponding to the columns B of A; the vector α is a column of A determined from the solution π to the first linear system. Based upon the solutions to the last two systems, the simplex method replaces a column from B with the column α and solves the systems again. Other details of the method are not required in the following discussion; we should note, however, that the system $Bw=b$ is solved easily from step to step, because the new solution after the basis change can be computed from the current solution w as $w-\theta y$, for a suitably chosen, and easily computed, scalar θ (given by the "simplex ratio test"). Consequently, we shall concentrate on methods for solving the first two systems.

In carrying out these computations, we desire an algorithm that is both numerically stable and efficient, in terms of the number of calculations performed. Unfortunately, it sometimes is difficult to achieve both of these objectives simultaneously, and often one of them must be sacrificed for the other.

Most commercial linear programming systems currently solve these systems using the basis inverse B^{-1} which is stored as a product of elementary matrices

$$B^{-1} = E_k E_{k-1} \cdots E_1 .$$

If α replaces the rth column of B, then the basis inverse is updated by adding a new elementary matrix E_{k+1} specified by the columns of the identity matrix I after pivoting on the rth row of y in the augmented matrix $[y:I]$.

These commercial codes strive to maintain the elementary matrices compactly with as few nonzero elements as possible, since less storage is then required and fewer computations are performed when solving the linear systems in (1). To achieve a compact representation, the codes reinvert the basis from scratch periodically, so that the list of elementary matrices does not become excessively long. Many of the codes also use heuristics to represent the product form compactly; for example Hellerman and Rarick [74] reorder the rows and columns of the basis to produce a near triangular form:

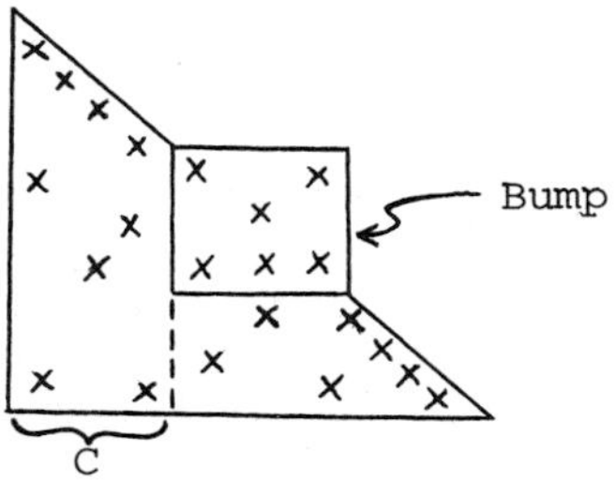

Bump

(x's are nonzero elements)

Fig. 1. Basis Sorted for Reinversion.

When computing the basis inverse, they pivot down the diagonal elements of the columns C producing elementary matrices as sparse as the basis itself and causing no fill in of nonzero elements in the matrix. Once the nontriangular bump is reached, pivots are selected from the remaining matrix in accordance with rating schemes aimed at giving sparse elementary matrices. In several experiments, Hellerman and Rarick found that this scheme produces a product form as much as 50% less dense than inversions that do not presort the basis.

Of course, any of the numerous procedures of linear algebra for solving systems of linear equations can be used in place of this basis inverse implementation as long as the method can be modified easily as the basis matrix B changes from step to step. Recognizing this, Markowitz [57] suggested Gaussian elimination, or LU factorization. Later Dantzig [63a] showed how LU factorization can exploit macro structure of the coefficient matrix A by maintaining a sparse representation for L and U; Bartels [71] and Bartels and Golub [69] showed how LU factorization, if implemented properly, provides numerical stability.

An LU factorization expresses the basis, with row permutations if necessary, as a product of a lower triangular matrix L and an upper triangular matrix U. When no row permutations are needed, B=LU and the systems in (1) can be written

$$\pi B=\pi(LU)=(\pi L)U=\gamma \text{ and } By=(LU)y=L(Uy)=\alpha.$$

In this form, the systems are solved by forward and backward substitution. In the π system, we first solve for $\bar{\pi}\equiv\pi L$ by backward substitution in $\bar{\pi}U=\gamma$; having $\bar{\pi}$, we solve for π from $\bar{\pi}=\pi L$ by forward substitution. Similarly, we solve the y sysem in two stages by computing $\bar{y}\equiv Uy$ first from $L\bar{y}=\alpha$ and then y from $Uy=\bar{y}$.

Although, we can, in principle, maintain both L and U explicitly (see Dantzig [63a]), the iterative nature of the simplex method and the required row permutations make it more attractive, particularly in terms of the logic of the resulting computer system, to store L^{-1} together with any row permutations in product form. Then

$$U = (E_m P_m E_{m-1} P_{m-1} \cdots E_1 P_1)B \tag{2}$$

where each P is a permutation matrix and each E is a lower triangular elementary matrix. We obtain U and the permutation and elementary matrices by familiar techniques from linear algebra. Starting with B, for each j=1,...,m we permute rows, if necessary, and then pivot on the diagonal element of column j of B, as updated so far, to produce zeroes below the diagonal.

Suppose now that we have a factorization $L^{-1}B=U$, where L^{-1} is used to denote all of the product form terms in parenthesis in (2) including the permutation matrices. After column α is introduced into B, we have to recover a new factorization. This updating can be performed in several ways depending upon how the columns of the new basis are arranged and what pivoting strategy is used. Figure 2 illustrates two of the most well known updating schemes.

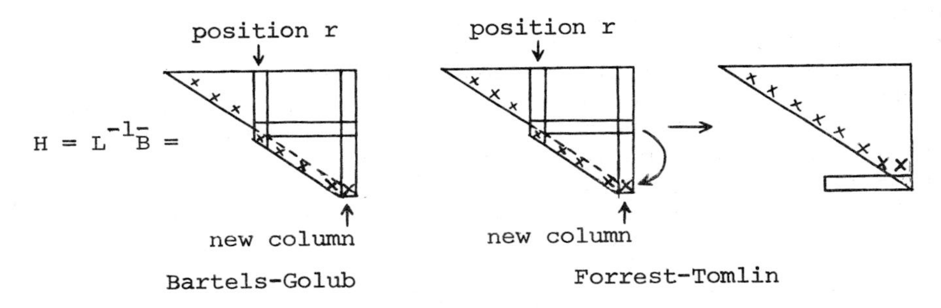

Fig. 2. Updating an LU factorization.

When column r leaves the basis, these procedures move columns (r+1),(r+2),...,(m) all one place to the left and append the incoming column α at the extreme right to give a new basis $\bar{B}$. The product form expression in (2) now becomes $L^{-1}\bar{B} = H$, where the Hessenberg matrix H is upper triangular except for the nonzero subdiagonal elements in columns r through m-1. Its columns are columns 1,2,...,r-1,r+1,...,m from U followed by $z=L^{-1}\alpha$.

To place H in upper triangular form, Bartels and Golub [69] (Bartels [71]) eliminate its subdiagonal elements in order from column r to column m-1. At each column j=r,r+1, ...,m-1 in the process, they pivot either on the diagonal element or subdiagonal element, selecting the element with largest absolute value. If the subdiagonal element is chosen, then they first permute rows j and j+1 so that this element appears on the diagonal. After these operations L^{-1} is premultiplied by the appropriate product of permutation and elementary matrices. Each elimination matrix is very simple, containing a single nondiagonal entry at one of the subdiagonal positions.

Bartels and Golub show that the simplex method is numerically stable if this updating procedure is used as long as the original LU factorization has been produced using maximal pivots; that is, when forming the LU factorization in (2), rows are permuted at each stage, so that the maximal absolute value element on or below the diagonal is moved to the diagonal position and chosen as the pivot element. Wilkinson [63] shows that this pivoting strategy gives an initial LU factorization that is numerically stable.

Forrest and Tomlin [72] have proposed a variant of the Bartels-Golub updating procedure to exploit sparsity. Before pivoting, the method moves row r to the bottom of the Hessenberg matrix H and moves rows (r+1),(r+2),...,(m) one position higher (see Fig. 2). Consequently, the first (m-1) rows are placed in upper triangular form. Pivoting in order on the rth through $(m-1)^{st}$ diagonal elements to eliminate the nonzero entries in the last row gives a new LU factorization.

Although the Forrest-Tomlin updating scheme does not ensure numerical stability, it is attractive for maintaining sparsity. Pivoting subtracts multiples of the rows from the last row only, adding no new nonzero entries to the upper triangular factor. Thus except for the newly added column, the upper triangular factor retains its sparsity at each step. In contrast, Bartels-Golub updating for the new upper triangular form can cause nonzero fill-in, in rows r through m.

Recently, Reid [76] and Saunders [76] have proposed modifications of Bartels-Golub updating for maintaining sparsity. Reid suggests row and column interchanges analogous to those used in basis inverse implementations to rearrange H as much as possible into triangular from to save pivoting operations and reduce fill-in. Saunders further suggests compact storage schemes for storing U. McCoy and Tomlin [74] compare Bartels-Golub and Forrest-Tomlin LU factorizations experimentally; Brayton, Gustavson, and Willoughby [69] discuss related issues for LU factorization.

Other variations of the LU approach can be used to exploit macro structure. As an illustration, consider the well known generalized upper bounding (GUB) problem in which certain equations, called the GUB rows, have the property that no variable appears in more than one GUB row. By exploiting this structure, we can construct a very sparse LU factorization. We simply place the GUB rows first and order the columns in any basis so that a nonzero coefficient of each GUB row appears on the diagonal. To obtain an LU factorization, we first pivot down the diagonal of the GUB rows to eliminate subdiagonal elements in the non-GUB rows and then pivot in the non-GUB rows to reduce these rows to upper triangular form, using any factorization procedure.

The resulting upper triangular factor looks like the

sparse matrices in Fig. 3. The x's are the coefficients in the original GUB rows; they are not altered by the elimination process which subtracts multiples of the GUB rows from the non-GUB rows. Note, also, that each elementary matrix used to express L^{-1} is sparse; its only off-diagonal elements appear in the non-GUB rows

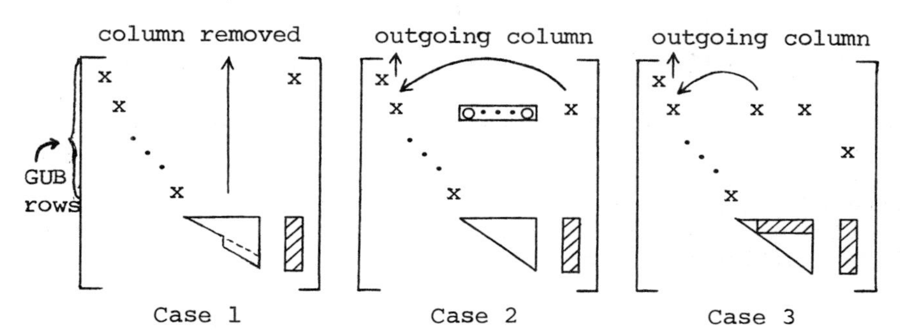

Fig. 3. Updating a GUB Factorization.

When applying the simplex method to the GUB problem, the Bartels-Golub and Forrest-Tomlin updates move the possibly dense triangular factor from the non-GUB rows into the GUB rows as columns of U are moved left. By altering the updating scheme this fill-in can be avoided.

The three cases shown above illustrate possiblilites for an updating procedure which preserves sparsity. In case 1, the column leaving the basis does not correspond to one of the leading GUB variables and any update, such as Bartels-Golub, can be made in the non-GUB constraints. In case 2, the basis contains only one variable from the GUB row corresponding to the outgoing column. Placing the entering column in place of the outgoing column and eliminating its elements from the non-GUB rows does not effect U; only L^{-1} must be updated. Finally, when a leading GUB variable leaves the basis and another variable from this row appears in the basis, the update can be made in two steps. First, rearrange the basic columns by exchanging the outgoing column with the next column to the right containing a variable in the same GUB row (see Fig 3). Putting U back into upper triangular form after the interchange only effects the shaded portion of U. After this exchange, we have reduced the update to case 1 and the update can be continued as in that case.

This updating procedure for the GUB problem was proposed,

and described in terms of both row and column transformations, by Dantzig and Van Slyke [67] for a basis inverse implementation. In this form, the procedure now is widely available on commercial codes. It has been used on problems with as many as 50,000 rows (all but 500 GUB rows) and 280,000 variables. Hax [67] seems first to have described the method solely in terms of row operations. Tomlin [74] suggested the LU version described here.

Procedures similar to this LU implementation of GUB apply to other linear programs with macro structure. For example, the method is modified in minor ways if the GUB rows are replaced by blocks of equations, with each variable appearing in at most one block. In this case, an LU factorization for each block replaces the leading GUB variables in the LU factorization for the GUB problem.

Figure 4 illustrates another example, a staircase system which typifies many time-phased applications:

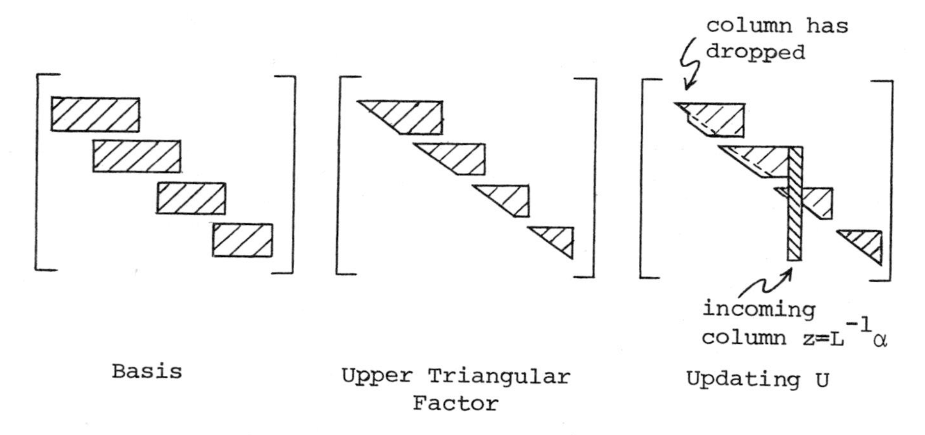

Fig. 4. Factorization for Staircase Systems.

If no row is moved from one of the four blocks of equations forming the staircase to another during the LU computations, then the nonzero elements of both L^{-1} and U lie within the staircase. Dantzig [63a] shows that if any incoming column is inserted within U in its original position relative to the other columns in the basis, then the same structure of L^{-1} and U can be recovered after making an update. Permutation and elementary matrices are applied as in Bartels-Golub updating to zero the subdiagonal elements, until the new column z is encountered. At this point, the elements of z below the diagonal are eliminated from the bottom up, so that only subdiagonal elements are generated in the columns to the right of z. That is, first a pivot is made on the second to last element of z to eliminate its last element. Then the third to last element is used to eliminate the second to last element, and so on until z fits into the triangular form (if the jth pivot is zero then rows j and j+1 are interchanged and row (j-1) is considered for the next pivot). Finally, the subdiagonal elements generated to the right of column z are eliminated. Given the importance of staircase systems in practice, it is surprising that this procedure has not been implemented and tested.

Another implementation of the simplex method, which is numerically stable, uses the factorization $B=QR$; the matrix Q is orthogonal ($Q^TQ=I$) and the matrix R is upper triangular. Like LU, this factorization can be computed by applying a sequence of matrices to reduce B, one column at a time, into triangular form. The matrices in this reduction use Householder transformations of the form $W=I-2uu^T$ where u is a column vector with a unit Euclidean norm. Wilkinson [65] shows that this reduction is very stable numerically.

Since Q is orthogonal, the QR factorization gives a Cholesky factorization $B^TB=R^TQ^TQR=R^TR$ of B^TB into a product of a lower triangular matrix and its transpose. The simplex method then can be implemented as follows. Define $\bar{\pi}$ by $\pi=\bar{\pi}B^T$. The systems $\pi B=\gamma$ and $By=\alpha$ in (1) become

$$(\bar{\pi}B^T)B=\bar{\pi}(B^TB)=\bar{\pi}(R^TR)=\gamma \text{ and } B^TBy=(R^TR)y=B^T\alpha$$

and can be solved by forward and backward substitution as in the LU factorization described previously.

The factorization $B=QR$ or $Q^TB=R$ can be updated in ways similar to the updating procedures used in LU factorization. For example, by placing the incoming column $z=Q^T\alpha$ at the end of the factor R and dropping the outgoing column, we can recover the factorization by eliminating subdiagonal elements. Now, however, orthogonal matrices must be used in place of elementary matrices to maintain a QR factorization. Plane rotations can be used for this purpose.

In a series of papers (Gill and Murray [70], Saunders [72a, 72b], Gill, Murray and Saunders [75] and Gill [76]), Gill, Murray and Saunders discuss many further details and variants for the QR method such as a product form implementation and several modifications aimed at reducing computations and saving storage. In particular, they propose factoring B^T, instead of B, into a QR form. As a consequence, the updating procedure must insert a row into R when the basis is changed.

This variant of the QR method does not require Q for any operations and saves storage; partial information about Q is generated easily when needed. Saunders also shows how the QR factorization exploits macro structure for GUB systems, staircase systems, and block angular generalizations of GUB systems. Our discussion follows Bartels, Golub, and Saunders' [70] description of QR factorization for linear systems.

The choice of a "best" implementation of the simplex method remains unsettled, and much further empirical testing and theoretical development is required before final judgement can be made. Nevertheless, the LU factorization provides flexibility, and the prospect of numerical stability, not given by a basis inversion implementation. Moreover, much of the "implementation art" for reinversion codes that has evolved over the years, such as multiple pricing (solving a subproblem restricted to a subset of the variables, with data often kept in core), remains valid for the factorization implementations. It is quite likely, then, that LU factorization will replace basis inversion in future commercial systems. (It has been used in some codes already.) QR methods seem to require more computational effort for problems without macro structure, but their ability to exploit macro structure and, at the same time, to provide numerical stability, suggests that they be tested and developed further.

Improvements on the simplex method may come from other sources as well. Kalan [71] suggests a data structure which exploits the observation that many of the nonzero coefficients in any linear programs are identical. Harris [72] and Goldfarb [76] have studied alternate rules for choosing the incoming column hoping to reduce the number of simplex iterations. Several researchers have investigated the question of what sequence of pivots provides the sparsest factorization representation. Bunch and Rose's [74] and Tarjan's [72] combinatorial investigations illustrate the nature of this work and provide additional references.

Research, such as Dantzig [73], Kallio [75] and Winkler [74], continues the search for new procedures for exploiting problem structure. In the context of ℓ_1 regression, Barrodale and Roberts [73] and Syropoulos, Kiountouzies, and Young [73] show that the steps of the simplex method, and not just its implementation, can be altered to exploit structure. For this application, their algorithm skips some of the bases changes required by the simplex method.

III. Minimum Cost Network Flow:

Most algorithms, and particularly those involving networks, can be improved by careful data management. Recent research for maximal flow (Edmonds and Karp [72]), for minimal spanning trees (Kershenbaum and Van Slyke [72]), for vehicle routing heuristics (Golden, Nguyen, and Magnanti [75]), and for shortest paths (Pierce [75] gives an extensive 431 source bibliography related to these problems), all illustrate this point.

Current and very active algorithmic research for minimum cost network flow models further demonstrates the importance of careful data management, and for an important problem with widespread application. Building upon work of Johnson [66], researchers centered at the University of Texas at Austin recently have developed a new simplex based code PNET for minimum cost network flow which improves significantly upon previous codes, and which has spurred much additional research for this class of problems. This work, which has been reported in a series of articles culminating with Glover, Klingman, Stutz [74], Glover, Karney, Klingman, and Napier [74] and the survey by Charnes et. al. [73], applies to the problem

$$\text{minimize } \{\sum_i \sum_j c_{ij}x_{ij} : \sum_j x_{ij} - \sum_k x_{ki} = b_i \quad (i=1,\dots,m)$$
$$0 \le x_{ij} \le u_{ij} \quad \text{all arcs } (i,j)\}.$$

Each summation in this formulation is limited to indices corresponding to arcs (i,j) and (k,i) in the network. We assume that the network is connected. Otherwise, we can solve the problem on each connected component separately.

When applied to this problem, the simplex method, as modified in minor ways for the upper bounds on variables, solves the systems $\pi B=\gamma$ and $By=\alpha$ with a basis B limited to the equality constraints (one of the equality constraints is redundant so that B contains $m-1$ linearly independent columns). In this case the systems become easy to solve because any basis is triangular (see Dantzig [63b]). Moreover, every basis corresponds to a spanning tree on the network. PNET takes advantage of this property by exploiting efficient list processing structures to store and manipulate trees.

Figure 5 shows a basis for a 12 node network flow problem. The tree has been rooted arbitrarily at node 4; any node can serve this purpose.

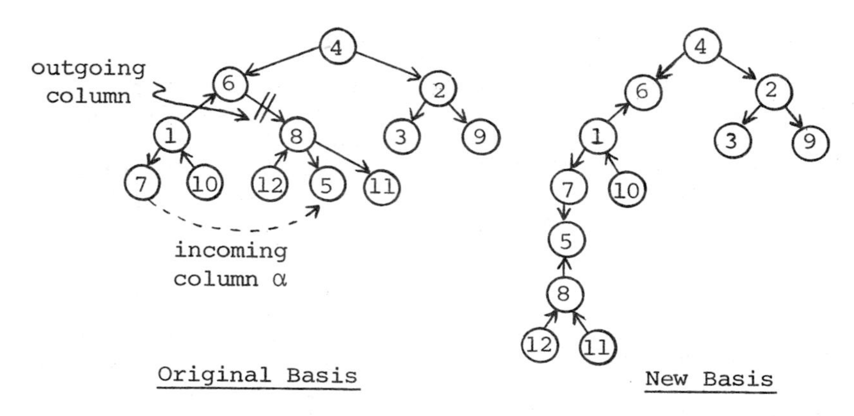

Fig. 5. Spanning Tree for a Network Basis.

Note that the tree can be specified, without left-right orientation, by knowing the predecessor (node above) each node. The left-right orientation is specified by knowing the brother of every node which is defined as the node, if any, to the immediate left of the node. Consequently, this basis can be stored on a computer, together with the problem data by keeping the following tables:

Tree (Basis) Data

Node i	Predecessor	Brother	Successor	Start i
1	6	-	10	1
2	-4	6	9	4
3	-2	-	-	6
4	-	-	2	14
•				
•				
•				

Arc Data (full network)

End j	x_{ij}	c_{ij}	u_{ij}
3	x_{13}	c_{13}	u_{13}
6	x_{16}	c_{16}	u_{16}
7	x_{17}	c_{17}	u_{17}
3	x_{23}	c_{23}	u_{23}
9	x_{29}	c_{29}	u_{29}
1	x_{31}	c_{31}	u_{31}
⋮			

Table 1. Partial Data for a 12 Node Network and a Basis

The sign in the predecessor column indicates the arc orientation from the node to its predecessor. The column entitled successor specifies the node immediately below node i in the tree which lies as much to the right as possible. For example, node 8, and not node 1, is the successor to node 6 since node 8 lies to the right of node 1. Although the successor data is redundant, it is convenient to have explicitly for any algorithm that moves down the tree frequently. Various applications involving trees will store part or all of the predecessor, brother, successor data depending upon how the tree is to be used.

The arc data is stored by grouping arcs with the same initial node; the final column in the tree data points to the arc table to indicate where the arcs starting with node i begin. Also, for simplex implementations the arc data x_{ij} only needs to be stored for arcs (i,j) in the tree. The flows x_{ij} on other arcs always will be 0 or u_{ij} and their values at any step can be specified by a flag.

To see how these list structures can be used, suppose that column α entering the basis corresponds to arc (7,5). Then α contains a +1 coefficient in row (equation) 7 and a -1 coefficient in row 5 because variable x_{75} appears only in these two equations and with these coefficients. The system $By=\alpha$ can be solved by finding the unique cycle that α determines when it is added to the spanning tree.

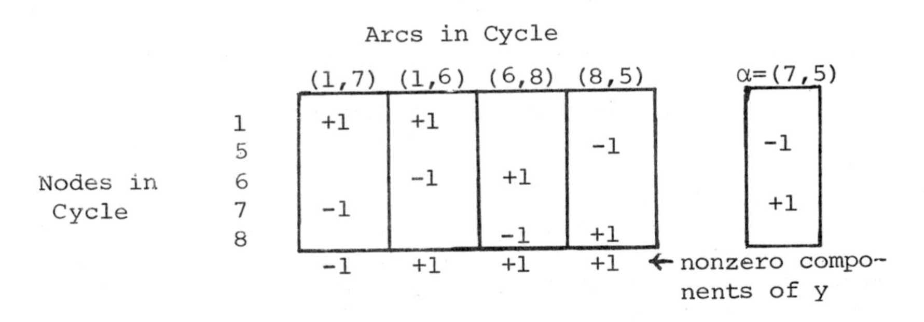

Nodes in Cycle	Arcs in Cycle (1,7)	(1,6)	(6,8)	(8,5)	α=(7,5)
1	+1	+1			
5				-1	-1
6		-1	+1		
7	-1				+1
8			-1	+1	
← nonzero components of y	-1	+1	+1	+1	

Table 2. Solving $By=\alpha$ for a Network.

As Figure 5 and Table 2 illustrate, the solution y to $By=\alpha$ has zero components for arcs not in this cycle, has -1 components for arcs in the cycle with the same orientation as α, and has +1 components for arcs in the cycle with orientation opposite to α. But these computations are easy to perform using the list structure in Table 1. Starting from node 7 and node 5 use the predecessor data to find the paths P_7 and P_5

from each of these nodes to the root. Deleting arcs common to these two paths specifies the arcs of the spanning tree in the cycle determined by α.

The list structure for the new spanning tree is also easy to compute. Since the arc dropping from the basis lies on path P_5 just above node 8, we reverse the path joining node 8 and 5, here a single arc, and add it below arc $\alpha=(7,5)$ in the new tree. All arcs attached to path P_5 in the previous spanning tree, here arcs (12,8) and (8,11), must be moved as well. The list structure describing the tree is modified accordingly.

The list structure permits the system $\pi B=\gamma$ to be solved easily, as well. Because each column of the node-arc incidence matrix B contains only a +1 and a -1, the equations $\pi B=\gamma$ read

$$\pi_i-\pi_j=\gamma_{ij} \tag{3}$$

and one component of π corresponds to each node in the network. Because one equation in the system is redundant, one component of π can be set arbitarily. Consequently, if π_4 is set to its previous value after the basis change, then the π components for every node not below node 8 in the previous spanning tree are unchanged. In particular, since π_7 does not change, the new value for π_5 is computed directly from $\pi_7-\pi_5=\gamma_{75}$. If Δ is the change in π_5, then every node below node 5 in the new tree also changes by Δ in order to maintain the conditions (3). Consequently, the values for π are easy to recover from step to step.

For further details about how this procedure is implemented most efficiently (for example with only two of the predecessor, brother, successor indices for each node) see the references cited above. Srinivasan and Thompson [73] report results related to PNET. Mulvey [75] has successfully modified the PNET code in several ways, by adding such facilities as multiple pricing of columns. Also, independent of the development of PNET, Graves and McBride [73] have studied data manipulation for network codes from a different viewpoint. Bradley, Brown, and Graves [75] have indicated the connection between the two approaches and have suggested additional mechanisms for exploiting data structure. They report computation times as much as twice as fast as the PNET results given in Klingman, Napier, Stutz [74], which were very fast themselves solving 1500 node, 4300-5700 arc problems in 15-20 seconds on a CDC 6600.

Aashtiani and Magnanti [75] have applied similar list

processing techniques to primal-dual, or out-of-kilter, algorithms for the minimum cost flow problem. These algorithms are tree constructing in the sense that they begin with the arc α and construct a tree around it until either a cycle is found containing α or a change can be made in the vector π. Their implementation stores forests (unconnected trees) instead of the spanning tree used in the simplex method, and "ties" the forests together when constructing the tree for α. This implementation gives significant improvements over previous codes. Barr, Glover, and Klingman [74] recently have suggested another implementation for the out-of-kilter algorithm.

IV. Dual Approaches for Sparse Systems

In the next two sections we consider the use of constructive duality for optimizing sparse systems. We consider the general optimization problem

$$v=\inf\{f(x):x\in X \text{ and } g(x)\leq 0\} \tag{4}$$

where X is a given subset of R^n, f is a real valued function and g is function with values in R^m, both defined on the set X. We call (4) the primal problem and assume that it contains at least one feasible solution. Since this primal problem can be specialized to be a linear program, the techniques that we discuss apply to each of the problems considered in the last two sections.

For any m-dimensional vector $u\geq 0$, define the Lagrangian function (also called the subproblem)

$$L(u)=\inf\{[f(x)+ug(x)]:x\in X\}.$$

It is well known (see Rockafellar [70]) that $L(u)\leq v$ and that $L(u)$ is upper semi-continuous and concave on its domain $D\equiv\{u:L(u)>-\infty\}$. The dual problem is to find the best Lagrangian lower bound to v given by

$$d=\sup\{L(u):u\in D \text{ and } u\geq 0\}. \tag{5}$$

If the primal problem is a linear program or a convex program (X convex, f and g convex on X) satisfying the Slater condition ($g(x)<0$ for some $x\in X$), then $v=d$ and solving the dual problem determines the optimal value, and frequently an optimal solution, to the primal problem. In general, however, duality gaps, $d<v$, exist for nonconvex problems.

The usual assumption when forming the dual problem is that the optimization problem involved in each Lagrangian evaluation is much simplier to solve than the original problem, so that it becomes attractive to iterate over the variables u in the dual problem. This notion has been recognized for some time, and the dual problem has been proposed and used in limited applications as a computational device for solving optimization problems. More recently, the fundamental importance of duality in optimization is becoming more apparent, not only qualitatively, but algorithmically. For example, penalty methods in nonlinear programming are being interpreted usefully as dual methods (see Luenberger [73]). Even in integer programming, where duality gaps are common, dual ap-approaches are being used in conjunction with branch and bound both to devise new algorithms and to unify and extend such concepts as surrogate constraints (Fisher and Shapiro [74], Geoffrion [74b]).

Because the Lagrangian function is generally nondifferentiable, its solution requires techniques extending classical differentiable procedures of nonlinear programming. The next section summarizes some recent research in this area. Before beginning this discussion, though, let us briefly recall some uses of duality for optimizing sparse systems. Further applications are given in the surveys, texts, and collections cited at the beginning of this paper.

(i) Inducing Seperability

Suppose that $X = X_1 \times X_2 \times \cdots X_k$ is the product of k sets so that any point $x \in X$ is expressed in terms of subvectors $x_j \in X_j$ as $x=(x_1,x_2,\cdots,x_k)$. Then if f and g are additively separable as $f(x) = f_1(x_1)+f_2(x_2)+\cdots+f_k(x_k)$ and $g(x) = g_1(x_1) + g_2(x_2)+\cdots+g_k(x_k)$, the Lagrangian subproblem

$$L(u)=\inf_{x \in X}[f(x)+ug(x)] = \sum_{j=1}^{n} \inf_{x_j \in X_j}[f_j(x_j)+ug_j(x_j)]$$

separates into several problems, one for each subvector of x. For each choice of u, the dual decomposes into several smaller optimization problems. In linear programming applications, each set X_j consists of linear constraints, such as GUB constraints, imposed upon only a subset of the problem variables.

(ii) Network subproblems:

Whenever the set X consists of network constraints, evaluation of the Lagrangian L(u) becomes a network problem. In multicommodity flow problems, for example, X contains a separate minimum cost flow constraints for each of several commodities. The constraints $g(x) \leq 0$ impose capacity limitations on the total flow of all commodity flows on each arc. The Lagrangian L(u) then separates, as above, but now as several minimum cost flow problems, one for each commodity. The algorithms of the next section can be used in its solution.

In their studies of the traveling salesman problem, Held and Karp [70,71] use a Lagrangian whose evaluation is a minimal spanning tree (or a slight variant called a 1-tree) computation. In his study of integer programs, Shapiro [71] shows how the usual group theoretic relaxation for an integer program arises in the context of dualizing integer programs. In this case, each Lagrangian evaluation involves a shortest path problem on the network induced by the group structure of the integer program. Fisher [72,73] uses network subproblems for solving machine sequencing problems.

(iii) Complicating variables:

In many applications, certain of the variables y in the primal problem are complicating in the sense that the problem becomes much easier to solve with only the remaining variables w when y is fixed. Assuming that the implicit constraint $x=(w,y)\in X$ separates into $w\in W$ and $y\in Y$ for given sets W and Y, it then becomes attractive to solve the problem by iterating over $y\in Y$ and solving:

$$v(y)=\inf\{f(w,y):g(w,y)\leq 0,w\in W\} \qquad (6)$$

which has the dual problem

$$d(y)=\sup_{u\geq 0} L(u;y) \text{ where } L(u;y)=\inf_{w\in W}\{f(w,y)+ug(w,y)\}$$

If there is no duality gap and $v(y)=d(y)$, then the original primal problem can be restated as

$$\inf\{d(y):y\in Y\cap\bar{Y}\}$$

where $\bar{Y}$ is the set of values of y for which problem (6) is feasible. The problem now can be solved by exploiting solution procedures for the dual problem.

This formulation was proposed initially by Benders [62] when (6) is a linear program and was generalized later by Geoffrion [72a]. Geoffrion further describes $\bar{Y}$ explicity as

$$\bar{Y}=\{y:\inf_{w\in W}[\lambda g(w,y)]\leq 0 \quad \text{for all } \lambda\geq 0\}$$

when X is a closed convex set and g(w,y) is lower semi-continuous and convex for each choice of y.

Benders originally applied this approach to mixed integer programming where fixing the vector y of integer variables gives a linear program. The method has been applied successfully to network distribution problems (Geoffrion and Graves [74]) where fixing the y variables gives a minimum cost network flow problem. Geoffrion's generalization of the method has been used to compute network equilibrium (Florian and Nguyen [74]).

V. Solving the Dual

In this section we summarize several techniques, each using an approximation scheme, for solving the dual problem (5). When applied to linear programs, these techniques compete with the simplex based methods discussed in sections II and III, much like iterative techniques for solving systems of linear equations compete with direct, or pivoting, methods for these problems.

The three methods that we consider might be called finite approximation, subgradient approximation, and dual ascent. They are illustrated in Figure 6.

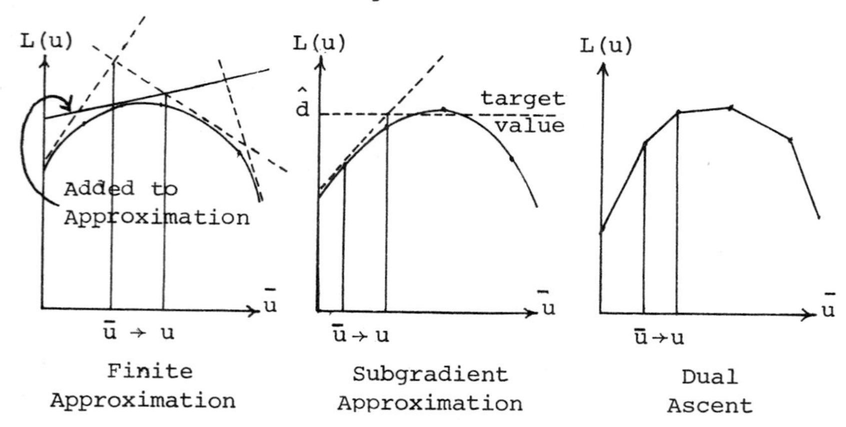

Fig. 6: Solving The Dual Problem

In the finite approximation method, the set X is replaced by a finite subset S of X to give an approximate Lagrangian $\bar{L}(u)$. $\bar{L}(u) \geq L(u)$ for any u and the dual approximation $\bar{d} = \sup\{\bar{L}(u): u \geq 0 \text{ and } u \in D\}$ satisfies $\bar{d} \geq d$. The dual approximation can be rewritten as

$$\bar{d} = \sup_{\substack{u \geq 0 \\ u \in D}} \{z: f(x) + ug(x) \geq z \text{ for all } x \in S\} \tag{7}$$

since the constraints and objective function imply that $z=\bar{L}(u)$ for any choice of u. This form of the problem shows the advantage of the approximation; (7) is a linear program with variables z and u and one constraint for each point x in S, as long as D is polyhedral. (In particular, if X is compact and f and g are lower semi-continuous, then $D=R^n$ and (7) is a linear program.)

After solving this linear program, the method next solves the Lagrangian subproblem to evaluate $L(\bar{u})$ at the solution $\bar{u}$ to (7). If $L(\bar{u})\geq\bar{d}$ then $\bar{u}$ solves the dual problem, since the inequalities $d\geq L(\bar{u})\geq\bar{d}\geq d$ on the value d of the dual problem are implied by the definition of d and the fact that $\bar{d}$ overapproximates d. If, on the other hand, $\bar{x}$ solves the Lagrangian subproblem at $\bar{u}$ and $L(\bar{u})=f(\bar{x})+ug(\bar{x})>\bar{d}$, then $\bar{x}$ corresponds to the most violated constraint in (7) when X replaces S. In this case; $\bar{x}$ is added to S and the procedure is repeated by solving the linear program (7) again.

The finite approximation method usually is called Dantzig-Wolfe decomposition or generalized programming. It was first proposed by Dantzig and Wolfe [61] when the primal problem is a linear program (they describe D explicitly as a cone generated by extreme rays of X and modify the method to approximate D as well as X). Dantzig [63b] proves that any limit point of the u's generated by the algorithm solves the dual whenever X is convex and compact and f(x) and g(x) are convex, and also shows that the algorithm can be used to find an optimal solution to the primal problem. The algorithm converges without any convexity assumptions when X is compact and f and g are lower semi-continuous even when there is a duality gap v>d. In this case the optimal dual value equals the value of an approximately defined convexified version of the primal problem (Magnanti, Shapiro, and Wagner [73]). The method also can be viewed as a cutting plane algorithm (Zangwill [69]).

Recently, Marsten, Hogan and Blankenship [75] have proposed a modification in which the value u* for the new solution is constrained at each step to lie in a box $\bar{u}_i-\beta_i\leq u_i\leq \bar{u}_i+\beta_i$ about the old point $\bar{u}$ for given parameters $\beta_i>0$. Having obtained u* as the solution to (7) with the added box constraints, they permit a one dimensional search on $L(u^*+\Theta[u^*-\bar{u}])$ for $\Theta\geq 0$ to determine the next point $\bar{u}$ to place the box about. For small values of β the BOXSTEP procedure tends to behave like steepest ascent, while for β sufficiently large it behaves like Dantzig-Wolfe decomposition. Computational experience indicates an "intermediate" value for β is best since the number of times the linear program (7) must be solved to become within a given $\varepsilon>0$ of d increases with β whereas the solution time to solve each linear program decreases with β.

Marsten [75] gives further computational experience with the method and introduces several refinements.

The decomposition procedure has been modified in other ways for staircase systems arising in time dependent problems. Glassey [73] and Manne and Ho [74] describe nested versions of the method in which decomposition is applied to the Lagrangian evaluations as well. The repeated use of decomposition for evaluating Lagrangians is carried out until one Lagrangian is identified for each time period. Manne and Ho's implementation is novel in that the Lagrangian problem is not solved to completion at each step.

Solving the linear programs (7) is the greatest computational burden when applying finite approximation to many problems. The subgradient procedure avoids these evaluations. Starting with a target value $\hat{d}<d$, the method attempts to find a value $\bar{u}$ for the dual variables satisfying $L(\bar{u})\geq\hat{d}$. It accomplishes this by making linear approximations to $L(u)$ at any potential solution $\bar{u}$ in terms of the concept of a subgradient of a concave function which is any vector γ satisfying.

$$L(u) \leq L(\bar{u})+\gamma(u-\bar{u}) \quad \text{for all } u. \tag{8}$$

When L is differentiable, the gradient is the unique subgradient at $\bar{u}$.

The subgradient method uses the right-hand-side in (8) as a linear approximation to $L(u)$ at $\bar{u}$ and finds the next potential solution from

$$L(\bar{u})+\gamma(u-\bar{u})=\hat{d}.$$

Setting $u=\bar{u}+\theta\gamma$ in this expression, the parameter θ is given by

$$\theta = \frac{\hat{d}-L(\bar{u})}{\|\gamma\|^2} \tag{9}$$

If the new point $\bar{u}+\theta\gamma$ has negative components or lies outside of D, it is projected onto the set $\{u\varepsilon D: u\geq o\}$, and the procedure is repeated.

This method satisfies the unusual property that the u's generated by the method converge, and, moreover, that the new point u generated at each step always lies closer to any solution $\hat{u}$ to $L(u)\leq\hat{d}$ than the previous point $\bar{u}$, i.e., $\|u-\hat{u}\| < \|\bar{u}-\hat{u}\|$. In fact, the same properties hold even if we use over or under relaxation by multplying the right-hand-side in (9) by a constant s_j at each step, as long as $0<\varepsilon<s_j<2$ and the resulting values θ_j of θ at each step satisfy $\theta_j\rightarrow 0$ and $\Sigma\theta_j=+\infty$. With these various modifications the procedure can be viewed as an analog of Newton's method (see figure 6) or of relaxation procedures for linear equality systems (Ortega and Reinboldt [70] provide a comprehensive discussion of these methods).

Finally, to implement the algorithm we must be able to generate a subgradient γ at each step. Fortunately evaluating the Lagrangian $L(\bar{u})$ provides a subgradient automatically. Suppose that $\bar{x}$ solves the Lagrangian subproblem so that $L(\bar{u})=f(x)+\bar{u}g(\bar{x})$. For any other choice at u, $L(u)\leq f(\bar{x})+ug(\bar{x})$ by the definition of L(u). Subtracting these two expressions gives $L(u)\leq L(\bar{u})+g(\bar{x})(u-\bar{u})$. Consequently, the function values $g(\bar{x})$ for any solution to the Lagrangian subproblem can be used as the subgradient γ. In fact, any convex combination of $g(\bar{x})$ evaluated at optimal solutions $\bar{x}$ also is a subgradient since the set of subgradients at any point forms a convex set.

Agmon [54] and Motzkin and Schoenberg [54] first proposed this subgradient approach for solving systems of linear inequalities. Held and Karp [70,71] have used the method successfully for the traveling salesman problem, and Held, Karp, and Wolfe [72] and Held, Wolfe, and Crowder [74] have applied the method to several optimization problems. They find that, in practice, the method seems to converge even if the target value is unattainable, that is, $\hat{d}>L(u)$ for all u. Polyak [69] shows convergence when the relaxation parameters s_j are introduced. Bertsekas and Mitter [73] introduce the concept of an ε-subgradient and an ε-subgradient optimization procedure in the context of Fenchel's equivalent to Lagrange duality (Magnanti [74]). The ε-subgradient concept is useful since any ε-solution $\bar{x}$ to the Lagrangian evaluation corresponds to an ε-subgradient $g(\bar{x})$ to L(u), and ε-solutions are all that we may be able to find when evaluating the Lagrangian in many applications. Polyak and Bertsekas and Mitter also give several sources to Soviet scientists who have studied nondifferentiable optimization. Our discussion follows Oettli's [72] highly recommended presentation of the subgradient method.

Subgradients have been used by Lemarechal [75] and Wolfe [75] to give conjugate direction algorithms for nondifferentiable optimization. Subgradient algorithms also have been proposed by Balas [74] and Marsten, Northrup, and Shapiro [74] for solving linear programs associated with set covering problems which are highly degenerate and difficult to solve by the simplex method. Bertsekas [75] and Madsen [75] are additional sources on nondifferentiable optimization.

The method of dual ascent is a feasible direction approach for solving the dual problem. We assume in this discussion that X is a finite set and initially that the domain D of L(u) is R^m. Given a current dual feasible point $\bar{u}\geq 0$, the method finds a direction of ascent h for L and chooses a new value for u as $\bar{u}+\theta h$ for a suitable parameter $\theta>0$.

Let S denote the set of optimal solutions to the Lagrangian subproblem at $\bar{u}$, i.e. $S=\{x\varepsilon X: f(x)+\bar{u}g(x)=L(\bar{u})\}$.

The method first solves a direction finding linear program:

$\bar{\sigma} = \max\{\sigma: h_i \geq 0 \text{ if } \bar{u}_i=0, -1 \leq h_i \leq 1, \text{ and } hg(x) \geq \sigma \text{ for all } x \in S\}$. The conditions $h_i \geq 0$ and the conditions on $hg(x)$ give, as we shall see, a feasible direction h of ascent for L(u).

This optimization can have two outcomes:

(i) If $\bar{\sigma}=0$, then $\bar{u}$ solves the dual problem. To see this, note that the optimal dual variables to the direction finding linear program are nonnegative multipliers $\lambda(x)$ defined for $x \in S$ which sum to one and satisfy (since $h_i=0$ is optimal)

$$\Sigma\lambda(x)g(x) \leq 0 \quad \text{and} \quad \bar{u}[\Sigma\lambda(x)g(x)]=0. \tag{10}$$

In these expressions, and those to follow, the indices of summation are limited to $x \in S$. These conditions imply that $u[\Sigma\lambda(x)g(x)] \leq 0$ for any $u \geq 0$ and thus, since $L(u) \leq f(x)+ug(x)$ for all $x \in X$, that

$$L(u) \leq \Sigma\lambda[f(x)+ug(x)] \leq \Sigma\lambda(x)f(x). \tag{11}$$

But, for $u=\bar{u}$, $L(\bar{u})=f(x)+\bar{u}g(x)$ for all $x \in S$ and $\bar{u}[\Sigma\lambda(x)g(x)]=0$ so that equality holds throughout (11). Consequently,

$$L(u) \leq L(\bar{u}) = \Sigma\lambda(x)f(x) \quad \text{for all } u \geq 0$$

and $\bar{u}$ solves the dual problem. (As we have noted before, $\gamma=\Sigma\lambda(x)g(x)$ is a subgradient to L(u) at $\bar{u}$; thus conditions (10) become $\gamma \leq 0$ and $\bar{u}\gamma=0$ which are the subgradient analogs of the optimality conditions $\nabla L(\bar{u}) \leq 0$ and $\bar{u}\nabla L(\bar{u})=0$ for maximizing a differentiable function L(u) over $u \geq 0$.)

(ii) If $\bar{\sigma}>0$, then any solution h is a direction of ascent for L(u) with $L(\bar{u}+\theta h)$ increasing by $\theta\bar{\sigma}$ as θ increases from 0. In fact, $L(\bar{u}+\theta h) \geq L(\bar{u})+\theta\bar{\sigma}$ whenever $f(x)+[\bar{u}+\theta h\; g(x) \geq L(\bar{u})+\theta\bar{\sigma}$ for all $x \in X$, or equivalently whenever 0 satisfies

$$0 \leq \theta \leq \bar{\theta} \equiv \min\{\frac{L(\bar{u})-f(x)-\bar{u}g(x)}{hg(x)-\bar{\sigma}} : x \in X \text{ and } hg(x)-\bar{\sigma}<0\}. \tag{12}$$

Since $hg(x) \geq \sigma$ for all $x \in S$, since X is finite, and since $L(\bar{u}) < f(x)+\bar{u}g(x)$ for all x in X, but not in S, the upper bound $\bar{\theta}$ on θ is strictly positive. We change $\bar{u}$ to $\bar{u}+\theta h$ where $\theta \leq \bar{\theta}$ is chosen as large as possible so that every component of $\bar{u}+\theta h$ remains nonnegative. The procedure then is repeated from the new point $\bar{u}+\theta h$.

Several papers have contributed ideas related to this dual ascent algorithm (Abadie and Williams [63], Balas [66], Bell [65], Bradley [67], Fisher and Shapiro [74], Geoffrion [70b], Gilmore and Gomory [63], Grinold [72], Lasdon [68], Silverman [68], and Zschau [67]). As Grinold notes, the direction finding view of the method presented here is equivalent to applying the primal-dual algorithm of linear programming to the linear programming dual problems

$$\max\{\Sigma\lambda(x)f(x): \Sigma\lambda(x)g(x)\leq 0,\ \Sigma\lambda(x)=1,\ \lambda(x)\geq 0\quad \forall x\varepsilon X\ \}.$$

and

$$\min\{\sigma:\ f(x)+ug(x)\geq\sigma \text{ for all } x\varepsilon X\}.$$

Several of these papers also note that (12) is a fractional programming problem which can be solved efficiently when f and g are linear-affine and the points in X are the extreme points of a polyhedral region. (The paper by Bitran and Magnanti [74] gives many references for fractional programming).

The dual ascent method described here can be modified easily when the domain D of L(u) is a polyhedral set. Grinold [72] specifies such an extension when the problem (5) is generated from a linear program. Fisher and Shapiro [74] describe D when applying dual approaches to integer programming problems. Both of these papers modify the basic procedure to use a subset of S, the set of optimal solutions to $L(\bar{u})$, and generate further points from S as needed. Fisher and Shapiro also note that the computational expense of solving the direction finding linear program, of generating points from S as needed, and of computing (12) may be offset by the fact that of the three methods describe here, only dual ascent provides monotonically increasing lower bounds $L(\bar{u})$ to v. These bounds are useful when applying branch and bound procedures to integer programs. Fisher, Northrup, and Shapiro [75] report computation experience with the method for integer programming applications.

Geoffrion [70b] shows how a variant of the dual problem (5) arises in resource directive approaches to optimization problems with "complicating constraints" and suggests several algorithms related to those presented in this section. In a related paper Geoffrion [72a] shows how a finite approximation method similar to the method presented here can be used to solve the optimization problems with complicating variables introduced in the last section.

In a promising new development, Bell and Shapiro [75] have shown how to resolve duality gaps in integer programming. They generate a sequence of integer programming dual problems which converge to the optimal primal value v to an integer program. The dual problems are large scale linear programs which decrease in size at each step, but at the expense of increasing computation for evaluating the Lagrangian.

In other recent paper of note, Cottle, Golub, and Sacher [74] propose a method for solving sparse linear complementarity problems that combines iterative overrelaxation tech niques with direct pivoting methods.

Acknowledgements

Preparation of this paper was supported in part by the Office of Naval Research (contract N00014-75-C-0556), by the Army Research Office, Durham (contract DAHC04-73-C-0032), and by the Department of Transportation (contract DOT-TSC-1058). I am grateful to Jerry Shapiro for many discussions and insights concerning constructive duality.

BIBLIOGRAPHY

AASHTIANI, H. and MAGNANTI, T.L., "Solving Large Scale Network Optimization Problems," presented at ORSA/TIMS National Meeting, Chicago, May 1975.

ABADIE, J.M. and WILLIAMS, A.C., "Dual and Parametric Methods in Decomposition," in Recent Advances in Mathematical Programming, (R. Graves and P. Wolfe, Eds.), McGraw-Hill, New York, 1963, pp. 149-158.

AGMON, S., "The Relaxation Method for Linear Inequalities," Can. J. of Math., 6, 1954, pp. 382-392.

BALAS, E., "An Infeasibility-Pricing Mechanism for Linear Programs," Opns. Res., 14(5), 1966, pp. 847-873.

———, Lectures on the Set Covering Problem, NATO Advanced Study Conference on Combinatorial Programming, Versailles, September 1974.

BARR, R.S., GLOVER, F. and KLINGMAN, D., "An Improved Version of the Out-of-Kilter Method and a Comparative Study of Computer Codes," Math. Prog., 7, 1974, pp. 60-86.

BARRODALE, I. and ROBERTS, F.D., "An Improved Algorithm for Discrete ℓ_1 Linear Approximation," SIAM J. on Num. Anal., 10(5), 1973, pp. 839-848.

BARTELS, R.H., "A Stabilization of the Simplex Method," Num. Math. 16, 1971, pp. 414-434.

——— and GOLUB, G.H., "The Simplex Method of Linear Programming Using LU Decomposition," Comm. ACM, 12, 1969, pp. 266-268.

———, ——— and SAUNDERS, M.A., "Numerical Techniques in Mathematical Programming," in Nonlinear Programming, (J. Rosen, O. Mangasarian, and K. Ritter, Eds.), Academic Press, New York, 1970, pp. 123-176.

BELL, E.J., "Primal-dual Decomposition Programming," Tech. Report ORC 65-23, Opers. Res. Center, Univ. of Cal., Berkeley, 1965.

BELL, D.E., and SHAPIRO, J.F., "A Finitely Convergent Theory For Zero-One Integer Programming," Tech. Report OR 043-75,

Opers. Res. Center, M.I.T., May, 1975.

BENDERS, J.F., "Partitioning Procedures for Solving Mixed Variables Programming Problems," Num. Math., 4, 1962, pp. 238-252.

BERTSEKAS, D.P., "Nondifferentiable Optimization via Approximation," Math. Prog., Study 3, 1975.

______, and MITTER, S.K., "A Descent Numerical Method for Optimization Problems with Nondifferential Cost Functionals," SIAM J. on Control, 11, 1973, pp. 637-652.

BITRAN,G., and MAGNANTI, T.L., "Duality and Sensitivity Analysis for Fractional Programs," Tech. Report OR-035-74. Opers. Res. Center, M.I.T., May 1974.

BRADLEY, G.H., BROWN, G.G. and GRAVES, G.W., "A Comparison of Storage Structures for Network Codes," presented at National ORSA/TIMS Meeting, Chicago, May 1975.

BRADLEY, S.P., "Decomposition Programming and Economic Planning," Tech. Report ORC 67-20, Opers. Res. Center, Univ. of Calif. Berkeley, June 1967.

BRAYTON, R.K., GUSTAVSON, F.G. and WILLOUGHBY, R.A., "Some Results on Sparse Matrices," RC-2332, IBM Research Center, Yorktown Heights, N.Y., February 1969.

BUNCH, J.R., and ROSE, D.J., "Partitioning, Tearing and Modification of Sparse Linear Systems," J. of Math. Anal. and Applic., 48, 1974, pp. 574-593.

CHARNES, A.F., GLOVER, F., KARNEY, D., KLINGMAN, D. and STUTZ, J., "Past, Present and Future of Development, Computational Efficiency, and Practical Use of Large-Scale Transportation and Transhipment Codes," Tech. Report C.S. 131, Center for Cybernetic Studies, Univ. of Texas, Austin, July 1973, to appear in Computers and Operations Research.

COTTLE, R.W., GOLUB, G.H. and SACHER, R.S., "On the Solution of Large, Structured Linear Complementarily Problems: III," Tech. Report 74-7, Dept. of Opers. Res., Stanford Univ., June 1974.

______ and KRARUP, J., Optimization Methods for Resource Allocation, The English Universities Press, London, 1974.

DANTZIG, G.B., "Compact Basis Triangularization for the Simplex Method," in Recent Advances in Mathematical Programming (R. Graves and P. Wolfe, Eds.), McGraw-Hill, New York, 1963a, pp. 125-132.

______, Linear Programming and Extensions, Princeton Univ. Press, Princeton, N.J., 1963b.

______, "Large Scale Linear Programming," in Mathematics of the Decision Sciences Part I, (G. Dantzig and A. Veinott, Eds.), Amer. Math. Soc., Providence, R.I., 1968, pp. 77-92.

______, "Solving Staircase Linear Programs by a Nested Block-

Angular Method," Tech. Report 73-1, Dept. of Opers. Res., Stanford Univ., January 1973.

______ and VAN SLYKE, R.M., "Generalized Upper Bounding Techniques," J. Computer System Science, 1, 1967, pp. 213-226.

______ and WOLFE, P., "Decomposition Principle for Linear Programs," Econometrica, 29(4), 1961, pp. 767-778. See also Opns. Res., 8(1), 1960, pp. 101-111.

EDMONDS, J. and KARP, R.M., "A Labeling Method for Maximal Network Flows which is Bounded by a Polynomial in the Number of Nodes," JACM, 19(2), 1972, pp. 248-264.

FISHER, M.L., "Optimal Solution of Scheduling Problems Using Lagrange Multipliers, Part I, Opns. Res., 21(5), 1973, pp. 1114-1127.

______, "Optimal Solution of Scheduling Problems Using Lagrange Multipliers, Part II," Symposium on Theory of Scheduling and Its Applications, North Carolina State Univ., Raleigh, 1972.

______, NORTHRUP, W. and SHAPIRO, J.F., "Using Duality to Solve Discrete Optimization Problems: Theory and Computational Experience," Math. Prog., Study 3, 1975.

______, and SHAPIRO, J.F., "Constructive Duality in Integer Programming," SIAM J. on Applied Math., 27, 1974, pp. 31-52.

FLORIAN, M. and NGUYEN, S., "A Method for Computing Network Equilibrium with Elastic Demands," Trans. Sci., 8, 1974, pp. 321-332.

FORREST, J.J.H. and TOMLIN, J.A., "Updating Triangular Factors of the Basis to Maintain Sparsity in the Product Form Simplex Method," Math. Prog., 2, 1972, pp. 263-278.

GEOFFRION, A.M., "Elements of Large Scale Mathematical Programming, Parts I and II," Man. Sci. 16, 1970a, pp. 652-691.

______, "Primal Resource-Directive Approaches for Optimizing Nonlinear Decomposable Systems," Opns. Res., 18(3), 1970b, pp. 375-403.

______, "Generalized Benders Decomposition," JOTA, 10(4), 1972a, pp. 237-260.

______ (Ed.), Perspectives in Optimization, Addison-Wesley, Reading, Mass., 1972b.

______, "A Guided Tour of Recent Practical Advances in Integer Linear Programming," Working Paper No. 220, Western Management Sci. Inst., UCLA, September 1974a.

______, "Lagrangian Relaxation and Its Uses in Ineger Programming;" Math. Prog., Study 2, 1974b, pp. 82-114.

______ and GRAVES, G., "Multicommodity Distribution System Design by Benders Decomposition," Man. Sci., 5, 1974c, pp. 822-844.

GILL, P., "The Use of Orthogonal Factorizations in Large-

Scale Linear Programming," this volume, 1976.
______ and MURRAY, W., "A Numerically Stable Form of the Simplex Algorithm," Tech. Report No. Maths. 87, National Physical Laboratory, Teddington, 1970.
______, ______, and SAUNDERS, M.A., "Methods for Computing and Modifying the LDV Factors of a Matrix," Math. Comp., 29 (32), 1975.
GILMORE, P.E., and GOMORY, R.E., "A Linear Programming Approach to The Cutting Stock Problem - Part II," Opns. Res., 11, 1963, pp. 863-888.
GLASSEY, R., "Nested Decomposition and Multi-Stage Linear Programs," Man. Sci., 20(3), 1973, pp. 282-292.
GLOVER, F., KARNEY, D., KLINGMAN, D. and NAPIER, A., "A Computations Study on Start Procedures, Basis Change Criteria, and Solution Algorithms for Transportation Problems," Man. Sci., 20(5), 1974, pp. 793-813.
______, KLINGMAN, D., and STUTZ, J., "An Augmented Threaded Index Method for Network Optimization," INFOR, 12, 1974, pp. 293-298.
GOLDEN, B., NGUYEN, H. and MAGNANTI, T.L., "Implementing Vehicle Routing Algorithms," Tech. Report TR-115, Opers. Res. Center, M.I.T., September, 1975.
GOLDFARB, D., "Using the Steepest-edge Simplex Method Algorithm to Solve Sparse Systems." this volume, 1976.
GRAVES, G.W. and MCBRIDE, R.D., "The Factorization Approach to Large-Scale Linear Programming," Working Paper No. 208, Western Management Sci. Inst., UCLA, August, 1973.
GRINOLD, R.C., "Lagrangian Subgradients," Man. Sci., 17, 1970, pp. 185-188.
______, "Steepest Ascent for Large Scale Linear Program," SIAM Review, 14, 1972, pp. 447-464.
HARRIS, P.M.J., "Pivot Selection Methods of the DEVEX, L.P. Code," British Petroleum Co., Ltd., London, 1972.
HAX, A., "On Generalized Upper Bounded Techniques for Linear Programming," Tech. Report 67-17, Opers. Res. Center, Univ. of Calif., Berkeley, May, 1967.
HELD, M. and KARP, R.M., "The Traveling-Salesman Problem and Minimum Spanning Trees," Opns. Res., 18, 1970, pp.1138-1162.
______ and ______, "The Traveling-Salesman Problem and Minimum Spanning Trees: Part, II," Math. Prog., 1, 1971, pp. 6-25.
______, ______ and WOLFE, P., "Large Scale Optimization and the Relaxation Methods," Proceeding of the 25th Annual ACM Cong., Boston, 1972, pp. 507-509.
______, WOLFE, P. and CROWDER, H., "Validation of Subgradient Optimization," Math. Prog., 6, 1974, pp. 62-88.
HELLERMAN, E. and RARICK, D., "Reinversion with the Preassigned Pivot Procedure," Math. Prog., 6, 1974, pp. 62-88.

HIMMELBLAU, D.M. (Ed.), Decomposition of Large-Scale Problems, North Holland, Armsterdan, 1973.

JOHNSON, E.L., "Networks and Basic Solutions, Opns. Res., 14, 1966, pp. 619-624.

KALAN, J., "Aspects of Large-Scale In-Core Linear Programming," ACM Proceedings, 1971, pp. 304-313.

KALLIO, M.J., "On Large-Scale Linear Programming," Tech. Report SOL 75-7, Dept. of Opers. Res., Stanford Univ., April 1975.

KERSHENBAUM, A. and VAN SLYKE, R.M., "Computing Minimum Spanning Trees Efficiently," Proc. 25th Annual Conf. of ACM, Boston, 1972, pp. 518-527.

KLINGMAN, D., NAPIER, A. and STUTZ, J., "NETGEN: A Program for Generating Large Scale Capacitated Assignment, Transportation and Minimum Cost Flow Network Problems," Man. Sci., 20, No. 5, 1974, pp. 814-820.

LASDON, L.S., "Duality and Decomposition in Mathematical Programming," IEEE Trans. System Sci. and Cybernetics, 4, 1968, pp. 86-100.

———, Optimization Theory for Large Systems, MacMillan, New York, 1970.

———, "A Survey of Large Scale Mathematical Programming," to appear in Handbook of Operations Research (Elmaghraby and Moler, Eds.), Van Nostrand Reinhold, New York, 1976.

LEMARECHAL, C., "An Extension of Davidon's Methods to Non-Differentiable Problems," Math. Prog., Study 3, 1975.

LUENBERGER, D.G., Introduction to Linear and Nonlinear Programming, Addison-Wesley, Reading, Mass., 1973.

MADSEN, K., "Minimax Solution of Non-Linear Equations without Calculating Derivatives," Math. Prog., Study 3, 1975.

MAGNANTI, T.L., "Fenchel and Lagrange Duality are Equivalent," Math. Prog. 7, 1974, pp. 253-258.

———, SHAPIRO J.F. and WAGNER, M.W., "Generalized Programming Solves the Dual," Tech. Report OR-019-73, Opers. Res. Center, M.I.T., September 1973, to appear in Man. Sci.

MANNE, A.S. and HO, J.K., "Nested Decomposition for Dynamic Models," Math. Prog., 6(2), 1974, pp. 121-140.

MARKOWITZ, H., "The Elimination Form of the Inverse and its Application to Linear Programming, Man. Sci. 3(3), 1957, pp. 255-269.

MARSTEN, R.E., "The Use of the BOXSTEP Method in Discrete Optimization," Math. Prog., Study 3, 1975.

______, HOGAN, W.W., and BLANKENSHIP, J.W., "The BOXSEP Method for Large-Scale Optimization," Opns. Res., 23, 1975, pp. 389-405.

______, NORTHRUP, R.W., and SHAPIRO, J.F., "Subgradient Optimization for Set Partitioning and Covering Problems," Presented at ORSA/TIMS National Meeting, San Juan, Oct. 1974.

McCOY, P.R., and TOMLIN, J.A., "Some Experiments on the Accuracy of Three Methods of Updating the Inverse in the Simplex Method," Tech. Report SOL 74-21, Dept. of Oper. Res., Stanford Univ., Dec. 1974.
MOTZKIN, T., and SCHOENBERG, I.J., "The Relaxation Method for Linear Inequalities," Can. J. of Math., 6, 1954, pp. 393-404.
MULVEY, J., "Developing and Testing a Truly Large Scale Network Optimization Code," presented at ORSA/TIMS National Meeting, Chicago, May 1975.
OETTLI, W., "An Iterative Method, Having Linear Rates Of Convergence, For Solving a Pair of Dual Linear Programs," Math. Prog., 3, 1972, PP. 302-311.
ORCHARD-HAYS, W., Advanced Linear Programming Computing Techniques, McGraw-Hill, New York, 1968.
ORTEGA, J., and REINBOLDT, W.C., Iterative Solution of Nonlinear Equations in Several Variables, Academic Press, New York, 1970.
PIERCE, A.R., "Bibliography on Algorithms For Shortest Path, Shortest Spanning Tree, and Related Circuit Routing Problems (1956-1974)," Networks, 5, 1975, pp. 129-149.
POLAK, E., "An Historical Survey of Computational Methods in Optimal Control," SIAM Review, 15(2), 1973, pp. 553-584.
POLYAK, B.T., "Minimization of Unsmooth Functionals," Zh. Uy̆chisl. Mat. Fiz., 9(3), 1969, pp. 509-521.
REID, J.K., (Ed.), Large Sparse Sets of Linear Equations, Academic Press, New York, 1971.
_______, "A Sparsity-Exploiting variant of the Bartels-Golub Decomposition For Linear Programming Bases," to appear, 1976.
ROCKAFELLAR, R.T., Convex Analysis, Princeton Univ. Press, Princeton, N.J., 1970.
ROSE, D.J., and WILLOUGHBY, R.A., (Eds.), Sparse Matrices and Their Applications, Plenum Press, New York, 1972.
SAUNDERS, M.A., "Large-Scale Programming Using the Cholesky Factorization," Report STAN-CS-72-252, Computer Science Dept., Stanford Univ., Jan. 1972a.
_______, "Product Form of the Cholesky Factorization for Large -Scale Linear Programming," Report STAN-CS-72-301, Computer Science Dept., Stanford Univ., Aug. 1972b.
_______, "A Fast Implementation of the Simplex Method Using Bartels-Golub Updating," this volume, 1976.
SHAPIRO, J.F., "Generalized Lagrange Multipliers in Integer Programming," Opns. Res., 19, 1971, pp.68-76.
SILVERMAN, G.J., "Primal Decomposition of Mathematical Programs by Resource Allocation," Tech. Memo 116, Opers. Res. Dept., Case Western Reserve Univ., 1968.
SPYROPOULOS, K., E. KIOUNTOUZIES, and YOUNG, A., "Discrete Approximations on the L_1 Norm," Computer Journal, 16, 1973,

pp. 180-186.

SRINIVASAN V. and THOMPSON G.L., "Benefit-Cost Analysis of Coding Techniques for the Primal Transportation Algorithm," J. of ACM, 20, 1973, pp. 194-213.

TARJAN, R.E., "Depth-first Search and Linear Graph Algorithms," SIAM J. on Computing, 1(2), 1972, pp. 146-160.

TOMLIN, J.A., "Survey of Computational Methods for Solving Large Scale Systems," Tech. Report N 72-25, Dept. of Opers. Res., Stanford, Univ., Oct. 1972.

______, "Generalized Upper Bounding and Triangular Decomposition in the Simplex Method," Opns. Res., pp. 664-668.

WETS, R.J.-B., "Stochastic Programs with Fixed Recourse: The Equivalent Deterministic Program," SIAM Review, 16(3), 1974, pp. 309-331.

WHITE, W.W., "A Status Report on Computing Algorithms for Mathematical Programming," Computer Surveys, 5, No. 3, Sept. 1973, pp. 135-166.

WILKINSON, J.H., Rounding Errors in Algebraic Processes, Prentice-Hall, Englewood Cliffs, N.J., 1963.

______, "Error Analysis of Transformations Based on the Use of Matrices of the Form $I-2ww^H$," in Error in Digital Computation, Vol. ii (L.B. Rall, Ed.), John Wiley and Sons, New York, 1965, pp. 77-101.

WILLOUGHBY, R.A., (Ed.), Sparse Matrix Proceedings, IBM, Yorktown Heights, N.Y., 1969.

WINKLER, C., "Bases Factorization for Block-Angular Linear Programs: Unified Theory of Partitioning and Decomposition Using the Simplex Method," Tech. Report SOL 74-19, Dept. of Opers. Res., Stanford Univ., 1974.

WISMER, D.A., (Ed.), Optimization Methods for Large Scale Systems, McGraw-Hill, New York, 1971.

WOLFE, P., "A Method of Conjugate Subgradients for Minimizing Nondifferentiable Functions," Math. Prog., Study 3, 1975.

ZANGWILL, W.I., Nonlinear Programming: A Unified Approach, Prentice-Hall, Englewood Cliffs, N.J., 1969.

ZSCHAU, E.U.W., "A Primal Decomposition Algorithm for Linear Programming," Working Paper No. 91, Grad. School of Business, Stanford Univ., 1967.

METHODS FOR SPARSE LINEAR LEAST SQUARES PROBLEMS

ÅKE BJÖRCK

Department of Mathematics
Linköping University, Linköping, Sweden

1. INTRODUCTION

The object of this paper is to survey direct and iterative methods for solving sparse least squares problems. These problems generally arise in the same contexts as sparse linear equations. Among the applications are geodesy (Ashkenazi [2]), photogrammetry (Glaser [32]), statistical computations (Gentleman [28]) and structural analysis (Lunde Johnsen [50]).

Since the subject is rapidly developing and so far not enough is known about the algorithms it has generally not been possible to make final assertions about the relative efficiency of different algorithms. It is still hoped that this survey will give some guidance to users and also stimulate further research.

2. THE LINEAR LEAST SQUARES PROBLEM

Let A be a given $m \times n$ matrix of rank r and b a given vector. The problem is

$$\text{minimize } ||b - Ax|| \,, \tag{2.1}$$

where $||\cdot||$ denotes the Euclidian norm, or equivalently to find x and r such that

$$r + Ax = b, \quad A^T r = 0 \,. \tag{2.2}$$

It follows immediately that a solution to (2.1) satisfies the normal system

$$A^T A x = A^T b. \tag{2.3}$$

The normal system is always consistent, with $x = A^+ b$ the unique solution of minimum norm. Here A^+ denotes the Moore-Penrose pseudo-inverse. When $r < n$ the problem is however ill-posed, since then $A^+ b$ is not a continous function of A and b (Wedin [74],[75]). In the following we assume unless otherwise stated that $r = n$. Then the matrix $A^T A$ of the normal system is positive definite, and (2.1) has a unique solution.

The sparse linear least squares problem can obviously be handled by techniques for sparse symmetric positive definite systems applied to (2.3). There are however two possible drawbacks with this approach, which give cause for a separate discussion of algorithms for the problem (2.1). For one thing there are sparse matrices A for which A^TA is dense. This will be discussed more in section 3. More important, the normal equations (2.3) are often too ill-conditioned to give a sufficiently accurate solution.

When rank(A) = n the condition number of A is defined to be

$$\kappa(A) = \sigma_1/\sigma_n \tag{2.4}$$

where $\sigma_1 \geq \sigma_2 \geq \ldots \geq \sigma_n > 0$ are the singular values of A. Unless we make some assumption about the vector b, the perturbation bounds for x contains a term proportional to $\kappa^2(A)$, see [36]. However, if the overdetermined system Ax = b is nearly consistent in the sense that

$$\kappa(A)||r|| \leq ||A|| \; ||x|| \;, \tag{2.5}$$

then (Björck [7]), the condition number for the problem (2.1) is essentially $\kappa(A)$. On the other hand, when A^TA is formed explicitly in (2.3) the condition number for computing x is always $\kappa(A^TA) = \kappa^2(A)$.

In practice an ill-conditioned matrix A often results from the following situation (Powell and Reid [58]). Suppose the first p < n equations are weighted heavily, so that

$$A = \begin{pmatrix} A_1 \\ \varepsilon A_2 \end{pmatrix}, \qquad b = \begin{pmatrix} b_1 \\ \varepsilon b_2 \end{pmatrix}, \qquad \varepsilon << 1. \tag{2.6}$$

If ε^2 is of the same order as the machine accuracy, then the matrix of the normal equations becomes

$$A^TA = A_1^TA_1 + \varepsilon^2 A_2^TA_2 \approx A_1^TA_1 \;.$$

Then all information from A_2 is lost, and the computed A^TA is singular. Note that for the particular right hand side in (2.6) the problem (2.1) is not necessarily ill-conditioned for small ε. Usually for small <u>relative</u> perturbations in the elements of A and b, the condition improves when $\varepsilon \to 0$.

We can write (2.2) as

$$\begin{pmatrix} I & A \\ A^T & 0 \end{pmatrix} \begin{pmatrix} r \\ x \end{pmatrix} = \begin{pmatrix} b \\ 0 \end{pmatrix} \qquad (2.7)$$

i.e. as an augmented system of (m+n) equations for the (m+n) unknown components of r and x. This system is symmetric and indefinite, and can be solved by sparse techniques for such systems. For ill-conditioned problems this approach may be more satisfactory than the use of the normal equations. If A is scaled so that $\sigma_1(A) = 2^{-1/2}$, then the condition number of the augmented matrix in (2.7) is approximately $\kappa^2(A)$. However, as was first pointed out to the author by Golub (private communication) if A is scaled so that $\sigma_n = 2^{-1/2}$, then the condition number is less than $2\kappa(A)$ (Björck [8]).

We end this section with the remark that since the Euclidian norm of a vector is unitarily invariant it follows that

$$||b - Ax|| = ||Qb - QAS(S^{-1}x)|| \qquad (2.8)$$

where Q is an arbitrary orthogonal m×m matrix and S an arbitrary n×n non-singular matrix. This completely describes the set of transformations which can be applied to transform A and b in the problem (2.1). In particular, it is possible to scale the columns of A but not the rows. If the columns of A are scaled to have equal Euclidian norm, then the matrix A^TA will have equal diagonal elements. It has been shown by van der Sluis [73], that if A^TA has at most q non-zero elements in any row, then for this scaling $\kappa(A^TA)$ is not more than a factor of q larger than the minimum under all symmetric diagonal scalings of A^TA. Note that it may be improper to scale a particular problem this way because this can cause inaccurate data to assume too much influence.

3. DIRECT METHODS BASED ON ELIMINATION

Direct methods have the general advantages over iterative methods that subsequent right hand sides can be treated efficiently and it is easier to obtain elements of related inverse matrices. With direct methods it is also possible to use the technique of iterative refinement. An excellent survey of direct methods for dense matrices has been given by Peters and Wilkinson [56].

The most straight forward method to solve the least squares problem is to form the normal equations (2.3) and then compute the Cholesky factorization

$$P_1 A^T A P_1^T = LL^T \quad , \tag{3.1}$$

where P_1 is a permutation matrix. The solution is then obtained by solving the two triangular systems $Ly = P_1A^Tb$ and $L^T(P_1x) = y$.

When A is a dense matrix, A^TA will always have fewer non-zero elements than A, and the computation of A^TA can be seen as a data reduction. For sparse matrices this is generally not true, and we now discuss the 'fill-in' when A^TA is formed. We have

$$(A^TA)_{jk} = a_j^Ta_k \ , \ A = (a_1,a_2,\ldots,a_n),$$

and obviously the row and column ordering of A will not influence the number of non-zero elements in A^TA. If we can ignore the possibility of cancellation in the inner product $a_j^Ta_k$, then $(A^TA)_{jk} \neq 0$ if and only if $a_{ij} \neq 0$ and $a_{ik} \neq 0$ for at least one row i.

If A is a random matrix such that $a_{ij} \neq 0$ with probability p, then, ignoring cancellation, $(A^TA)_{jk} \neq 0$ with probability

$$q = 1 - (1-p^2)^m \approx (mp)p \ .$$

The average number of non-zero elements in half the symmetric matrix A^TA is approximately $n^2mp^2 = mnp(1/2np)$. Thus, if np, the average number of non-zero elements in a row of A, is small, then the 'fill-in' when forming the normal equations will be small.

If A has the property that

$$a_{ij} \neq 0 \text{ and } a_{ik} \neq 0 \Rightarrow |j - k| \leq w, \tag{3.2}$$

then A is said to be a rectangular band matrix of band width w. It follows immediately that

$$(A^TA)_{jk} = 0 \quad \text{if} \quad |j - k| > w \ ,$$

i.e. A^TA is a symmetric band matrix of the same band width as A. Also during the formation of A^TA we need only (w+1) columns at a time.

Although the situation is favourable for random matrices and band matrices there are sparse matrices A for which A^TA is completely filled but where the Cholesky factor L again is sparse. A simple example of such a matrix is

$$\begin{pmatrix} x & x & x & x \\ & x & & \\ & & x & \\ & & & x \\ & & & x \end{pmatrix} .$$

This illustrates the difficulty of making theoretical comparisons with other methods, where A^TA is not explicitly computed.

We remarked earlier that by the transformation of variables $y = S^{-1}x$ we get a least squares problem with matrix AS. In the method of Peters and Wilkinson [56], A is reduced by elementary row operations to an upper triangular $n \times n$ matrix U. In this reduction row and column interchanges should be used to preserve sparseness and insure numerical stability. The resulting decomposition becomes

$$P_1AP_2 = L\,U\ , \tag{3.3}$$

where L is a unit lower trapezoidal $m \times n$ matrix and where by our assumption U is nonsingular. We then solve the least squares problem

$$\text{minimize}\ ||P_1b - Ly|| \tag{3.4}$$

and compute x from the triangular system

$$UP_2^Tx = y\ . \tag{3.5}$$

Experience at NPL indicates, that if row interchanges are used to limit the size of the off-diagonal elements of L, then L is usually a well-conditioned matrix. Thus (3.4) can be solved using the normal equations $L^TLy = L^TP_1b$.

In particular the method of Peters and Wilkinson works well for systems of the form (2.6). For this case it is easily seen that the last (n-p) rows of U and y will only have elements of magnitude ε, and after equilibration of rows (3.5) is a well-conditioned system.

A variation of the method of Peters and Wilkinson results if we instead use elementary column operations to reduce A to unit lower trapezoidal form. If completed, this reduction gives the same decomposition (3.3). However, an important difference is, that it is now possible to stop after $p < n$ steps of the reduction. We then have a factorization

$$P_1 A P_2 = L_p U_p, \tag{3.6}$$

where the first p rows of the m×n matrix L_p has non-zeros only in a unit lower triangle and the last (n-p) rows of U_p equals the last (n-p) rows of the unit matrix I_n. For systems of type (2.6) we can now avoid ill-conditioned normal equations by only doing $p = \text{rank}(A_1)$ steps of the reduction.

A promising new approach suggested by Hachtel (IBM, Yorktown Heights) is to apply sparse elimination techniques to the symmetric indefinite system (2.7). Note that if we pivot down the diagonal nothing new results, since after the first m steps the last n equations are just the normal equations $-A^TAx = -A^Tb$.

Duff and Reid [20] investigate two different schemes. In the first a symmetric decomposition is obtained by following Bunch [10] in using both 1×1 and 2×2 pivots,

$$P_1 \begin{pmatrix} \alpha I & A \\ A^T & 0 \end{pmatrix} P_1^T = L\, D\, L^T . \tag{3.7}$$

Here D is block diagonal with 1×1 and 2×2 blocks, and α is a scaling factor, which by the analysis in section 2 should be chosen as an approximation to the smallest singular value of A.

In the second approach one ignores symmetry and computes an unsymmetric decomposition

$$P_1 \begin{pmatrix} \alpha I & A \\ A^T & 0 \end{pmatrix} P_2 = L\, U , \tag{3.8}$$

where L is unit lower triangular.

Gentleman [30] has applied a generalized row elimination technique to the system (2.7). He uses elementary stabilized row transformations, where not necessarily the same pivot row is used for elimination in any particular column. This seems to be a promising area for future research.

When computing the factorizations (3.1), (3.3), (3.7) and (3.8) the pivots should be chosen to maintain sparsity in the factors and ensure numerical stability. Similar criteria apply as for the linear equation case, see Reid [62]. Usually only a rather mild numerical test is advocated (Curtis and Reid [16]).

For the band matrix case the columns of A should be ordered to minimize e.g. the profile of A^TA (Cuthill [17]). The Cholesky decomposition can then be carried out very efficiently even when backing store is needed by using the technique described by Jennings and Tuff [44].

For ordering the columns in the general sparse case one can use either a static scheme, where the columns are reordered initially before the elimination, or a dynamic scheme where the reordering is done as the elimination proceeds. Static ordering allows a simpler data structure to be used and has, especially for problems requiring backing store, substantial implementation advantages.

Duff and Reid [20] use dynamic reorderings in their tests. For the Cholesky decomposition the pivot is chosen at each step to minimize the number of non-zeros in the pivotal row. For the Peters-Wilkinson's decomposition (3.3) the Markowitz criterion is used, i.e. minimize $(r_i-1)(c_j-1)$ where r_i and c_j are the number of non-zeros in the pivotal row and column respectively. For the unsymmetric decomposition (3.8) Duff and Reid note that if early pivots occur in late rows and late pivots in late columns, then not all of L and U need to be retained, and use a modified Markowitz criterion to achieve this. For the symmetric Hachtel decomposition (3.7) a generalization of the Markowitz cost to 2×2 pivots is used.

We note that in the Peters-Wilkinson decomposition one wants to maintain sparsity not only in L and U but also in L^TL. Duff and Reid give the following simple example, where the Markowitz criteria fails to give a good choice of pivots,

$$A = \begin{pmatrix} x & x & & & \\ & x & x & & \\ & & x & x & \\ & & & x & x \\ x & & & & x \end{pmatrix}.$$

Here the choice of pivots will be down the diagonal, which leads to a factorization for which L^TL is full. If the rows are reordered 5,1,2,3,4 then L becomes bidiagonal and L^TL retains the sparsity of A.

A systematic comparison of the methods discussed in this section with respect to storage and number of operations can be found in Duff and Reid [20]. In their tests the normal equations and the two Hachtel schemes performed best, with the Peters-Wilkinson method close behind.

4. DIRECT METHODS BASED ON ORTHOGONALIZATION

It is well known that for any m×n matrix A of rank n, there exists an essentially unique factorization

$$A = Q\,U\,, \tag{4.1}$$

where Q is an m×n orthogonal matrix and U is upper triangular. Note, that since $P_1A = (P_1Q)U$, the factor U is independent of the row ordering of A. It is easily seen that $U = L^T$ where L is the lower triangle in the Cholesky decomposition of A^TA. Since we have $\kappa(U) = \kappa(A)$, the problem of ill-conditioning can to a great extent be avoided by using methods based on the decomposition (4.1).

Perhaps the simplest of the methods based on orthogonality is the modified Gram-Schmidt (MGS) method, Bauer [4], Björck [7]. In this method we explicitly determine a set of orthogonal vectors $q_1,q_2,\ldots,q_n$, which span R(A), the range of A. Note that an advantage of this method is that it is possible to stop after $p < n$ steps of the orthogonalization. If we put $Q_p = (q_1,\ldots,q_p)$, we have then

$$A = (Q_p,A_p)\begin{pmatrix} U_p & S_p \\ 0 & I \end{pmatrix}, \qquad b = b_p + Q_pc\,, \tag{4.2}$$

where $Q_p^TA_p= 0$ and $Q_p^Tb_p = 0$. The least squares problem (2.1) then decomposes into

$$\text{minimize } ||b_p - A_px_2|| \,, \quad U_px_1 = c - S_px_2\,, \tag{4.3}$$

where $x^T = (x_1^T,x_2^T)$ has been partitioned conformely. If all n steps are carried out we have $A = Q_nU_n$, which is the decomposition (4.1).

Unfortunately the MGS method performs badly from a sparsity point of view. Bauer [4] has shown that this method can be seen as a row elimination method, where a weighted combination of rows is used as pivot row. This weighted pivot row will of course be much denser than the original rows.

A second method for computing the decomposition (4.1) is due to Golub [33] and uses a sequence of Householder transformations

$$P_n \dots P_2 P_1 \, A = \begin{pmatrix} U \\ 0 \end{pmatrix}, \quad P_k = I - 2w_k w_k^T, \quad ||w_k|| = 1. \quad (4.4)$$

Note that in this method Q is not explicitytly computed, and only the vectors $w_1, w_2, \dots w_n$ are stored. This saves storage even in the dense case. Tewarson [72] has compared theoretically the fill-in for the MGS method and Golub's method. He concludes that the factored form in (4.4) generally is more sparse than the explicit matrix Q. There is however a close connection between the two methods. It has been pointed out by Golub (private communication) that the MGS method numerically gives the same matrix U as Golub's method applied to the problem

$$\text{minimize } ||\binom{0}{b} - \binom{0}{A} x|| ,$$

where A and b have been augmented by n zero rows. Also the vectors w_k in (4.4) are then exactly equal to $1/\sqrt{2}(e_k^T \; w_k^T)^T$, where e_k is the k:th unit vector. When m>>n, then we can thus expect the fill-in in the factored form of Q in Golub's method to be about the same as the fill-in in Q for MGS.

The orthogonal decomposition (4.4) can also be performed by using a sequence of Givens transformations (plane rotations), Gentleman [27]. For full matrices using the original expression for the Givens transformation this requires twice the number of multiplications and many more square roots than if Householder transformations are used. This disadvantage can however largely be eliminated by using one of the modifications proposed by Gentleman [27] and Hammarling [38]. Duff and Reid [20] point out that the Givens procedure is likely to show advantage for sparse problems "because each Householder transformation results in all rows with non-zeros in the pivotal column taking the sparsity pattern of their union, whereas the sequence of Givens' rotations performing the same task results in each row taking the sparsity pattern of the union of those rows that are involved in the earlier rotations in the sequence".

The potential advantage of Givens' method over the straightforward use of Householder transformations is well demonstrated by the case when A is a band matrix (3.2). Reid [60] has shown that Golub's method requires about mwn multiplications compared to about $\frac{1}{2}mw(w+1)$ for the method of normal equations. Givens' method however will take full advantage of the band-structure, provided that the rows of A are ordered so that, if the first non-zero in the i:th row is in column k_i, then $k_{i+1} \geq k_i$; the increase in number of multiplications over the method of normal equations will be similar to that for full matrices. Reid [60] shows that by performing Householder transformations blockwise in a certain way, a similar reduction in operation count can be obtained.

As for elimination methods, the pivotal columns should be chosen to maintain sparsity and ensure numerical stability. Chen and Tewarson [13] have described a method to minimize the local fill-in at each step in MGS and Golub's method. The computational cost of this strategy however seems to be prohibitive. Tewarson [72] describes a cheaper algorithm for the a priori arrangement of columns for these two methods. Rudolphi [63] reports on some experiments with MGS using the dynamic strategy of taking at each step the column with fewest non-zero elements as pivot column. Duff [18] recommends the same dynamic strategy for Givens' method. In his tests this strategy performed much better than ordering the columns a priori after ascending column count.

We now discuss the selection of pivot row. The MGS method is easily seen to be numerically invariant under row permutation. Golub's method on the other hand may need row pivoting to preserve numerical stability in the presence of widely different row scalings (Powell and Reid [59]). The fill-in is essentially not affected by the choice of pivot row, provided only that a non-zero pivot is chosen. In Givens' method, however, an appropriate choice of pivot row and row ordering is essential. This is evident from the band matrix case where, although the final matrix R is uniquely defined, the intermediate fill-in and the number of Givens transformations in the reduction depends critically on the row ordering.

Duff [18] reports on extensive tests of row orderings in the Givens' reduction of sparse matrices. He recommends choosing as pivot the non-zero element in the pivot column with least non-zeros in its row. A row ordering is then used, which for each rotation minimizes the fill-in in the pivot row. Note that this leads to the desired row ordering for the band matrix case.

It has been suggested by Gentleman [29],[30] that in order to postpone fill-in, one should allow the pivot row to vary at each minor step in the Givens' reduction. In [30] an error analysis is given for this generalized reduction method. One possibility is to rotate at each step the two rows with non-zeros in the pivot column, which have the lowest number of non-zeros in them. Duff [18] has tested this strategy, and reports results comparable with the best fixed pivot strategy. These generalized reduction methods are however still new, and it seems to be too early to answer the question how they compare with the more classical ones.

Duff and Reid [20] reach from their experimental results the conclusion, that except for matrices with very few columns, Givens' method is the best on operation counts among the methods based on orthogonality. On storage, when subsequent solutions are wanted, Golub's method usually showed a slight advantage. The Peters-Wilkinson algorithm was however usually superior, sometimes with a large factor, and in these tests was the best of the more stable methods.

Methods which start with the LU-decomposition (3.3) and then solve (3.4) with Golub's method (Cline [14]) or MGS (Plemmons [58]) have been suggested. Cline shows that for dense matrices his method is more efficient than all other standard methods when $m < 4n/3$.

Methods for updating the QU-factorization (4.1) when A is modified by a matrix of low rank have been devloped by Gill et al [31]. These methods may be used e.g. when the given matrix can be represented as $A + uv^T$, where A is sparse.

For large problems, which exceed the capacity of the high speed storage, efficient retrieval of elements from the backing storage is essential. One way to handle this is to use algorithms for partitioned matrices. Such methods for computing the QU-decomposition are discussed in [9] and [64]. See also [37], where a method using recursive partitioning for computing A^+ is given.

5. ITERATIVE REFINEMENT

For dense problems iterative refinement is a cheap and simple way to improve the accuracy of a computed solution. It also has the advantage that it gives useful information about the condition of the problem [8],[36]. For sparse problems this is not always as attractive, since then the cost of processing a subsequent right hand side may often not be much smaller than the cost for the initial decomposition. On the other hand it is often true, that the extra storage needed and the cost of computing residuals involving A and A^T is very small.

The simplest schemes for the refinement of an approximate least squares solution x_0 start by computing an accurate residual vector $r_0 = b - Ax_0$, e.g. by using double precision accumulation. The correction is then computed in single precision from

$$LL^T(x_1-x_0) = A^T r_0, \qquad (5.1)$$

where L is the Cholesky factor of A^TA. The numerical properties of this process are discussed in [36] and they turn out to be not quite satisfactory.

By instead computing accurate residuals to the augmented system (2.7)

$$f_0 = b - r_0 - Ax_0, \quad g_0 = -A^T r_0 \qquad (5.2)$$

it is possible to obtain much better results. Golub [34] solves the correction from

$$U^T U(x_1-x_0) = A^T f_0 - g_0, \qquad (5.3)$$

with U computed by (4.4). This method has the important advantage that the matrix Q need not to be retained. However, the rate of convergence will only be about $\kappa^2(A)\,2^{-t}$, and there is no assurance that the final solution will be correct.

Much better results are possible if Q is retained. Björck [8] has given algorithms for MGS, Golub's and Given's methods, for which the rate of convergence is $\kappa(A)\,2^{-t}$ for all right hand sides b, and which will practically always produce x to full single precision accuracy.

We finally remark that the two Hachtel methods seem very well suited for use in an iterative refinement scheme. This is especially true for the unsymmetric version (3.8), which is the more efficient when subsequent right hand sides are to be processed (Duff and Reid [20]).

6. ITERATIVE METHODS

In this section we consider iterative methods, which are not finitely terminating. The conjugate gradient method and related methods are surveyed in section 7 . Following Householder [43] we distinguish in this section between norm reducing methods and methods of projection.

The general linear stationary method of first degree for solving the non-singular system $Ax = b$,

$$x_{i+1} = Gx_i + k, \qquad (6.1)$$

can be obtained by splitting A into the form $A = M - N$, where M is non-singular, and taking $G = M^{-1}N$. When $r = n$, then A^TA is positive definite and these methods can be applied to the normal system (2.3). When $r < n$, then the normal system is consistent but singular. For this case the convergence of the iteration (6.1) has been investigated by Keller [46], Joshi [45] and Young [78]. Whitney and Meany [76] proved that Richardson's first order method for the normal system,

$$x_{i+1} = x_i + \alpha A^T(b - Ax_i) \qquad (6.2)$$

converges to the least squares solution A^+b if $x \in R(A^T)$, the range of A^T, and $0 < \alpha < 2/\lambda_{max}(A^TA)$.

The concept of splitting has been extended to rectangular matrices by Plemmons [57]. Berman and Plemmons [6] define $A = M - N$ to be a proper splitting if the ranges and null-spaces of A and M are equal. They show that for a proper splitting, the iteration

$$x_{i+1} = M^+Nx_i + M^+b \qquad (6.3)$$

converges to A^+b for every x_0, if and only if the spectral radius $\rho(M^+N) < 1$. The iterative method (6.3) avoids the explicite recourse to the normal system. Necessary and sufficient conditions for $\rho(M^+N) < 1$ are given, extending known results for the non-singular case.

Tanabe [69] considers iterative methods of the form (6.1) for computing the more general solution $x = A^-b$, where A^- is any generalized inverse of A, $(AA^-A = A)$. He shows that the iterations can always be written

$$x_{i+1} = x_i + B(b - Ax_i)$$

for some $n\times m$ matrix B, and characterizes the solution in terms of the range of AB and the nullspace of BA.

Splittings of rectangular matrices and corresponding iterative methods have also been investigated by Chen [12]. Chen shows that if $r = n$, then for any consistent iterative method (6.1) there exists a splitting such that the method can be written in the form (6.3). Also, for every such method we have $G = I - C^{-T}A^TA$ for some nonsingular matrix C. Thus, the most general iterative method of the form (6.1) for the least squares problem is equivalent to Richardson's first order method applied to the preconditioned normal system $C^{-T}(A^TAx - A^Tb) = 0$.

A survey of iterative methods for positive definite systems is given in the book by Young [78]. If we assume that the columns of A have been scaled so that the diagonal elements in A^TA equals unity, then in the SOR method the matrix A^TA is split into

$$A^TA = L + I + L^T, \qquad (6.4)$$

where L is a strictly lower triangular matrix. The SOR iteration

$$x_{i+1} = x_i + \omega[A^Tb - Lx_{i+1} - (I + L^T)x_i] \qquad (6.5)$$

always converges, provided that the relaxation parameter ω satisfies $0 < \omega < 2$. This method has the advantage of simplicity and small storage requirements. Ashkenazi [2] has applied it to very large systems of geodetic normal equations.

Young [78] notes that when A^TA does not have property 'A' or is not an L-matrix, then the SOR method may not be effective for any choice of ω. The semi-iterative method based on Richardson's method may however still be effective. If the eigenvalues of A^TA satisfy

$$0 < \bar{a} \leq \lambda(A^TA) \leq \bar{b} ,$$

then this method can be written in the stable form

$$x_{i+1} = x_{i-1} + \rho_i(x_i - x_{i-1}) + 2\rho_i/(\bar{b}+\bar{a})\ A^T(b - Ax_i), \qquad (6.6)$$

where the parameters ρ_i are recursively computed by

$$\rho_0 = 2, \quad \rho_i = (1 - \frac{\mu^2}{4}\rho_{i-1})^{-1}, \quad \mu = (\bar{b}-\bar{a})/(\bar{b}+\bar{a}).$$

Note that in contrast to the SOR-iteration, (6.6) may be performed without explicitly forming the normal equations. This can be expected to improve the numerical properties of the method and we can also avoid possible fill-in when forming A^TA.

To improve the rate of convergence, we can apply (6.6) to the preconditioned least squares problem

$$\text{minimize } ||b - (AC^{-1})z|| \quad , \quad Cx = z. \tag{6.7}$$

Note that we have the choice of either computing AC^{-1} explicitly, or solving two linear systems with matrix C and C^T respectively in each iteration step. Instead of iterating for z using the residuals $C^{-T}A^T(b - AC^{-1}z_i)$ and finally solving $Cx = z$ for x, it is also possible to iterate for x directly using the residuals $C^{-1}C^{-T}A^T(b - Ax_i)$.

Several different choices of the matrix C are possible. In the SSOR preconditioning one takes $C = I + \omega L^T$, where L is defined by (6.4) and ω is a relaxation parameter. This method has been discussed by Axelsson [3], Evans [24] and Andersson [1]. In this method C is not explicitly inverted. Note also that at the expense of storing both L and A, the iterations can be performed with the original matrix A and not A^TA.

Other choices of C result from using elimination methods on A. This seems to have been first suggested by Läuchli [52], who applies Jordan elimination with complete pivoting to A. If we for simplicity assume that the corresponding permutations have been carried out in advance on A and write $A^T = (A_1^T \; A_2^T)$, where A_1 is n×n and non-singular, then in the Jordan step one computes

$$A_1^{-1} \; , \; AA_1^{-1} = \begin{pmatrix} I \\ B \end{pmatrix} \; , \quad B = A_2A_1^{-1} \; . \tag{6.8}$$

Läuchli notes that the parameters for the semi-iteration can easily be computed from the elements of B since

$$\bar{a} = 1 \leq \lambda(I+B^TB) \leq 1+\text{trace}(B^TB) = \bar{b} \; .$$

We remarked earlier that the least squares problem (3.4), which is solved in the method of Peters and Wilkinson, usually is well-conditioned. By choosing the preconditioning matrix $C = UP_2^T$ from (3.3) we get a method which closely resembles Läuchli's method. It has the advantage that the matrix L in (3.3) usually is more sparse than the matrix B in (6.8).

Chen [12] has described another closely related method, which is due to Gentleman. He assumes that the partitioning $A^T = (A_1^T \; A_2^T)$ can be found a priori and applies semi-iteration to the method

$$x_{i+1} = x_i + A_1^{-1}(I, A_1^{-T}A_2^T)\binom{r_1}{r_2}, \quad r = b - Ax_i, \qquad (6.9)$$

without forming A_1^{-1} or $A_1^{-T}A_2^T$ explicitly.

A unified treatment of methods of projection have been given by Householder and Bauer [4]. Although these methods usually are defined only for square and nonsingular systems, they directly generalize to rectangular systems. Two different classes of methods result. In the first we let $p_1, p_2, \ldots$ be a sequence of somehow selected n-vectors and compute for $i = 1,2,\ldots$

$$x_{i+1} = x_i + a_i p_i, \quad a_i = p_i^T A^T r_i / ||Ap_i||^2 \qquad (6.10)$$

For this class of methods the residual will be orthogonal to Ap_i, and therefore (Householder [41]) we have $||r_{i+1}|| \leq ||r_i||$. This class of iteration methods is also discussed by Chen [12], who calls them residual reducing methods.

The second class of projection methods results from selecting a sequence of m-vectors $q_1, q_2, \ldots$ and computing for $i = 1,2,\ldots$

$$x_{i+1} = x_i + a_i A^T q_i, \quad a_i = q_i^T r_i / ||A^T q_i||^2 \qquad (6.11)$$

For this class of methods the error $||x_i - x||$ is monotonically decreasing. Taking $q_i = r_i$ in (5.12) leads to the method of steepest descent for the normal equations. If we choose q_i to be the unit vectors in cyclic order, we get the Kaczmarz method. Tanabe [68] has proved that for a suitable choice of starting vector the Kaczmarz method converges to A^+b for all matrices A with non-zero rows. The Kaczmarz method ordinarily converges slowly and Dyer [21] has studied methods to accelerate the convergence.

A more general projection method for the least squares problem has been suggested by Tewarson [71]. He considers iterations of the form

$$x_{i+1} = x_i + Q(A^TA)A^T(b - Ax_i),$$

where Q is a polynomial of degree p, the coefficient of which are determined from a least squares problem of size m×p.

We finally mention that iterative methods for computing the generalized inverse A^+ have been given by Ben-Israel and Cohen [5] and by Garnett, Ben-Israel and Yau [25]. Söderström and Stewart [67] give a careful discussion of the numerical properties of such methods.

7. METHODS OF CONJUGATE GRADIENT TYPE

The prototype for this class of methods is the conjugate gradient method of Hestenes and Stiefel [40]. They gave the following algorithm for solving $Ax = b$, where A is n×n and non-singular

$$r_0 = b - Ax_0, \quad p_0 = s_0 = A^T r_0, \tag{7.1a}$$

$$q_i = Ap_i, \quad a_i = ||s_i||^2/||q_i||^2,$$

$$x_{i+1} = x_i + a_i p_i, \; r_{i+1} = r_i - a_i q_i, \; s_{i+1} = A^T r_{i+1}, \tag{7.1b}$$

$$b_i = ||s_{i+1}||^2/||s_i||^2, \quad p_{i+1} = s_{i+1} + b_i p_i.$$

This algorithm also applies to the least squares problem, even when rank (A) < n. If $x_0 \in R(A^T)$ and in the absense of rounding errors (7.1) will give the exact solution $x = A^+ b$ in t iterations, where t equals the number of distinct non-zero singular values of A. When the singular values of A are clustered, good estimates of x may be quickly obtained. The algorithm also has the property that both $||x - x_i||$ and $||r_i||$ are monotonically decreased.

Hestenes and Stiefel [40] point out, that although the algorithm (7.1) is theoretically equivalent to the conjugate gradient method applied to the normal equations, it can be expected to be numerically superior since $A^T A$ is not explicitly computed. Stiefel [66] and Läuchli [51] have discussed the application of (7.1) to the linear least squares problem. Hestenes [39] gives a second least squares conjugate gradient algorithm and shows how this and (7.1) can be used for computing A^+.

The conjugate gradient method requires the storage of three n-vectors and two m-vectors, and each iteration step is more complicated than for the semi-iterative method (6.6). A great advantage is that it does not depend on any a priori information about the eigenvalues of $A^T A$. Indeed the parameters for (6.6) may be estimated from a_i and b_i in (7.1).

Paige [54] has derived two different algorithms for the least squares problem from the bidiagonalization algorithm of Golub and Kahan [35] applied to A and A^T respectively. For a consistent system Paige's first method is equivalent to the method of Craig [15]. For a non-consistent least squares problem Paige's second method is the more straight forward.

It requires storage of one m-vector less than (7.1). Paige and Saunders [55] give a modification of Paige's second method. This algorithm, denoted LSLQ, takes a little more storage and computation per step, but is conjectured to be numerically more stable.

An extensive survey of algorithms related to the conjugate gradient algorithm has been given by Lawson [47]. He shows that Paige's second method mathematically generates the same sequence of approximate solutions as the conjugate gradient method (7.1).

Chen [12] considers several different methods related to the conjugate gradient method. He derives two methods, denoted RRLS and RRLSL, from the bidiagonalization algorithm. The method RRLS is mathematically equivalent to (7.1) and therefore also to Paige's second method. Chen's numerical examples show that, although there is little difference for well conditioned problems, RRLS gives more accurate results than both Paige's second method and LSLQ for ill conditioned problems. On the other hand RRLS and (7.1) behaved similarly in all cases, as did also RRLSL.

Elfving [22] has compared several numerically different formulations of the conjugate gradient method on a set of compatible and ill-conditioned least squares problems. He recommends the use of the original algorithm (7.1), which choice is also supported by the recommendation of Reid [61] for the linear equation case.

Stewart [65] has generalized the notion of a set of directions conjugate to a matrix, which leads to a variety of finitely terminating iteration methods. Some of these can be used for solving the linear least squares problem.

The convergence of the methods considered in this section can be accelerated by applying them to the preconditioned problem (6.7), i.e. by substituting AC^{-1} for A and $z = Cx$ for x. The same choices of matrix C as in section 6 can be used. Läuchli [52] analysed the conjugate gradient method for the choice $C = A_1$ (6.8). He comments that this gave superior results compared to the corresponding semi-iterative method. Chen [12] studies a variant of the same method. The use of SSOR preconditioning combined with the conjugate gradient method has been studied by Axelsson [3] and Andersson [1].

ACKNOWLEDGEMENTS

Many people have helped in preparing this survey by submitting advice and preprints. In particular I would like to thank I.S. Duff, W.M. Gentleman, G.H. Golub and J.K. Reid. I would also like to thank T. Elfving for reading a draft of this paper and making a number of very helpful suggestions.

REFERENCES

1. L. Andersson, SSOR preconditioning of Toeplitz matrices with application to elliptic equations, Department of Computer Sciences, 75.02, Chalmers University of Tech.
2. V. Ashkenazi, Geodetic normal equations, 57-74, in Large sparse sets of linear equations, J.K. Reid (Ed.) Academic Press (1971).
3. O. Axelsson, On preconditioning and convergence acceleration in sparse matrix problems, CERN 74-10, Geneva.
4. F.L. Bauer, Elimination with weighted row combinations for solving linear equations and least squares problems, Num. Math. 7, 338-352 (1965).
5. A. Ben-Israel and D. Cohen, On iterative computation of generalized inverses and associated projections, SIAM J. Numer. Anal. 3, 410-419 (1966).
6. A. Berman and R.J. Plemmons, Cones and interacitve methods for best least squares solutions of linear systems, SIAM J. Numer. Anal. 11, 145-154 (1974).
7. Å. Björck, Solving linear least squares problems by Gram-Schmidt orthogonalization, BIT 7, 1-21 (1967).
8. Å. Björck, Iterative refinement of linear least squares solutions I, BIT 7, 257-278 (1967).
9. O.E. Brönlund and T.L. Johnsen, QR-factorization of partitioned matrices, Computer Methods in Appl. Mech. and Eng. 3, 153-172 (1974).
10. J.R. Bunch, Partial pivoting strategies of symmetric matrices, SIAM J. Numer. Anal. 11, 521-528 (1974).
11. P. Businger and G.H. Golub, Linear least squares solutions by Householder transformations, Numer. Math. 7, 269-276 (1965).
12. Y.T. Chen, Iterative methods for linear least squares problems, Research Report CS-75-04, University of Waterloo, Canada, Feb. 75.
13. Y.T. Chen and R.P Tewarson, On the fill-in when sparse vectors are orthonormalized, Computing 9, 53-56 (1972).
14. A.K. Cline, An elimination method for the solution of linear least squares problems, SIAM J. Numer. Anal. 10, 283-289 (1973).

15. E.J. Craig, The N-step iteration procedures, J. Math. Phys. 34, 65-73 (1955).
16. A.R. Curtis and J.K. Reid, The solution of large sparse unsymmetric systems of linear equations, J. IMA 8, 344-353 (1971).
17. E. Cuthill, Several strategies for reducing the bandwidth of matrices, in Sparse matrices and their applications, D.J. Rose and R.A. Willoughby (Eds.) Plenum Press (1972).
18. I.S. Duff, Pivot selection and row ordering in Givens reduction on sparse matrices, Computing 13, 239-248 (1974).
19. I.S. Duff and J.K. Reid, A comparison of sparsity orderings for obtaining a pivotal sequence in Gaussian elimination, J. Inst. Maths. Applics. 14, 281-291 (1974).
20. I.S. Duff and J.K. Reid, A comparison of some methods for the solution of sparse overdetermined systems of linear equations, C.S.S.12, A.E.R.E., Harwell, Jan. 75.
21. J. Dyer, Acceleration of the convergence of the Kaczmarz method and iterated homogenous transformations, Ph.D. thesis, University of California, Los Angeles (1965).
22. T. Elfving, Some numerical results obtained with two gradient methods for solving the linear least square problem, LiH-MAT-R-75-5.
23. M. Engeli, T. Ginsburg, H. Rutishauser and E. Stiefel, Refined iterative methods for computation of the solution and the eigenvalues of self-adjoint boundary value problems, Birkhäuser, Basel (1959).
24. D.J. Evans, Iterative sparse matrix algorithms, 49-83, in Software for numerical mathematics, D.J. Evans (Ed.) Academic Press (1974).
25. J.M. Garnett, A. Ben-Israel and S.S. Yau, A hyperpower iterative method for computing matrix products involving the generalized inverse, SIAM J. Numer. Anal. 8, 104-109 (1971).
26. W.M. Gentleman, Basic procedures for large, sparse, or weighted linear least squares problems, Research Report CSRR 2068, University of Waterloo, Canada, July 1972.
27. W.M. Gentleman, Least squares computation by Givens transformations without square roots, J. Inst. Math. Applics. 12, 329-336 (1973).
28. W.M. Gentleman, Regression problems and the QR-decomposition, Bulletin of the IMA 10, 195-197 (1974).
29. W.M. Gentleman, Error analysis of QR decompositions by Givens transformations, Linear Algebra and Appl. 10, 189-197 (1975).
30. W.M. Gentleman, Row elimination for solving sparse linear systems and least squares problems, Proceedings Conf. on Numerical Analysis, Dundee, July 1975.

31. P.E. Gill, G.H. Golub, W. Murray and M.A. Saunders, Methods for modifying matrix factorizations, Math. of Comp. 28, 505-535 (1974).
32. G.H. Glaser and M.S. Saliba, Applications of sparse matrices to analytical photogrammetry, in Sparse matrices and their applications, D.J. Rose and R.A. Willoughby (Eds.) Plenum Press (1972).
33. G.H. Golub, Numerical methods for solving linear least squares problems, Numer. Math. 7, 206-216 (1965).
34. G.H. Golub, Matrix decompositions and statistical calculations, in Statistical Computation, R.C. Milton and J.A. Nelder (Eds.) Academic Press (1969).
35. G.H. Golub and W. Kahan, Calculating the singular values and pseudo-inverse of a matrix, J. SIAM Numer. Anal. Ser.B 2, 205-224 (1965).
36. G.H. Golub and J.H. Wilkinson, Note on the iterative refinement of least squares solutions, Numer. Math. 9, 139-148 (1966).
37. C.R. Hallum and M.D. Pore, Computational aspects of matrix generalized inversion for the computer with applications, Comp. & Maths. with Appls. 1, 145-150 (1975).
38. S. Hammarling, A note on modifications to the Givens plane rotation, J. Inst. Math. Applics. 13, 215-218 (1974).
39. M.R. Hestenes, Pseudoinverses and conjugate gradients, Comm. of the ACM 18, 40-43 (1975).
40. M.R. Hestenes and E. Stiefel, Methods of conjugate gradients for solving linear systems, J. Res. Nat. Bur. Standards Sect. B 49, 409-436 (1952).
41. A.S. Householder, Terminating and nonterminating iterations for solving linear systems, J. SIAM 10, 67-72 (1955).
42. A.S. Householder and F.L. Bauer, On certain iterative methods for solving linear systems, Numer. Math. 2, 55-59 (1960).
43. A.S. Householder, The theory of matrices in numerical analysis, Blaisdell, Boston, Mass. (1964).
44. A. Jennings and A.D. Tuff, A direct method for the solution of large sparse symmetric simultaneous equations, in Large sparse sets of linear equations, J.K. Reid (Ed.) Academic Press (1971).
45. V.N. Joshi, A note on the solution of rectangular linear systems by iterations, SIAM Review 12, 463-466 (1970).
46. H.B. Keller, On the solution of singular and semidefinite linear systems by iterations, J. SIAM Ser.B 2, 281-290 (1965).
47. C.L. Lawson, Sparse matrix methods based on orthogonality and conjugacy, Tech. Memo. 33-627, Jet Propulsion Laboratory, Pasadena, Calif. (1973).

48. C.L. Lawson and R.J. Hanson, Solving least squares problems, Prentice-Hall, Englewood Cliffs, N.J. (1974).
49. D.G. Luenberger, The conjugate residual method for constrained minimization problems, SIAM J. Numer. Anal. 7, 390-398 (1970).
50. T. Lunde Johnsen and J.R. Roy, On systems of linear equations of the form $a^tAx = b$ error analysis and certain consequences for structural applications, Computer methods in Applied Mech. and Eng. 3, 357-374 (1974).
51. P. Läuchli, Iterative Lösung und Fehlerabschätzung in der Ausgleichsrechnung, Z.f. angew. Math. u. Phys. 10, 245-280 (1959).
52. P. Läuchli, Jordan-Elimination und Ausgleichung nach kleinsten Quadraten, Numer. Math. 3, 226-240 (1961).
53. O. Nerdal, Sparse matrix techniques to solve geodetical network problems, Report UMINF-39. Umeå University (1973).
54. C.C. Paige, Bidiagonalization of matrices and solution of linear equations, SIAM J. Numer. Anal. 11, 197-209 (1974).
55. C.C. Paige and M.A. Saunders, Solution of sparse indefinite systems of equations and least squares problems, STAN-CS-73-399, Nov. 1973.
56. G. Peters and J.H. Wilkinson, The least squares problem and pseudo-inverses, Computer J. 13, 309-316 (1970).
57. R.J. Plemmons, Monotonicity and iterative approximations involving rectangular matrices, Math. Comp. 26, 853-858 (1972).
58. R.J. Plemmons, Linear least squares by elimination and MGS, J. ACM 21, 581-585 (1974).
59. M.J.D. Powell and J.K. Reid, On applying Householders method to linear least squares problems, Proc. IFIP Congress 1968.
60. J.K. Reid, A note on the least squares solution of a band system of linear equations by Householder reductions, Computer J. 10, 188-189 (1967-1968).
61. J.K. Reid, On the method of conjugate gradients for the solution of large sparse systems of linear equations 231-254, in Large sparse sets of linear equations, J.K. Reid (Ed.) Academic Press (1971).
62. J.K. Reid, Direct methods for sparse matrices, 29-47, in Software for numerical mathematics, D.J. Evans (Ed.) Academic Press (1974).
63. I. Rudolphi, Orthogonalization of sparse vectors, Report UMINF-38. Umeå University (1973).
64. A. Schönhage, Unitary transformations of large matrices, Numer. Math. 20, 409-417 (1973).

65. G.W. Stewart, Conjugate direction methods for solving systems of linear equations, Numer. Math. 21, 285-297 (1973).
66. E. Stiefel, Ausgleichung ohne Aufstellung der Gausschen Normalgleichungen, Wiss. Z. Technische Hochschule Dresden 2, 441-442 (1952/53).
67. T. Söderström and G.W. Stewart, On the numerical properties of an iterative method for computing the Moore-Penrose generalized inverse, SIAM J. Numer. Anal. 11, 61-74 (1974).
68. K. Tanabe, Projection method for solving a singular system of linear equations and its applications, Numer. Math. 17, 203-214 (1971).
69. K. Tanabe, Characterization of linear stationary iterative processes for solving a singular system of linear equations, Numer. Math. 22, 349-359 (1974).
70. R.P. Tewarson, On the orthonormalization of sparse vectors, Computing 3, 268-279 (1968).
71. R.P. Tewarson, A least squares iterative method for singular equations, Comput. J. 12, 388-392 (1969).
72. R.P. Tewarson, Sparse matrices, Academic Press (1973).
73. A. van der Sluis, Condition numbers and equilibration of matrices, Numer. Math. 14, 14-23 (1969).
74. P.Å. Wedin, Perturbation theory for pseudo-inverses, BIT 13, 217-232 (1973).
75. P.Å. Wedin, On the almost rank deficient case of the least squares problem, BIT 13, 344-354 (1973).
76. T.M. Whitney and R.K. Meany, Two algorithms related to the method of steepest descent, SIAM J. Numer. Anal. 4, 109-118 (1967).
77. D.M. Young, Iterative solution of large linear systems, Academic Press, New York-London (1971).
78. D.M. Young, On the consistency of linear stationary iterative methods, SIAM J. Numer. Anal. 9, 89-96 (1972).

THE ORTHOGONAL FACTORIZATION OF A LARGE SPARSE MATRIX

Philip E. Gill and Walter Murray

Division of Numerical Analysis and Computing
National Physical Laboratory, Teddington, Middlesex.

1. INTRODUCTION

By and large, methods based upon Gaussian elimination have been used to solve large sparse systems of equations more extensively than methods based upon orthogonalization. Although orthogonalization techniques give superior numerical stability, this is offset by the increase in storage requirements caused by fill-in. However, the use of orthogonalization can be beneficial in some applications, for example, in linear least-squares and linear programming. In the first case the orthogonal factorization can give a significant amount of statistical information (see Golub and Styan, 1971). In the linear programming case, a factorization is repeatedly modified and an algorithm with a more stable method of modifying the factors may require fewer reinversions.

In this paper we shall discuss the application of the orthogonal factorization to both the large sparse linear least-squares problem and the large-scale linear programming problem.

2. LINEAR LEAST-SQUARES PROBLEMS

Consider the problem

$$\text{minimize } ||b - Ax||_2 , \tag{1}$$

where A is a large sparse $m \times n$ matrix with $m \geq n$ and b is an m vector. The method of Businger and Golub (1965) requires the computation of the orthogonal factorization

$$Q\,A = \begin{bmatrix} R \\ 0 \end{bmatrix} , \tag{2}$$

where Q is an orthogonal matrix such that $QQ^T = Q^TQ = I$

and R is a non singular $n \times n$ upper-triangular matrix. (We shall not consider the rank-deficient case here.) The vector x which minimizes (1) is given by the solution of

$$\begin{bmatrix} R \\ 0 \end{bmatrix} x = Q\, b \,. \tag{3}$$

We shall consider the Householder and Given's methods for obtaining the orthogonal factorization (2). These methods proceed by forming a sequence of orthogonal matrices $\{Q_j\}$ and a sequence of rectangular matrices $\{A_j\}$ such that

$$A_1 = A \,, \quad A_{j+1} = Q_{j+1} A_j \,, \tag{4}$$

with A_τ, the last member of $\{A_j\}$, equal to $\begin{bmatrix} R \\ 0 \end{bmatrix}$. The matrix Q is then given by $Q = Q_\tau Q_{\tau-1} \ldots\ldots Q_2$.

2.1 Householder's method

With the method of Householder, the matrix Q_{j+1} of the sequence (4) reduces the last $m-j$ elements of the jth column of A_j to zero, giving the matrix A_{j+1} in the form

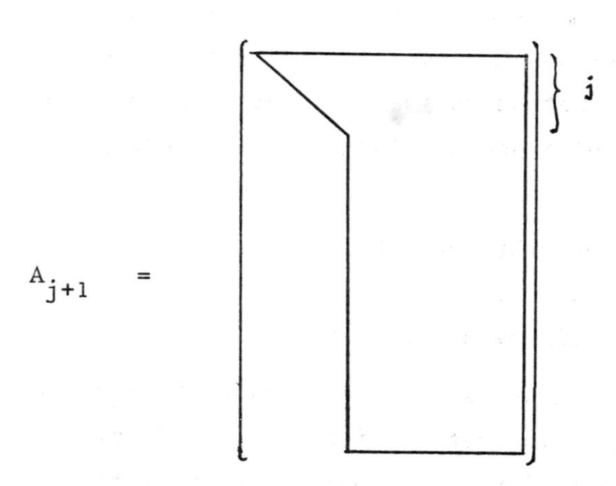

If a denotes the jth column of A_j, then Q_{j+1} is of the form

$$Q_{j+1} = I - \frac{1}{\gamma} w w^T \,,$$

where $\quad w^T = (0,\ldots,a_j + \rho, a_{j+1},\ldots,a_m),$

$$\gamma = -\rho w_j \text{ and } \rho = \text{sign}(a_j)\left(\sum_{i=j}^{m} a_i^2\right)^{\frac{1}{2}}.$$

(In order to simplify the notation a little, we have omitted the subscript denoting the particular Householder matrix being used.) The new diagonal element of R is $-\rho$. At the jth stage, the jth column of R and the matrix Q_{j+1} are completely defined by the jth column of A_j and the number ρ defined above.

When A is large and sparse, a convenient method of implementing the Householder method is to store, in sequence, the jth columns of A_j as they are formed. At each stage of the reduction the (j+1)th column can be found from the recursion

$$A_{j+1}\, e_j = Q_{j+1} \ldots\ldots Q_2\, Ae_j$$

and stored in packed form. When the m columns formed in this way are stored, the construction of the matrices Q_j for the computation of the right hand side of (3) can be effected in a straightforward manner.

The solution of the upper-triangular system in (3) can be performed with the storage scheme just suggested by taking advantage of the well-known fact that

$$R = R_n\, R_{n-1} \ldots R_1\ ,$$

where R_k is an $n\times n$ identity matrix with its kth column replaced by the kth column of R. Then

$$R^{-1} = R_1^{-1}\, R_2^{-1} \ldots R_n^{-1}\ ,$$

where R_k^{-1} is an $n\times n$ identity matrix with its kth column replaced by

$$(-r_{1k}/r_{kk},\ -\ r_{2k}/r_{kk},\ldots\ldots,\ 1/r_{kk},0,0,\ldots.0)^T.$$

We note that solving (3) involves access to the sequence of stored vectors in forward and reverse order. This implies that if an external storage device is used, no rewinding is necessary.

2.2 Givens' method

Unlike Householder's method, where a column of A is reduced to zero at each step, the Givens' method used here involves the reduction of a row to zero.

Consider the sequence (4) where A_j is of the form

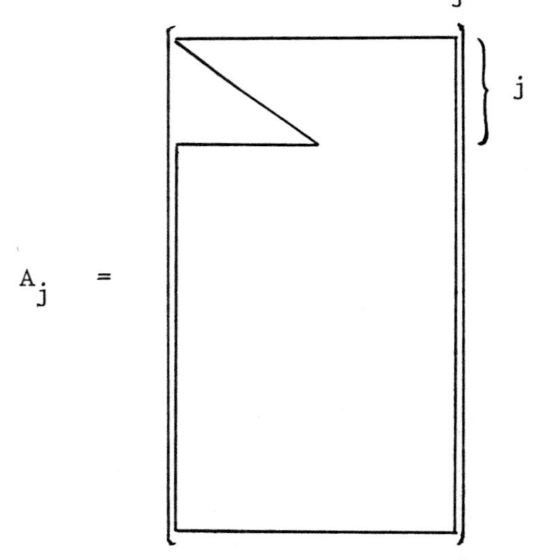

The matrix Q_{j+1} is the product of orthogonal matrices

$$Q_{j+1} = P^i_{j+1}\ P^{i-1}_{j+1} \ldots P^1_{j+1}$$

where each P^i_{j+1} is a symmetric orthogonal matrix. The matrix P^i_{j+1} is an identity matrix except for the ith and (j+1)th rows and columns whose elements make up the 2×2 submatrix

$$\begin{bmatrix} c_i & s_i \\ s_i & -c_i \end{bmatrix} .$$

The elements c_i and s_i are chosen such that the (j+1)th diagonal submatrix of A_{j+1} is upper triangular.

An immediate disadvantage with this method is that a square root is required for each non-zero element which needs to be eliminated. Also, during the reduction, the matrix R needs to be stored in-core so that newly created elements can be inserted. These difficulties can be overcome by computing a slightly different orthogonal factorization using a method which can be viewed as a generalization of the method of Givens reduction. The technique requires no square roots and both factors can be stored in product form using the same information.

Let a diagonal matrix $D = \text{diag}(d_1,d_2,d_3,\ldots,d_m)$ be defined by

$$d_i = \begin{cases} r_{ii}^2 , & i = 1,2,\ldots, n, \\ 1 , & i = n+1,\ldots, m. \end{cases}$$

A VDR factorization of A may then be written in the form

$$DVA = D\begin{bmatrix} \hat{R} \\ 0 \end{bmatrix} \qquad (5)$$

where $\hat{R}$ and V are defined in terms of R, Q and D by the equations

$$\begin{bmatrix} R \\ 0 \end{bmatrix} = D^{\frac{1}{2}}\begin{bmatrix} \hat{R} \\ 0 \end{bmatrix} , \quad D^{\frac{1}{2}}V = Q .$$

The diagonals of $\hat{R}$ are unity and the rows of V are orthogonal. The relations $VV^T = D^{-1}$ and $V^TDV = I$ are easily proved. Henceforth we shall use the notation R for both R and $\hat{R}$ above, since it will always be clear from the context whether or not R has a unit diagonal. In the following we shall outline the method of computing the VDR factorization; for further details the reader is referred to the original reference (Gill, Murray and Saunders, 1975).

The method for computing the VDR factorization is based upon the observation that if a row is added to a matrix with known VDR factors V and R, say, the new VDR factors are $\hat{V}V$ and $\hat{R}R$ where $\hat{V}$ and $\hat{R}$ are defined by two sparse vectors which can be stored in packed form. If the rows of A are added one by one and the initial factors $V_o = I$, $D_o = 0$ and $R_o = I$ are updated accordingly we will finally obtain V and R such that

$$V = V_m V_{m-1} \ldots . V_1 ,$$

and

$$R = R_m R_{m-1} \ldots . R_1 .$$

A particular pair of matrices R_k and V_k, computed when a row a_k is added to A are completely defined by two sparse vectors p and β. The vector p is found from the equations $R_{k-1}^T \ldots . R_1^T p = a_k$ and β from the recurrence relations

$$\left.\begin{array}{ll} \text{(i)} & \text{Set } t \leftarrow 1 ; \\ \text{(ii)} & \text{for } j = 1,2,\ldots, n \text{ define} \\ & s \leftarrow t + p_j^2/d_j , \\ & \beta_j \leftarrow p_j/(d_j s) , \\ & d_j \leftarrow d_j s/t , \\ & t \leftarrow s ; \\ \text{(iii)} & d_k = 1/t . \end{array}\right\} \qquad (6)$$

The special treatment of the case $d_j = 0$ is described in the original reference. After all the rows of A have been added the d_j are the diagonal elements of D defined in (5).

As in Householder's method, the solution of (3) requires a forward and backward access to the numbers defining V and

R (that is, the vectors p and β). We can utilize this fact and implement the algorithm by sequentially storing the sparse p-vectors only. This is because at any stage of evaluating the recurrence relations (6), β_j can be constructed from d_j and p_j. During a forward pass through the data the iteration is commenced with $d_j = 0$, $j = 1,2,\ldots., m$, and before a backward pass the d_j are set to the diagonals of D. The recurrence relations during the backward transformations are:

$$\left.\begin{array}{ll} \text{(i)} & \text{Set } t \leftarrow 1 \ ; \\ \text{(ii)} & \text{for } j = 1,2,\ldots., n \text{ define} \\ & d_j \leftarrow d_j - p_j^2/t \ , \\ & s \leftarrow t + p_j^2/d_j \ , \\ & \beta_j \leftarrow p_j/(d_j s) \ , \\ & t \leftarrow s \ ; \\ \text{(iii)} & d_k \leftarrow 1 \ . \end{array}\right\} \quad (7)$$

As a numerical check we should have $d_k = 1/t$ before step (iii).

The method can be generalized further by choosing the matrix D in (5) such that the upper-triangular matrix R does not have unit diagonal elements (see Hammarling, 1974).

The recurrence relations (6) can be rearranged to give those of Gentleman (1973). This is to be expected since we are effectively computing the same factorization. However, at the kth stage Gentleman recurs the vector $y_{k-1} = V_{k-1} V_{k-2} \ldots . V_1 b$ and implicitly forms $y_k = V_k y_{k-1}$. In contrast, the product form of Givens' method presented here enables problems with multiple right-hand sides to be solved. Our method has the added advantage that it is not necessary to maintain the upper-triangular factor R in-core as Gentleman's requires. We note that "product-form" algorithms have significant advantages when implemented in-core on virtual-memory machines since the scattered nature of information stored using linked-list-type data structures often leads to the phenomenon of "thrashing" on a machine with a paged memory.

2.3 Sparsity considerations

With the Householder and Givens' methods as we have described them, the factors are obtained by taking linear combinations of rows of A. If these methods are used on a matrix with the "dual-angular" structure:

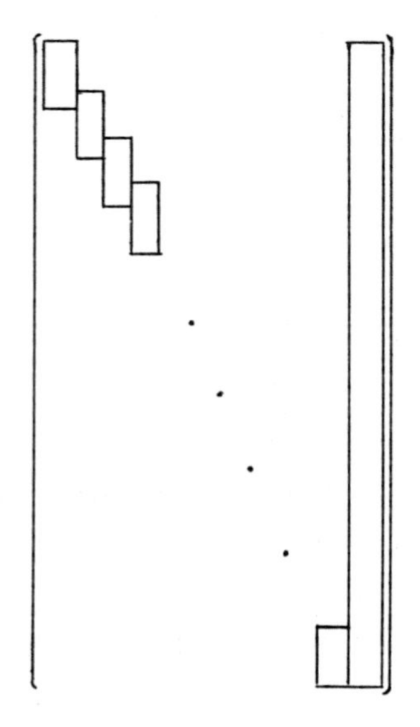

fill-in will be restricted to the diagonal and vertical blocks. When a matrix is arbitrarily sparse, the rows and columns can be reordered so that it is in approximate dual-angular form. Such an algorithm has been given by Weil and Kettler (1971). Unfortunately in some statistical applications (Gentleman, private comm.) it may not always be valid to permute the columns of A and as a result, the sparsity may not always be exploited to full advantage. In this case, when Givens' method is being used, the rows must be added in such an order that the worst fill-in is confined to the last stages of the reduction (see Gentleman, 1975). Similarly, it is difficult to apply preprocessing algorithms to least-squares problems where a row of A is generated as it is required. However, the advantages of preprocessing can be so significant that it may be worthwhile computing and storing the complete matrix A *ab initio*.

2.4 Orthogonal factorization by rows or columns?

The fundamental difference between the Householder and Givens' methods we have described is that the first obtains the orthogonal factorization by reducing columns to zero and the second by reducing rows. Special Householder transformations with two off-diagonal elements could be used to reduce a row to zero but it can be shown that apart from sign, they are identical to the Givens' transformations defined in Section 2.2 (see Gill, Golub, Murray and Saunders, 1974).

Gentleman (1973) has pointed out that an advantage of his method over the conventional Householder when computing the orthogonal factorization of a matrix by rows is that at any one time only the upper-triangular matrix and row being processed need to be held in-core. This argument does not apply to the two methods described here since each requires only a single sparse vector (defining R_k and Q_k) in-core at any one time. However, as we mentioned in the last section, it may be possible to generate only a single row of the matrix A at each step. In this case we believe that the product form of the VDR factorization has significant advantages over Householder reduction.

3. LARGE-SCALE LINEAR PROGRAMMING

The Simplex method for linear programming involves the repeated solution of a set of linear equations when a single column of the $m \times m$ basis matrix B is replaced at each iteration. Gill, Murray and Saunders (1973) have suggested the implementation of the Simplex Method by recurring the VDR factors of B^T or equivalently, the LDV factors of B. When a column of B is replaced, the effect on the LDV factorization is to add two more updates (L_k, V_k) and (L_{k+1}, V_{k+1}) to the product form. If $B = LDV$, the computation required for a single iteration is as follows.

(i)	solve	$Lp = a_s$,	(FTRANL)
(ii)	form	$y = V^T p$,	(BTRANV)
(iii)	form	$q = Ve_r$,	(FTRANV)
(iv)	solve	$Lw = q$.	(BTRANL)

For the full significance of these equations the reader should refer to Gill (1974) and Gill, Murray and Saunders (1975), to supplement the following outline comments. The vector a_s is the incoming column, a_r is the outgoing column and e_r is the rth column of the identity matrix. The direction of access to the vectors defining the LDV factorization is shown in parenthesis (FTRANV = Forward TRANsformation of the V-updates, etc). The vectors p and q are used to update the factors, the matrices (L_k, V_k) being defined by a pair of vectors (p, β) and (L_{k+1}, V_{k+1}) by (q, σ). The recurrence relations for β and σ are similar to (6). In the original form of the algorithm (p, β) and (q, σ) were stored during each iteration. However, a similar technique, which was suggested in Section 2.2 for the linear least-squares problem, can be used to compute σ and β as they are required. In this form the algorithm requires the storage of two sparse vectors p and

q for each iteration. Note that a forward transformation is always followed by a backward transformation, ensuring that the vector of diagonal elements of D is always available when required to initiate the recurrence relations.

Extensive numerical comparisons of this algorithm with methods based upon the LU factorization will appear in a forthcoming paper.

4. CONCLUSIONS

We have demonstrated that it is possible to take advantage of sparsity of a large matrix by arranging the orthogonal factorization of a matrix to be in product form. The resulting algorithm can be applied directly to the linear-least-squares problem and the linear-programming problem.

A thorough comparison with alternative techniques on linear programming problems has yet to be made, but the algorithm outlined in this paper has been implemented and preliminary tests are encouraging.

Acknowledgements

The authors would like to thank M. G. Cox and D. W. Martin for their helpful comments in discussion.

References

BUSINGER, P., and GOLUB, G.H., (1965). Linear least-squares solutions by Householder transformations, Numer. Math. 7, 269-276.

GENTLEMAN, W.M. (1973). Least-squares computations by Givens transformations without square roots. J. Inst. Maths. Applics. 12, 329-336.

GENTLEMAN, W.M. (1975). Error Analysis of QR Decompositions by Givens' Transformations. Linear Algebra and its Applications, 10, 189-197.

GILL, P.E. (1974). Recent developments in numerically stable methods for linear programming. Bull. Inst. Maths Applics. 10, 180-186.

GILL, P.E., GOLUB, G.H., MURRAY, W. and SAUNDERS, M.A. (1974). Methods for modifying matrix factorizations. Math. Comput. 28, 505-535.

GILL, P.E., MURRAY, W., and SAUNDERS, M.A. (1973). The Simplex Method Using the LQ Factorization in Product Form. Paper presented at VIII International Symposium on Mathematical Programming, Stanford University, Stanford, California.

GILL, P.E., MURRAY, W. and SAUNDERS, M.A. (1975). Methods for computing and modifying the LDV factors of a matrix. Math. Comput. 29, Nr. 132.

GOLUB, G.H. and STYAN (1971). Numerical Computations for univariate linear models, Computer Science Department Report Number CS-236-71, Stanford University, Stanford, California.

HAMMARLING, S. (1974). A note on Modifications to the Givens Plane Rotation. J. Inst. Maths. Applics. 13, 215-218.

A FAST, STABLE IMPLEMENTATION OF THE SIMPLEX METHOD USING BARTELS-GOLUB UPDATING[†]

Michael A. Saunders

Systems Optimization Laboratory
Department of Operations Research
Stanford University
Stanford, California

Summary

The only stable method for updating LU factors of the basis matrix in linear programming is that due to Bartels and Golub. Two in-core implementations of the method have been described by J.K. Reid and shown to be very efficient. Here we describe an alternative approach which allows L and most of U to reside on disk. It is shown that the remainder of U (the part which is modified during updating) is typically small enough to be stored in-core as a dense matrix. The efficiency of this scheme is confirmed by experimental results on several practical problems.

[†] Research for this paper was supported by the Energy Research and Development Administration Contract AT(04-3)-326 PA #18, at Stanford University.

1. Introduction

Large-scale linear programming problems of the form

$$\text{minimize} \quad c^T x \qquad \text{subject to} \quad Ax = b, \quad \ell \le x \le u$$

continue to be solved by the simplex method (Dantzig (1963)). Here we consider relatively recent attempts to implement the simplex method in a way which preserves numerical stability throughout all iterations. In chronological order the main theoretical proposals are as follows. They are all based on some factorization of the usual $m \times m$ basis matrix B.

1. Bartels and Golub (1969), $B = LU$, L in product form, U explicit.

2. Gill and Murray (1973), $B = LQ$, L explicit, $Q^TQ = I$, Q not stored.

3. Gill, Murray and Saunders (1974), $B = LDV$, L and V in product form, D diagonal, $V^TDV = I$.

The second method as implemented by Saunders (1972) showed promise for certain classes of problem. However the explicit lower triangular L can in some cases suffer catastrophic fill-in. This difficulty is overcome in the third method by keeping L and V in product form. The method has been implemented by the author but as yet requires further practical evaluation. At present it seems most suited to cases where B has more columns than rows.

In this paper we restrict our attention to the method of Bartels and Golub (Method BG) as there seems to be more chance of maintaining sparsity with LU factors. The only difficulty with the method is that it must allow for fill-in in the matrix U. (Here we mean both creation of nonzeros and modification to existing nonzeros.) This is a serious difficulty when weighed against the closely related method of Forrest and Tomlin (1972). The latter scheme (Method FT) is potentially unstable but requires only deletion of nonzeros from U. It can therefore be implemented very efficiently using disk storage for U.

It was long thought that the Bartels-Golub approach could cause fill-in throughout U, in unpredictable locations and to a rather arbitrary degree. The experimental results of Tomlin (1972a) provided first evidence that the total number of nonzeros in L and U may in fact grow very slowly, at a rate comparable to Method FT. Hence one could deduce that the actual fill-in in existing columns of U must be quite moderate. In spite of this success the question of _where_ such fill-in occurs remained uninvestigated.

More recently, Reid (1973 and 1975) has given two implementations of Method BG which confirm that the growth of L and U is slight and can even be less than with Method FT. Indeed for some problems Reid's 1975 approach maintains $L = I$ and $U =$ (some permutation of B) throughout all iterations. In terms of any attempt to maintain sparsity in a fully general way, this must surely be the ultimate in success.

In this paper we offer an alternative to Reid's in-core implementations. Our aim is to demonstrate that most of U can be stored on disk exactly as in the Forrest-Tomlin scheme, and that the remaining part of U is small enough to be held in-core. This should open the way to solution of arbitrarily large problems in a practical, stable manner. The method has been implemented and tested on several medium-scale problems. It follows an earlier theoretical proposal given in Saunders (1975).

2. Bumps and Spikes

Here we assume the reader to be familiar with the "bump and spike" structure of a typical basis matrix, as described by Hellerman and Rarick (1971, 1972) in connection with their preassigned pivot procedures P^3 and P^4. To summarize, the rows and columns of a sparse basis can be permuted so that the resulting matrix B is block lower triangular, and furthermore each block (_bump_) is almost strictly lower triangular except for a few columns (_spikes_) which have some nonzeros above the diagonal.

The implementation discussed in this paper depends critically on empirical evidence that the total number of spikes in a basis is invariably quite small. For example, certain classes of linear programming problems produce bases which are strictly lower triangular (0 spikes). More generally it appears that anything from 1 to 100 spikes is typical and that 150 or 200 spikes would be extreme even for the largest of problems.

We shall refer several times to the matrix B sketched in Figure 1. This has 3 bumps, each containing 3 spikes.

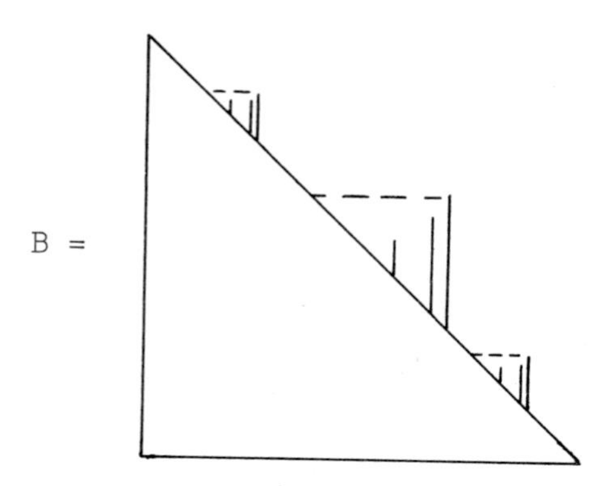

Fig. 1: Bumps and spikes in B.

Non-spike columns in B will sometimes be called triangle columns.

2.1. LU factors of B

Suppose that the triangular factorization $B = LU$ is computed for the matrix B in Figure 1. (The need for column interchanges to preserve stability is discussed in Section 6.3.) A certain amount of fill-in occurs but this is restricted to spike columns. Hence the bump and spike structure of B is transmitted to L and U as illustrated in Figure 2.

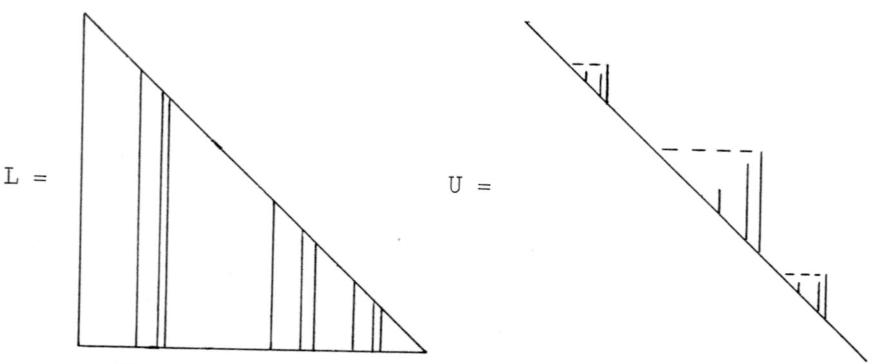

Fig. 2: The spike structure in L and U.

3. A Partitioning of U

Here we introduce the main simplifying observation. Let P be the column permutation which moves all spikes to the end of U without changing their order. If the same permutation is applied to the rows of U, it happens that the resulting matrix P^TUP is of the form

$$\tilde{U} \equiv P^TUP = \left[\begin{array}{c|c} I & R \\ \hline & F \end{array}\right]$$

where F is upper triangular.

The structure and relative sizes of R and F are important. For example, if B were of order 1000 with 50 bumps and 50 spikes (a conceivable situation), then R would be a 950×50 sparse matrix with at most one nonzero per row, while F would be a 50×50 diagonal matrix.

More typically, the matrices in Figures 1 and 2 would produce a $\tilde{U}$ of the form shown in Figure 3.

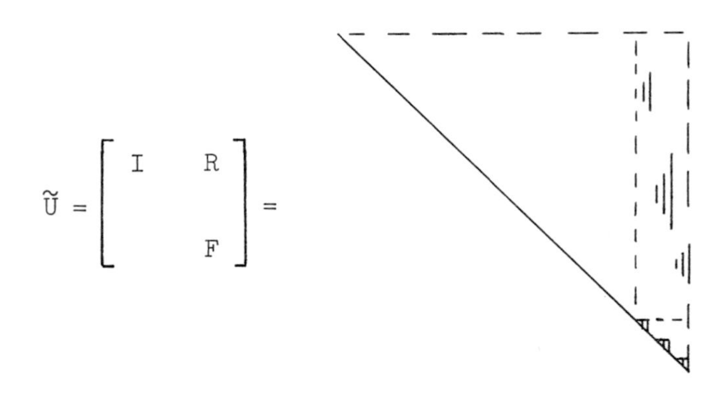

Fig. 3: R and F before updating.

<u>Note</u>: The structure shown in R and F would <u>not</u> be present if all spikes were moved to the end of B before the LU factorization.

From now on we shall regard $\tilde{U}$ as being U itself.

4. Updating

When the p-th column of B is replaced by some other vector a_q, the following steps are required to update L and U.

1. Delete the p-th column of U (call this u_p).
2. Add a sparse vector $v = L^{-1}a_q$ to the end of U (a new spike).
3. Extract the p-th row of U (call this r^T) and permute it to the bottom.
4. Reduce r^T to a multiple of the unit vector e_m^T using Gaussian elimination.

Figure 4 shows the intermediate state of U just before the last step. Note that r^T may come from either F or R depending on whether the deleted column u_p was a spike or not. In both cases, however, r^T is a short vector as depicted in Figure 4. Hence when Gaussian elimination is applied, only F and the bottom part of v are involved and <u>no further change occurs to</u> R.

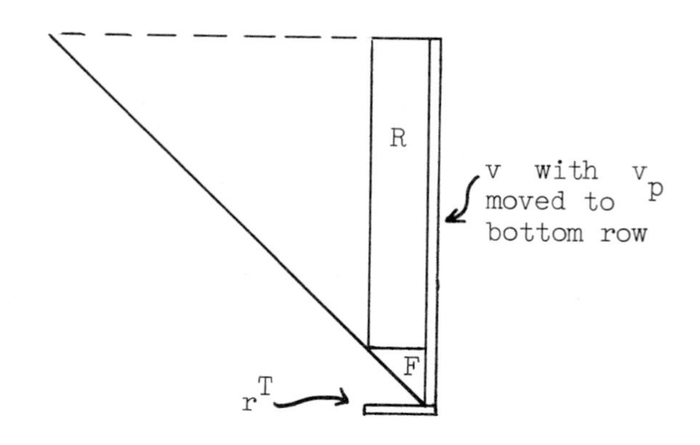

Fig. 4: Intermediate state of U prior to elimination.

The reason for distinguishing between R and F now becomes clear. The matrix R has exactly the same properties as U in Method FT. It is essentially "read-only" and can therefore be stored on disk. Furthermore, since F is always a relatively small matrix it may be stored in-core and potential fill-in presents no difficulty. Hence row interchanges may be incorporated during the elimination. It follows that the stable updating scheme envisaged by Bartels and Golub can now be realized in a practical sense.

5. Implementation Details

In order to investigate the typical size and density of F, an in-core implementation of the simplex method has been constructed as follows:

1. The constraint matrix A is stored column-wise as a list of row indices and pointers into a table of distinct coefficients. Two separate arrays specify upper and lower bounds on all variables.

2. An implementation of P^3 as described by Hellerman and Rarick (1971) is used to locate bumps and spikes in B.

3. An LU factorization of B is computed using column interchanges where necessary to preserve stability.

4. The columns of L corresponding to non-spike columns of B are represented by pointers into the matrix A.

5. U is stored in two sections as R and F. R is packed column-wise as a general sparse matrix. The upper-triangular part of F is stored column-wise as a dense matrix.

6. During updating, deletion of a row from R is accomplished by over-writing any existing nonzeros. Columns deleted from R are simply flagged as such (i.e., no attempt is made to recover storage). On the other hand storage <u>is</u> recovered when rows and columns are deleted from F, since F is confined to a small area of core.

7. Refactorization occurs at a user-specified frequency k, or whenever there is insufficient core to update L, R or F. If P^3 detects s spikes in B then $t \times (t+1)/2$ words of core are allocated to F (where $t = s+k$ is an upper bound on the number of spikes there could be before the next refactorization). The remaining available core is allocated to L and R.

The test program is written in Fortran IV to run on a machine with a paged memory, in this case an IBM 370/168 operating under OS/VS2. In such an environment the strategy of imbedding L in A is tolerable only if the amount of core required for A is moderate relative to the total real core available to the program. Our main concern here however is with access to U, and in fact the storage scheme chosen for R and F will avoid frequent page-swapping regardless of problem size.

6. Experimental Results

Three medium-scale LP models were chosen as test problems. Their physical characteristics are summarized in Table 1.

	Rows	Columns	Elements	Density	Iterations performed	Optimal?
STAIR	357	468	3927	2.35	700	Yes
SHELL	537	1776	4902	0.51	582	Yes
BP	822	1572	11414	0.88	5200	Not quite

Table 1: Test problem statistics

6.1. The density of F

One of our primary aims was to investigate the size and density of F in order to determine how it should best be stored.

In terms of computational effort, the dense storage scheme as implemented is very efficient for any problem, regardless of the density of F. This is obvious if F should happen to be nearly full, but in any case only a few rows or columns of F are accessed during updating (since in Figure 4 the vector r^T is usually very sparse) or during solution of $Uy = z$ or $U^T y = z$ (since y is normally sparse also).

The density of F is shown in Table 2 for various iterations. In the case of STAIR we see that treating F as a dense matrix does not imply much work or storage overhead (50% is far from what is commonly regarded as sparse). However to allow for densities of 10 or 20% as exhibited by the other problems, F could well be stored row-wise using a linked list. This was the data structure used by Reid (1973) for storing the whole matrix U. Reid (1975) subsequently switched to ordered row and column lists with "elbow room", in order to avoid frequent page-swapping on a machine with virtual memory. In our case there should be no paging problem as F is always of moderate size.

	Iteration	Iterations since LU	No. of spikes (dimension of F)	Density of F (%)
STAIR	500	0	60	50.7
	550	50	87	45.4
	560	0	56	53.8
	620	60	82	56.4
SHELL	80	0	0	-
	155	75	71	7.9
	310	70	62	16.3
	465	65	56	9.9
BP	1500	3	86	9.7
	2200	20	116	16.8
	2700	48	116	24.5
	3350	49	136	31.1
	4150	2	108	19.8
	4900	25	126	22.6

Table 2: The density of F

6.2. The initial LU

The success of any implementation depends very much on starting each series of iterations with a sparse yet accurate factorization of the current basis. In our case it is vital that a stable LU factorization be computed with a minimal number of spikes in U.

Sparsity

First of all this means that the "below-bump" part of B must be stored in L, as shown in Figure 2. (It is now common practice to split off this part of B and store it in U, thereby avoiding fill-in in the bottom part of the spike columns.) A more dense initial L probably means that columns added to U during updates will be more dense, but two things offset this disadvantage:

(a) The triangle columns of B are more easily imbedded in A.

(b) The subsequent growth of L is extremely low (typically 0 to 20 nonzeros per iteration). This is discussed further in Section 6.3.

Stability

Some rearrangement of a preassigned pivot order is usually necessary to guarantee stability in the initial LU. With bump and spike ordering it is clear that row interchanges could increase the total number of spikes by an arbitrary amount. Instead we are prepared to re-order the columns within any one bump.

Some relative pivot tolerance u $(0 < u \leq 1)$ is used to decide whether each column should be accepted in the preassigned order. If all columns pass the test the resulting LU is equivalent to Gaussian elimination with row interchanges with tolerance u (except that no row interchanges are performed). If a column a_k fails the test the best possible substitute a_s is chosen by solving $L^T_p = e_k$ and finding $|p^T a_s| = \max |p^T a_j|$ amongst all remaining spikes a_j in the current bump. An interchange of a_k and a_s is called a spike swap or a triangle swap according to whether a_k was a spike or not. It is equivalent to Gaussian elimination with column interchanges with pivot tolerance 1.0.

The computation of p involves a backward pass through the columns of L. It may be stopped at the beginning of the current bump but can be expensive if incurred too often by a large value of u. Note that a triangle swap usually increases the total number of spikes by one, so the choice of u also affects the final density of L and U. Table 3 shows this effect on a particular basis from problem STAIR. The number of triangle swaps is clearly excessive when $u = 0.1$, and as a compromise between stability and sparsity we should probably use $u = 0.01$ as recommended by Tomlin (1972b). In actual practice we normally use $u_1 = 0.001$ and make allowance for recomputing LU with $u_{j+1} = 10 \times u_j$ if the computed residual vector $b - B\tilde{x}$ is too large after the j-th factorization. (It has never yet been necessary to refactorize.)

6.3. Growth of Nonzeros During Updating

The one useful theorem here is that the number of nonzeros added to L each iteration is bounded by the current number of spikes in U. This is easily seen from Figure 4. In practice the vector r^T being eliminated is normally so

Basis Statistics	Problem:	STAIR	Iteration:	560
	Slacks:	22	Structurals:	335
	Elements:	3478	Density:	2.7
	Bumps:	1	Spikes:	52
	Estimate of condition number:			10^5

u	Triangle swaps	Spike swaps	Final no. of spikes	A*	L**	R	F	LU =A+L+R+F
0.001	3	8	56	2688	2243	3051	858	8840
0.01	11	7	63	2602	2764	2883	1003	9252
0.1	55	36	108	2099	7348	1997	1805	13249

Table 3: Nonzeros in L and U for various pivot tolerances u.

* A = number of nonzeros in L (triangle columns) that are imbedded in A.

** L = other nonzeros in L (spike columns).

sparse that the actual growth is considerably less than this (already low) bound. For example, near the end of the three test runs the average growth of L was respectively 15.6, 1.4 and 13.8 nonzeros per iteration. This remarkably low growth helps to offset the high initial density of L.

Corresponding figures for the growth of R are 70, 11 and 136 nonzeros per iteration, measured over 60, 80 and 50 iterations. This includes rows and columns that have been deleted from R since this storage is not recovered.

A more detailed summary is given in Figure 5 for problem STAIR. Note that L includes a constant 2688 nonzeros that are imbedded in A. These are not shown in the curve for L. Note also that the number of spikes grows at about 0.5 per iteration, so the growth rate of meaningful numbers in R is about half that shown.

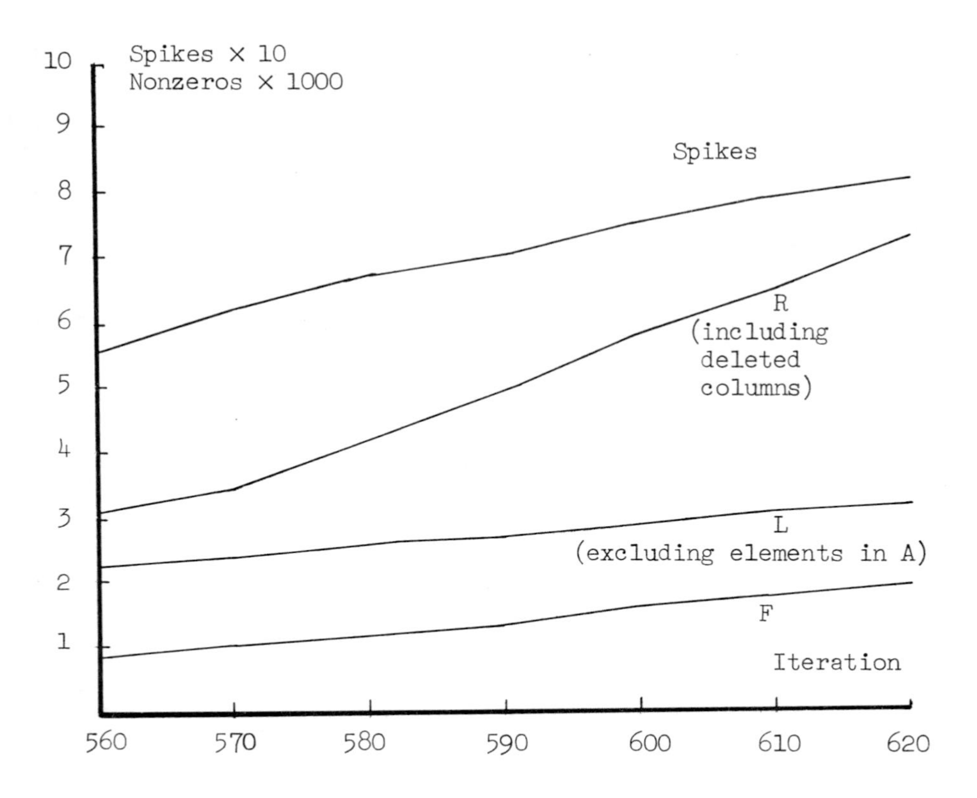

Fig. 5: Problem STAIR -- growth of spikes and nonzeros in L, R and F.

6.4. Discussion

1. The triangle swaps discussed in Section 6.2 are as important as spike swaps for maintaining stability in the initial LU. The scaling algorithm of Curtis and Reid (1972) should help here as it makes the elements of A as close in size as possible.

2. The initial LU factors in our experiments are considerably more dense than those reported by Reid (1973,1975). This is largely due to our use of Hellerman and Rarick's P^3 rather than their P^4 which is known to produce the maximal number of bumps and fewer spikes. (On problem BP for example, P^4 typically gives 8 to 10 bumps and about 90 spikes. On advanced bases, our version of P^3 usually gives only 1 bump and about 110 spikes.)

3. At this stage the growth rates during updating are more representative of the performance of our method. The results shown in Figure 5 are reasonably comparable with those given by Reid and are vastly better than those for the standard product form of inverse (e.g., see Figure 3 in Saunders (1972)).

7. Conclusion

We have shown that it is possible to implement the simplex method using a stable update of LU factors, in a way which allows most of U to be stored in sequential, read-only form either on disk or in virtual memory. The scheme used would involve a relatively minor change to any existing implementation of the method of Forrest and Tomlin. (In fact that is how the present implementation was developed.)

The experimental results given here must be regarded as preliminary, but they illustrate that the major difficulties associated with the Bartels-Golub approach can be overcome. We hope to report considerably lower densities in L and U when the preassigned pivot procedure P^3 is used in conjunction with a maximal bump-finding algorithm.

Acknowledgment

I wish to thank J.A. Tomlin for numerous discussions during the course of this research.

References

R.H. Bartels and G.H. Golub (1969). The Simplex Method of Linear Programming Using LU Decomposition. Comm. ACM 12, 266-268.

A.R. Curtis and J.K. Reid (1972). On the Automatic Scaling of Matrices for Gaussian Elimination. J. Inst. Maths. Applics. 10, 118-124.

G.B. Dantzig (1963). Linear Programming and Extensions, Princeton University Press, Princeton, New Jersey.

J.J.H. Forrest and J.A. Tomlin (1972). Updating Triangular Factors of the Basis to Maintain Sparsity in the Product Form Simplex Method. Mathematical Programming 2, 263-278.

P.E. Gill and W. Murray (1973). A Numerically Stable Form of the Simplex Algorithm. J. Linear Algebra Applics. 7, 99-138.

P.E. Gill, W. Murray and M.A. Saunders (1974). Methods for Computing and Modifying the LDV Factors of a Matrix. Report NAC 56, National Physical Laboratory, Teddington.

E. Hellerman and D. Rarick (1971). Reinversion with the Preassigned Pivot Procedure. Mathematical Programming 1, 195-216.

E. Hellerman and D. Rarick (1972). The Partitioned Preassigned Pivot Procedure (P^4). In Sparse Matrices and Their Applications, D.J. Rose and R.A. Willoughby (eds.), Plenum Press, New York.

J.K. Reid (1973). Sparse Linear Programming using the Bartels-Golub Decomposition. Presented at the VIII International Symposium on Mathematical Programming, Stanford University, California.

J.K. Reid (1975). A Sparsity-Exploiting Variant of the Bartels-Golub Decomposition for Linear Programming Bases. AERE Report CSS , Harwell, Didcot, Oxford.

M.A. Saunders (1972). Large-Scale Linear Programming Using the Cholesky Factorization. Report STAN-CS-72-252, Stanford University, California.

M.A. Saunders (1975). The Complexity of LU Updating in the Simplex Method. In The Complexity of Computational Problem Solving, R.S. Anderssen and R.P. Brent (eds.), Queensland University Press, Queensland.

J.A. Tomlin (1972a). Modifying Triangular Factors of the Basis in the Simplex Method. In Sparse Matrices and Their Applications, D.J. Rose and R.A. Willoughby (eds.), Plenum Press, New York.

J.A. Tomlin (1972b). Pivoting for Size and Sparsity in Linear Programming Inversion Routines. J. Inst. Maths. Applics. 10, 289-295.

USING THE STEEPEST-EDGE SIMPLEX ALGORITHM TO SOLVE SPARSE LINEAR PROGRAMS.

By D. Goldfarb
The City College of The City University of New York

Abstract:

An algorithm for solving sparse linear programs in core is described. This algorithm selects pivot columns according to a "steepest-edge" strategy. Although this approach is non-standard and requires some additional storage and increased computing time per iteration, computational results are given which show it to be competative with other approaches. A variant of the Bartels-Golub algorithm developed by Reid which creates a minimal amount of fill-in while maintaining numerical stability is used to update the factorization of the basis matrix after each simplex step. Combining the steepest-edge approach with partial pricing and applying it to obtain a dual algorithm are also discussed.

1. Introduction:

We consider in this paper the steepest-edge simplex algorithm applied to the standard linear programming problem:

$$\text{minimize} \quad c^T x \tag{1.1}$$

$$\text{subject to} \quad Ax = b \tag{1.2}$$

$$\text{and} \quad x \geq 0, \tag{1.3}$$

where A has dimension m x n. This algorithm has been known for some time to be very effective in terms of the number of simplex iterations required to solve (1.1) - (1.3); (e.g. see Kuhn & Quandt[1] and Cutler & Wolfe [2]. However, until the development of the practicable algorithm of Goldfarb and Reid [3], the steepest-edge algorithm was of only theoretical interest because of the excessive amount of computation that appeared to be necessary at each iteration on other than very small problems.

Our attention here is primarily focused on moderately large - say 1,000 to 10,000 variable - sparse problems. After briefly outlining the steepest-edge algorithm of Goldfarb and Reid [3] in the next section, we describe some algorithmic details embodied in its implementation as a Harwell library subroutine in section 3 . In section 4 test results obtained on genuine problems with as many as 3333 variables are given. Included are comparisons with other algorithms and actual computer timings.

Finally, practicable variants of the steepest-edge algorithm other than the one described in section 3 are discussed in section 5. In particular, multiple pricing, partial pricing and dual algorithms are considered.

2. A practicable steepest-edge algorithm

Consider a single simplex step in which variable p in the basis is replaced by variable q. If we assume, without loss of generality, that the first m variables are basic at the start of the step, then $p \leq m$, $q > m$ and the old basic solution is given by

$$x = \begin{bmatrix} B^{-1}b \\ 0 \end{bmatrix}, \qquad (2.1)$$

where B, the basis matrix, consists of the first m columns of A. Moreover, after the step, which we indicate by the use of bars, the basic solution is given by

$$\bar{x} = x + \theta\, \eta_q \qquad (2.2)$$

where

$$\eta_q = \begin{bmatrix} -w \\ e_{q-m} \end{bmatrix} = \begin{bmatrix} -B^{-1}a_q \\ e_{q-m} \end{bmatrix}, \qquad (2.3)$$

$a_q = Ae_q$, e_q is the q-th column of the n x n identity matrix,

and θ is a scalar chosen so that $\bar{x}_p=0$. The old basic solution x is a vertex of the polytope $P = \{x \mid Ax=b, x \geq 0\}$ and the n-m vectors $\eta_{m+1}, \ldots, \eta_n$ point down the n-m edges of P that emanate from x.

In the simplex method one chooses a pivot column q with negative "reduced cost",

$$z_q = c^T \eta_q, \tag{2.4}$$

(i.e., a direction η_q that is downhill). If all $z_i \geq 0$, x is optimal. In the standard column selection algorithm q is chosen so that

$$z_q = \min_{i>m} \{z_i\}. \tag{2.5}$$

In the steepest-edge algorithm q is chosen so that

$$\frac{z_q}{||\eta_q||_2} = \frac{c^T \eta_q}{||\eta_q||_2} = \min_{i>m} \left\{ \frac{c^T \eta_i}{||\eta_i||_2} \right\}. \tag{2.6}$$

That is, the direction η_q chosen is most downhill.

Explicitly computing all of the norms $||\eta_i||_2$, $i>m$ at each step is clearly impracticable for all but very small problems. Fortunately, the following updating formulas for the vectors η_i and the numbers $\gamma_i = ||\eta_i||_2^2 = \eta_i^T \eta_i$ can be derived, (see Goldfarb and Reid [3]).

$$\bar{\eta}_p = -\eta_q / \alpha_q \tag{2.7a}$$

$$\bar{\eta}_i = \eta_i - \eta_q \bar{\alpha}_i \qquad i > m, \ i \neq q \tag{2.7b}$$

and

$$\bar{\gamma}_p = \gamma_q / \alpha_q^2 \tag{2.8a}$$

$$\bar{\gamma}_i = \gamma_i - 2\bar{\alpha}_i a_i^T B^{-T} w + \bar{\alpha}_i^2 \gamma_q, \ i>m, \ i \neq q \tag{2.8b}$$

where α_q is the p-th component of $w = B^{-1}a_q$ and the numbers $\bar{\alpha}_i$ are the components of the p-th row of $\overline{B^{-1}}$ A. Premultiplying both sides of (2.7a) and (2.7b) by c^T, and recalling the definition (2.4) we observe that the numbers $\bar{\alpha}_i$ can also be used to update the "reduced costs".

Updating the γ_i requires some extra computation. In addition to an extra backward transformation operation $B^{-T}w$, inner products of this vector with non-basic columns, a_i, corresponding to $\bar{\alpha}_i$ that are nonzero are needed. Fortunately, in sparse problems, most of the $\bar{\alpha}_i$ are zero. On the basis of the computational experience reported in section 4, the increase in work per step over the standard simplex algorithm is roughly 25 percent. Extra storage for the γ_i's is, of course, also needed. Time must also be spent computing the γ_i's initially.

3. The Harwell library subroutine

The steepest-edge simplex algorithm described in the preceeding section has been successfully implemented by the author and John Reid as a Harwell library subroutine. Before presenting test results obtained with this code we describe here some of its more important details. A full description including a listing are given in [4].

The basis matrix B is kept in factorized form as LU where U is a permutation of an upper triangular matrix and L is a product of matrices which differ from I in just one off-diagonal element (and therefore represent elementary row operations). This is updated after a simplex pivot step by an ingenious variant of the Bartels-Golub algorithm developed by Reid [5] which is designed to create a minimal amount of fill-in while maintaining numerical stability, and which whenever it is possible updates the factorization of B by row and column interchanges alone.

If for simplicity we assume that U is an upper triangular matrix, then after a simplex pivot step the basis matrix

$$\bar{B} = B + (a_q - a_p)e_p^T$$

can be written as $\bar{B} = LS$ where S is an upper triangular matrix with a spike in column p. If this spike has its last non-zero in row k then the submatrix of S consisting of rows and columns p through k is called the "bump". In Reid's algorithm, the rows and columns of this bump except for the spike are scanned for singletons. Whenever a row (column) singleton is found, a symmetric permutation is applied to the bump to bring this singleton to the last (first) row and column of the bump, thereby reducing the dimension of the bump by one and the scan is continued. If the spike in the bump that remains after the column sweep and the row sweep is a column singleton, additional permutations are applied before resorting to eliminations to reduce the permuted S to upper triangular form. This updating, factorizing (or refactorizing) a given basis matrix B and both backward and forward transformations using B are all handled by a subroutine written by Reid, (see [6].)

The numbers $\bar{\alpha}_i$ are obtained by first computing the vector

$$y = \bar{B}^{-T} e_p. \qquad (3.1)$$

Surprisingly, it can be shown that this requires less work than computing

$$y = \frac{1}{\alpha_q} B^{-T} e_p$$

when variants of the Bartels-Golub algorithm are used [7]. In fact, if either the original Bartels-Golub [12] or the Forrest-Tomlin [13] algorithm is used to update B, then (3.1) involves only a backward solution using $\bar{L}^T$.

In addition to the method of column selection, our code is non-standard in its procedure for selecting the pivot row, using the method of Harris [8] . This relaxes the usual restriction that x be nonnegative to

$$x_i \geq -\delta \qquad i = 1,\ldots,m$$

where δ is some small tolerance.

By doing this, near degeneracy is treated as degeneracy, with ties resolved by choosing the largest pivot. Numerically, this is good as it tends to prevent the basis matrix from becoming ill-conditioned.

Unless an initial basis is specified by the user, the code determines one with a basis matrix which is a permutation of an upper triangular matrix by suitably adding artificial variables. Moreover, negative infeasibilities are allowed and are handled in phase 1 by a variant of the composite method; (see [9] for example). This necessitates some minor modification of the Harris' row selection procedure and the formulas used to update the reduced costs.

4. Test Results

In this section we report the results of using a slightly modified version of the Harwell steepest-edge code [4] on six test problems made available by P. Gill (Blend), S. Powell (Powell) and M. A. Saunders (Stair, Shell, GUB, and BP). The modifications were incorporated so that the Harris algorithm, (a variant of steepest-edge in which the weights γ_i are calculated approximately), and the original Dantzig algorithm could be run as well, and realistic comparisons made. Because of sparsity it is impossible to predict just how expensive the recurrences (2.8) will be and it seems best to rely on actual machine times.

All algorithms had the same initial basis for phase 1 generated by the code. In order that they should all have the same starting basis for phase 2, the feasible point generated by the steepest-edge algorithm was used to restart all three algorithms.

Our results are summarized in Table 1. In the case of Blend the initial basis generated was feasible so comparisons could be made only on phase 2. In the case of Powell, the problem had no feasible solution so we could only compare algorithms in phase 1. The others had two phases and we ran each algorithm on each phase, apart from the original Dantzig algorithm on BP which was curtailed because of the high computing cost. All runs were carried out on the IBM 370/168.

We have given separately the time taken to calculate γ_i, $i>m$, initially for the steepest-edge algorithm. This is an overhead only for phase 1 and it can be seen that it is usually quite slight compared with the rest of the time taken in phase 1 and 2, the worst overhead being 18% on Shell and the average being 5%.

We also show in Table 1 the amount of storage required for the arrays used by the algorithms in order to indicate the practicality of working in core with medium-scale problems. We held the basis factorization and the vectors x,w,y in double-length (8 bytes) but A,b,c,z and γ in single-length (4 bytes) except that we found double length necessary for γ in BP.

In terms of numbers of iterations, the steepest-edge algorithm was best except for Powell (14% more iterations than Harris) and phase 2 of Shell (1% more iterations than Harris) and the average percentage gain over the Harris algorithm was 19%. On the other hand the time per iteration was on average 22% greater, so that overall the two algorithms are very comparable. In fact the total time taken by the steepest-edge algorithm was less than that taken by the Harris algorithm on Blend, Stair, GUB and BP and greater on Powell and Shell.

In the above averages, equal weight was given to all six test problems. As three of these problems (Blend, Powell and Shell) are much easier to solve than the other three problems, both in terms of iterations and computing time required, the total number of iterations and total time required to solve all six problems is probably a better measure of the relative performance of the algorithms. On this basis, the Harris algorithm required 33% more iterations (5698 versus 4297) and 7% more time (771 versus 719) than the steepest-edge algorithm.

Similar relative iteration counts were obtained by Crowder and Hattingh [10] on a different set of test problems, using a code which explicitly calculated the weights γ_i. The superiority of the Harris algorithm over the original Dantzig algorithm in terms of iterations is further supported by the results reported in [8] and [10] .

In Table 2 we present some results obtained using the steepest-edge code that relate to the basis matrix factorization. This table should be self-explanatory.

Table 1 Summary of Results

Problem		Blend	Powell	Stair		Shell		GUB		BP	
Number of rows, m		74	548	362		662		929		821	
Number of cols, n		114	1076	544		1653		3333		1876	
Number of non-zeros		560	7131	4013		5033		14158		11719	
Phase		2	1	1	2	1	2	1	2	1	2
Iterations	Steepest edge	48	226	210	157	48	183	475	644	1124	1182
	Harris	65	198	313	163	77	181	672	796	1414	1819
	Dantzig	70	285	244	387	78	181	854	1102	Not run	>3976
Time (secs)	Steepest edge Set up	0.02	1.1	0.4	-	2.4	-	7.5	-	3.5	-
	Steepest edge run	0.90	16.6	26	28	2.9	10.2	57	109	190	263
	Harris	1.04	11.7	33	24	3.8	8.6	78	111	195	305
	Dantzig	1.06	16.9	26	51	3.8	8.5	96	146	Not run	>596
Time per iteration	Steepest edge	.019	.073	.12	.18	.060	.056	.12	.17	.17	.22
	Harris	.016	.059	.11	.15	.049	.048	.12	.14	.14	.17
	Dantzig	.015	.059	.11	.13	.049	.047	.11	.13	Not run	.15
Array storage (370 k-bytes)	Steepest edge	20	135	190		126		268		297	
	Harris	20	135	190		126		268		289	
	Dantzig	19	127	185		114		248		275	

Table 2 Results for Steepest-edge Algorithm

Problem			Powell	Stair	Shell	GUB	BP
Basis dimension,		m	548	362	662	929	821
Non-zeros in initial basis factorization		L	0	0	0	0	0
		U	2144	2472	1267	2713	3011
Non-zeros in basis factorization before and after last refractorization	before	L	629	5232		22	4197
		U	3551	4099		3306	5780
	after	L	427	2584	0	0	2141
		U	2134	3051	1288	2716	3391
Refactorizations +			2	7	0	10*	36*
Iterations			226	367	231	1119	2306

* One refactorization due to roundoff.

\+ The initial factorization and refactorizations after phase 1 and phase 2 as a precaution against roundoff are not included in these numbers.

5. Other variants of the steepest-edge algorithm

Many commercially available LP codes incorporate "multiple pricing" (i.e., suboptimization) and "partial pricing" options for choosing the pivot column in order to reduce the average computing cost of an iteration by eliminating the need to price out all nonbasic columns at each iteration. By partial pricing, we mean the type of heuristic where at each iteration a sequential search is made of the nonbasic columns, starting from where the last iteration left off, until some number of columns, say k, either set by the user or dynamically determined by the code, has been found with negative reduced costs, or all columns have been considered. Because, the steepest-edge algorithm requires the updating of the column weights, γ_i, on every iteration, neither of these options would appear to be applicable to it. Although it is possible to delay the updating of the γ_i's until after a multiple pricing pass, followed by a "multiple" update of them, this does not result in a computational savings.

We propose here a partial pricing variant of the steepest-edge algorithm in which only approximations to the weights $\gamma_i=\eta_i^T\eta_i$ are stored and updated. During a partial pricing iteration, only those columns searched with negative reduced costs whould be updated by

$$\bar{\gamma}_i=\max(\gamma_i-2\bar{\alpha}_i a_i^T B^{-T} w+\bar{\alpha}_i^2\gamma_q, 1+\bar{\alpha}_i^2) \qquad (5.1)$$

Although, this formula is used in the Harwell subroutine for numerical reasons, it is absolutely essential in the partial pricing variant. If γ_i is an under-estimate of $\eta_i^T\eta_i$ then (2.8b) can yield a zero or negative $\bar{\gamma}_i$. Also, instead of storing reduced costs z_i for all nonbasic columns and updating them by the formula

$$\bar{z}_i = z_i - \bar{\alpha}_i z_q \qquad i > m,\ i\neq q \qquad (5.2)$$

as is done in the Harwell subroutine, one should store the pricing vector

$$\pi = B^{-T}\hat{c}$$

and update it by

$$\bar{\pi} = \pi - z_q y,$$

(e.g., see [11]),

where $\hat{c}$ is a vector of the first m components of c and

$$y = \bar{B}^{-T}e_p.$$

Storing π is usually preferable to storing the possibly very long vector of z_i's. If $\hat{k}$ is the number of columns searched and θ is the fraction of elements in the pivot row that are nonzero then the number of inner products with nonbasic columns of A (i.e. $a_i^T\bar{\pi}$, a_i^Ty, and $a_i^TB^{-T}w$) required by the partial search algorithm would be approximately $\hat{k}+(1+\theta)k$. In the Harwell steepest-edge code $(1+\theta)(n-m)$ inner products are needed.

Initially all γ_i's should be set equal to one. Although the test results reported in the last section indicate that the cost of initializing the γ_i's can be substantial it may be of some value to compute those corresponding to negative z_i on an initial pass through all nonbasic columns of A. Note that

$$\gamma_q = 1 + w^T w \qquad (5.3)$$

and, hence this formula and (2.8a) provide a correct value of γ_p.

Along with the tests reported in section 4, John Reid and the author, ran the steepest-edge algorithm with all weights initially set equal to one. In terms of iterations the performance of this variant fell between that of the steepest-edge algorithm with correct initial weights and the Harris algorithm. Although the cost of these extra iterations was not compensated by the time saved in **not** initializing the weights, these results bode well for a partial **pricing** steepest-edge algorithm. Also for reasons similar to ones that apply to standard LP codes, it may be profitable to do multiple pricing with the above partial **pricing variant.**

Sometimes the solutions of the LP problem (1.1)-(1.3) are required for several different vectors b in (1.2). In such a situation or when a constraint is added to an LP problem whose solution is already known, it is advantageous to use a dual feasible algorithm. Thus we consider briefly a dual steepest-edge simplex algorithm. One might better call such an algorithm a maximal distance algorithm since at each step the pivot row selected is the one which in the transformed set of equations

$$B^{-1}Ax = B^{-1}b, \qquad (5.4)$$

has a negative right hand side and whose corresponding hyperplane is furthest from the origin. Algebraically, one considers the elements of $B^{-1}b$ weighted by the norms of the corresponding rows.

Under the assumptions of section 2 it is simple to show that the following recurrences hold for the rows of $B^{-1}A$, $\rho_i = e_i^T B^{-1} A$, and the square of their norms $\beta_i = \rho_i^T \rho_i$

$$\bar{\rho}_q = \rho_p / w_p \qquad (5.5a)$$

$$\bar{\rho}_i = \rho_i - w_i \bar{\rho}_p \qquad i \lesssim m,\ i \neq p \qquad (5.5b)$$

and

$$\bar{\beta}_q = \beta_p / w_p^{\,2} \qquad (5.6a)$$

$$\bar{\beta}_i = \beta_i - 2(w_i/w_p) e_i^T B^{-1} \hat{A}\hat{A}^T B^{-T} e_p + w_i^{\,2} \bar{\beta}_p \ \left\{ \begin{matrix} i \leqslant m \\ i \neq p \end{matrix} \right. \qquad (5.6b)$$

where $A = [B : \hat{A}]$.

As in the primal algorithm the pivot column $w = B^{-1} a_q$ and all elements of the pivot row $\alpha_i = e_p^T B^{-1} a_i$ must be computed. Note that $\bar{\beta}_i \geqslant 1 + w_i^{\,2}/w_p^{\,2}$. In the primal algorithm $B^{-T} w$ is needed whereas here we require $B^{-1} y$ where $y = \sum_{i>m} \alpha_i a_i$. y can be computed in the same loop as the α_i's. As in the primal case most of these will be zero in sparse problems with a consequent reduction in the work required to compute y. Small savings also result from zeros in the pivot column.

The vector of row weights β can also be initialized economically by setting all of its components to one and then adding to these the square of the respective components of $B^{-1} a_j$ for $j > m$. Thus it is clear that it is possible to implement a practicable dual steepest-edge algorithm for large sparse LP problems.

Acknowledgements

The author wishes to thank John Reid for his invaluable contributions to this work, while the author was visiting Harwell, the AERE Harwell and the U.S. National Science Foundation under Grant GJ 36472 for their support, P. Gill, S. Powell, and M. A. Saunders for kindly providing the test problems and Joel Cord of IBM for providing useful information about the MPSX code.

References

1. H. W. Kuhn and R. E. Quandt, "An experimental study of the simplex method", in: Metropolis et al., ed., Proc. of Symposia in Applied Maths, Vol. XV (A.M.S.1963).

2. P. Wolfe and L. Cutler, "Experiments in linear programming", in: Graves and Wolfe, ed., Recent advance in mathematical programming (McGraw-Hill, 1963).

3. D. Goldfarb and J. K. Reid, "A practicable steepest-edge simplex algorithm", to appear.

4. D. Goldfarb and J. K. Reid, "Fortran subroutines for sparse in-core linear programming", A.E.R.E. Report to appear.

5. J. K. Reid, "A sparsity-exploiting variant of the Bartels-Golub decomposition for linear programming bases", to appear.

6. J. K. Reid, "Fortran subroutines for handling sparse linear programming bases", A.E.R.E.Report, to appear.

7. D. Goldfarb, "On the Bartels-Golub decomposition for linear programming bases", Harwell C.S.S. Report 18, (1975).

8. P.M.J. Harris, "Pivot selection methods of the Devex LP code", Mathematical programming, 5 (1973), 1-28.

9. P. Wolfe, "The composite simplex algorithm", SIAM Review, 7, (1965).

10. H. Crowder and J. M. Hattingh, "Partially normalized pivot selection in linear programming", IBM research report RC 4918 (Yorktown Heights, 1974).

11. J. A. Tomlin, "On pricing and backward transformation in linear programming", Mathematical programming, 6, (1974), 42-47.

12. R.H.Bartels, G.H. Golub, and M.A. Saunders, "Numerical techniques in mathematical programming". Rosen, Mangasarian, and Ritter, eds., in: Nonlinear Programming (Academic Press, 1970).

13. J. J. H. Forrest and J. A. Tomlin, "Updating triangular factors of the Basis to maintain sparsity in the product-form Simplex method", Mathematical Programming, 2, (1972), 263-278.

IV
Mathematical Software

SPARSE MATRIX SOFTWARE

W. Morven Gentleman
and
Alan George

Department of Computer Science
University of Waterloo
Waterloo, Ontario, N2L 3G1
Canada

ABSTRACT

In this paper we consider the problem of designing and implementing computer software for sparse matrix computations. We consider the data access requirements for typical sparse matrix computations, and review some of the main data structures used to meet these demands. We also describe some tools and techniques we have found useful for developing sparse matrix software.

§1 INTRODUCTION

This paper is concerned with the design and development of computer software for sparse matrix computations. The particular computation we use as an illustration is the problem of solving the sparse linear algebraic system

$$(1.1) \qquad Ax = b,$$

where A is $n \times n$ and nonsingular. The several Gaussian elimination variants we consider for solving this problem serve to exhibit many of the data structure requirements found in other algorithms involving sparse matrix computations.

In the numerical methods we consider, and which are discussed in section 2, the sparse matrix A is transformed by a series of elementary transformation matrices which yield a sequence of successively "simpler" matrices, typically more nearly upper triangular or diagonal. These transformation matrices achieve this by introducing <u>zeros</u> in appropriate positions, and these zeros are preserved by subsequent transformations. Unfortunately, except in some rather exceptional circumstances, these transformations also cause "fill"; that is, they also introduce <u>nonzeros</u> in positions in the transformed matrix which contain zeros in the original matrix. In addition, in some contexts, the transformations may introduce some zeros fortuitously ("accidental" cancellation), and these should also be exploited, if possible or convenient, in subsequent transformations.

The point of these observations is that if we are to exploit the sparsity in this matrix sequence, we must provide a data structure which either "knows" a priori where fill will occur, and provide space for the generated nonzeros, or else the data structure must be designed to dynamically allocate space for fill during the computation. Various pivoting strategies for reducing fill and preserving numerical stability also make demands on the data structure. In section 3 we review some of the main data structure ideas employed in sparse linear equation packages, and try to show how their choice depends upon the problem characteristics and user requirements.

A sparse matrix package is typically a rather large and complex piece of software. This observation suggests at least three things. First, the package is probably expensive to develop, which in turn means that it should be portable, so it can be made widely available. Second, the package should be sensibly modularized to facilitate understanding its operation, and to ease maintenance and enhancement. Third, the sparse matrix package should provide a "friendly" interface to the user. Of course, this is a desirable feature of any software package intended for general use, but it is particularly critical in connection with sparse matrix packages because they typically employ sophisticated data structures. The matrix data usually must be mapped from the user representation onto that used by the sparse matrix package. Finally, the package should be well tested, and its applicability clearly defined. Input data should be checked for consistency, and the package should exhibit graceful degradation for problems outside its range of applicability.

All these are "motherhood" attributes, accepted as desirable, and well known. However, as anyone who has been involved in large software projects well knows, these attributes do not appear automatically, and are not easily achieved. Good performance for such a complex piece of software is not even easily defined, and certainly not easily achieved. Thus, the construction of the software itself will be a nontrivial process. Paper design and theoretical analysis are desirable and important, but not enough; the abstractions they represent are often too far from the realities of actual software. To produce a quality product, as in any field of engineering, it is often necessary to build prototypes, instrument them, measure their performance in various situations, study their response to specific inputs, compare them with alternatives, and expose them to field testing to reveal design deficiencies and to suggest enhancements. From this experience, the next generation of prototypes can be constructed, the process repeating until a satisfactory product is obtained. If a comprehensive development is undertaken to produce and compare methods and implementations, we need several fundamentally different packages, and much of the work must be repeated for each one. The scale of the enterprise may be indicated by the fact that the current codes of the authors' total perhaps 15,000 lines of source; development coding over the past few years would probably double that number.

The human labour involved in such development would be prohibitive unless modern tools and techniques of software development are employed. Some of these tools and techniques are common to any kind of software development, while others are specific to sparse matrix software development. In section 4 we review some of these tools and techniques which we have found to be particularly helpful in our software development.

Section 5 contains our concluding remarks.

§2 GAUSSIAN ELIMINATION

Ignoring row and/or column interchanges for the moment, Gaussian elimination can be described by the following equations:

$$(2.1)\quad A = A^{(0)} = \begin{bmatrix} d_1 & r_1^T \\ c_1 & B^{(1)} \end{bmatrix} = \begin{bmatrix} 1 & 0 \\ \frac{c_1}{d_1} & I_{n-1} \end{bmatrix} \begin{bmatrix} d_1 & r_1^T \\ 0 & B^{(1)} - \frac{c_1 r_1^T}{d_1} \end{bmatrix}$$

$$= L^{(1)} A^{(1)}$$

$$(2.1) \quad A^{(2)} = \begin{bmatrix} d_1 & 0 & 0 \\ 0 & d_2 & r_2^T \\ 0 & c_2 & B^{(2)} \end{bmatrix} = \begin{bmatrix} 1 & & \\ 0 & 1 & \\ 0 & \frac{c_2}{d_2} & I_{n-2} \end{bmatrix} \begin{bmatrix} d_1 & r_1^T & \\ 0 & d_2 & r_2^T \\ 0 & 0 & B^{(2)} - \frac{c_2 r_2^T}{d_2} \end{bmatrix}$$

$$= L^{(2)} A^{(2)}$$

$$\vdots$$

$$A^{(n-1)} = U.$$

Here U is upper triangular, and it is easy to verify that the unit lower triangular matrix L is given by

$$L = \left(\sum_{i=1}^{n-1} L^{(i)}\right) - (n-2)I,$$

and that $A = LU$.

In general, to maintain numerical stability, some pivoting strategy (row and/or column interchanges) is required. A second reason for wishing to perform row and/or column interchanges is to preserve sparsity. It is well known that by a judicious choice of pivotal strategy, the amount of fill suffered during Gaussian elimination can often be reduced substantially. Unfortunately, it is usually the case that pivoting for numerical stability and pivoting to preserve sparsity yield conflicting pivot selections, and some compromise between the desiderata must be made.

Some common pivoting strategies are as follows:

1) Partial row{column} pivoting: at the i-th step, row{column} i and row{column} j, $j \geq i$ are interchanged so that $d_i \geq \|c_i\|_\infty$ $\{\|r_i\|_\infty\}$.

2) Complete pivoting: at the i-th step, row and column j and k, $j \geq i$, $k \geq i$, are interchanged so that

$$|d_i| \geq \max\{\|c_i\|_\infty, \|r_i\|_\infty, \max_{\ell,m} |B_{\ell m}^{(i)}|\}.$$

3) Markowitz criterion: in its "pure form", at the i-th step, row and column j and k, $j \geq i$, $k \geq i$, are interchanged so as to minimize $N(c_i)N(r_i)$, where $N(v)$ denotes the number of nonzeros in the vector v. The rationale is that since this strategy minimizes the number of nonzeros in the rank-one matrix $c_i r_i^T/d_i$, it should tend to minimize fill. For symmetric matrices, this algorithm is often referred to as the minimum degree algorithm [26].

For general matrices, this latter strategy can lead to serious numerical stability or fail, since d_i can be unacceptably small or zero. Thus, this pivoting strategy is usually applied along with some constraint which preserves (to varying degrees) numerical stability. One such constraint is to require that the pivot not lead to an increase in the component of greatest magnitude in the matrix by more than a prescribed factor. This essentially represents a balance between strategies (2) and (3) above. There are also various threshold embellishments of (1) which attempt to preserve sparsity as well as stability, and minrow-within-mincolumn strategies [7]. For the purposes of this paper, strategies (1) and (3) make sufficient demands on the data structures to illustrate most of the points we wish to make in section 3.

We have now described the algorithms which will serve as the examples of sparse matrix algorithms whose data structure, storage management needs and general implementation are to be considered in the next sections. Our main concern is to communicate the basic considerations which go into the data structure and program design of a sparse matrix package, rather than to show that any particular scheme or design is necessarily better than any other, except in a restricted and perhaps unimportant sense.

§3 STORAGE METHODS AND STORAGE MANAGEMENT FOR SPARSE ELIMINATION

In this section we consider the issues relevant to the selection of a storage scheme appropriate for the application of Gaussian elimination to a sparse matrix. The storage schemes we discuss are only examples; space does not permit us to be exhaustive in either scope or detail. We assume the reader has some familiarity with standard implementations of lists, and with some methods for allocation and deallocation of fixed and variable length records [2,20].

Obviously, one of the prime objectives in the design or selection of a storage scheme for sparse elimination is to achieve a net reduction in storage requirements over that required if we simply ignore sparsity altogether. We emphasize the word net here; normally, there are two computer storage components associated with a storage scheme for sparse matrices, one part used for the actual matrix components (primary storage), and a second part required for pointers, indexing information, and other "structural" information (overhead storage). As we increase the sophistication of our storage scheme, exploiting more and more zeros, the primary storage obviously decreases, but overhead storage generally increases. Usually it makes little sense to reduce primary storage unless there is also a net reduction in total storage requirements.

In this connection, it is unfortunate that the majority of complexity analyses of sparse elimination are done in terms of arithmetic requirements and/or number of non-zero components (i.e., primary storage). We shall see later in this section that in some circumstances one must be very careful in the design of the data structure if execution time is to be reliably predicted by the arithmetic operation count. Also, comparing competing methods in terms of primary storage needs may be grossly unfair if the appropriate data structures for the respective methods have very different overhead storage needs. Such comparisons are further complicated by differences in machine architecture. What is easy to do on one machine may be difficult to do on another. Different word lengths can mean different ratios of overhead to primary storage if packing of overhead information is done. Ideally, claims about the practical usefulness of a new method or idea should be accompanied by careful numerical experiments and comparisons with competitors, solving non-trivial problems.

The way we store a matrix for sparse elimination is intimately associated with the general problem of ordering and/or pivoting, and the two issues cannot reasonably be discussed in isolation. Storage methods are usually chosen with a particular ordering and/or pivoting strategy in mind. Conversely, ordering algorithms are (or should be) designed with a particular data structure in mind. Obviously, the choice of an ordering or pivoting strategy which preserves sparsity makes little sense if the data structure used does not exploit sparsity! (Note that it may very well be sensible to ignore some zeros, if the overhead storage incurred by exploiting them is more than the corresponding decrease in primary storage.)

In addition to distinguishing among different methods for storing a sparse matrix, it is perhaps just as important to consider the way a storage scheme is utilized. For example, suppose we represent our sparse matrix as a set of linked lists, one per row, with each record (list element) containing space for a matrix component, a column subscript, and a pointer to the next list element. Suppose first that our lists only contain the nonzero components of the matrix. If we then apply Gaussian elimination to the matrix, we generally must add components to the appropriate lists to accommodate fill. On the other hand, if we can predict before the computation where fill will occur, the lists can already contain members corresponding to the fill. The actual elimination then proceeds with the storage structure remaining static. In either case, the method of storing the matrix is the same, but its utilization is different. It is clear that use of a storage method in a static manner is only appropriate when the location of fill can be predicted before the computation begins.

When does this situation prevail? One very common situation is when A is symmetric and positive definite, or diagonally dominant. For these matrices it is well known that Gaussian elimination is stable when applied to PAP^T, for any permutation matrix P [29]. Thus, P can be chosen without regard to numerical stability; i.e., solely on the basis of the zero/nonzero structure of A or its factors.

A second situation where a static data structure can be used (and which may coincide with the above) is when A can be permuted so that it has a small bandwidth. If no permutations are performed during the decomposition, then all fill is confined within the band. If row permutations are performed, the lower band remains the same, and the upper band can at most double [21]. General row and column permutations as suggested by the Markowitz criterion seem to be inappropriate with band orderings, since they can lead to an unacceptable increase in both the upper and lower band. That is, using a storage method which would accommodate the worst fill situation is unacceptably wasteful.

In either of the above situations, it is a simple matter to exploit the profile of the matrix and its factors; that is, the set of elements between the first and last nonzeros in each row (or column) [17]. A static storage scheme for profile is clearly adequate to store the matrix, and the profile for the factors of a matrix can readily be obtained from the profile of the matrix.

Linear programming applications typically produce basis matrices which can be permuted so as to be block lower triangular, with each diagonal block ("bump") almost lower triangular, except for a few nonzero columns ("spikes"). Gaussian elimination with column pivoting preserves numerical stability. Moreover, fill can occur only in the relatively few spike columns [18,28].

Frequently, a matrix can be ordered so that the nonzeros of its factors fall in blocks; that is, there are rectangular submatrices which are quite dense, whereas the matrix as a whole is very sparse [12,28]. This suggests representing the matrix as a two level structure, a matrix of matrices, where the outer structure is represented in perhaps some linked manner, but the nonzero blocks of the inner structure are represented as conventional dense matrices. In engineering applications this idea, recursively applied, is termed the "hypermatrix" method [10].

Why is the use of a static storage scheme desirable? First, in terms of software organization, it promotes modularity. The three problems of finding a suitable ordering, setting up the appropriate storage structure for the matrix factors, and the actual numerical computation can be isolated in separate modules. Since these tasks are performed sequentially, storage used for one task can be reused by the next.

Second, storage methods can be specifically tailored for the job at hand. For example, the use of bit matrices or lists may be most appropriate for finding a good ordering, but decidedly inappropriate in connection with actually storing the matrix or its factors.

Third, in many engineering design applications numerous different matrix problems having the same structure must be solved, and for which a prior ordering is numerically stable. If we use a static storage scheme, the ordering and storage structure initialization need be done only once. (In some situations, it may be appropriate to generate straight line (loop-free) code which only performs those eliminations actually required [14,15].)

Finally, given a specific storage structure, not having to modify it during the elimination inevitably leads to improved execution efficiency. In addition, knowing that we can use a storage scheme in a static manner often allows us to select a method which is more efficient than any that would be appropriate if the storage structure had to be altered during the elimination.

Unfortunately, not all matrices are positive definite or diagonally dominant, and not all can be permuted so that they have a small bandwidth. For a general indefinite sparse matrix A, some form of pivoting is necessary in general to ensure numerical stability. Thus, given A, one normally ends up with a factorization of PA or PAQ, where P and Q are permitation matrices of the appropriate size. They are determined during the decomposition by a combination of numerical stability and sparsity considerations. Different matrices, even though they may have the same zero/nonzero structure, will normally yield different P and Q, and therefore have factors with different sparsity patterns. We are then obliged to use some form of dynamic storage structure which allocates storage for fill as the computation proceeds.

How crucial is this? As we explain more fully in the next few paragraphs, in the general (undefinite, non-diagonally dominant) case, a sophisticated data structure and a fairly elaborate code appear to be necessary to avoid requiring at least $0(n^2)$ execution time for the program, regardless of the actual arithmetic operation count or fill count. On the other hand, if P is known beforehand, and the data structure set up accordingly, it is possible to arrange that program execution time is proportional to the arithmetic operation count [15,12].

An $n \times n$ matrix A is generally considered to be sparse if the number of nonzeros is $0(n)$. In many cases, if A is suitably permuted, the number of nonzeros in its factors is also $0(n)$, as is the cost in arithmetic operations of performing the factorization. In situations such as this, one should expect that the execution time of any sparse matrix package for solving the associated linear equations problem should also be proportional to n.

Now consider what this expectation implies for ordinary Gaussian elimination with partial pivoting, as implied by equations (2.1). At the i-th step we must determine in which row j, $j \geq i$, the largest component of column i resides, and then perform the interchange, either implicitly or explicitly. If our code is to execute linearly in time with n, our data structure must provide us with the positions of the nonzero components of column i below the diagonal; if we are obliged to perform i comparisons to determine which are nonzero, our code cannot execute faster than $0(n^2)$. This suggests that we need a representation which allows us to find the components in the pivotal column in time independent of n.

Suppose for the moment then that the matrix is stored as a set of columns, each of which is stored as a linked list. This meets the requirements above. At the i-th step, after having determined which row is to be the pivotal row, we must determine which columns have nonzero components in the pivotal row. Unless we also have some linkage along rows telling us explicitly where the nonzero components are in each row, we will have to examine each column r, $r > i$, which again implies that our code cannot execute in less than $0(n^2)$ time.

This argument suggests that if our problem is such that linear execution time is possible (in theory), a representation of the matrix involving both rows and columns is necessary. Knuth [20, pp.299-302] suggests storing, for each $a_{ij} \neq 0$, the quantities a_{ij}, i, j, and the pointers NEXTI and NEXTJ, which point to records containing the next nonzero component in column j and row i respectively. A schematic example appears in Figure 3.1.

Does this storage scheme solve our problem of $0(n^2)$ execution time? For this discussion, suppose the matrix L+U has no more than γ nonzeros in any row or column, where γ is assumed to be small with respect to n, and independent of n. Consider the first step of Gaussian elimination with partial pivoting applied to the example in Figure 3.1. Scanning the first row and column lists, we determine that no interchange is necessary, and that either fill or modification of an existing component will occur in positions (4,3) and (4,5). If order to carry out these operations and to make any updates to the data structure due to fill, it appears necessary to scan the row 4 list, and column lists 3 and 5. As observed and exploited by Rheinboldt et al. [24], since we

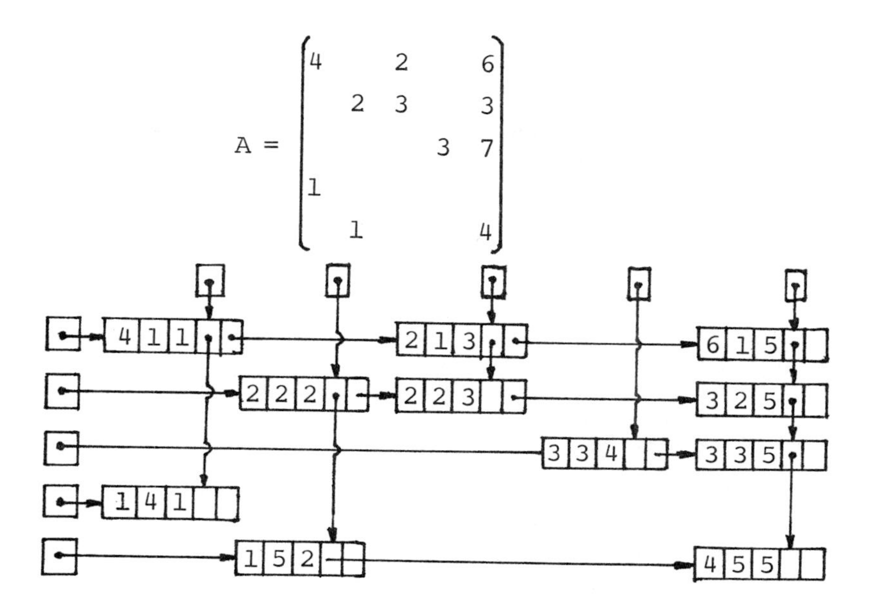

Figure 3.1 Orthogonal linked list representation of a sparse matrix

have to scan these rows and columns anyway, we can make the lists circular, include the row and column numbers in the listhead, and avoid storing the subscripts in each record. In general, it is obvious that at the i-th step, the number of data accesses and arithmetic operations is at most $0(\gamma^2)$, which implies that the execution time for this partial pivoting Gaussian elimination algorithm is bounded by $0(\gamma^2 n)$.

Notice that when fill occurs, new (fixed length) data structure records must be created, and inserted into the data structure. In addition, the elimination creates zeros in the pivotal column, and in some cases the storage used by the previously nonzero component must be returned to a storage pool, to be used for subsequent fills. Thus, a convenient mechanism for allocating and deallocating fixed length records is necessary. One such standard scheme can be found in [2,pp.44-46]. (For paged machines, however, this storage management scheme has the potential danger of causing a page fault for each successive member of a list, so more sophisticated schemes must be employed [8,22].)

Of course this preoccupation with $0(n^2)$ time may be practically irrelevant, for reasons we will discuss below. However, we have included it to emphasize the influence that data structure design has on the potential performance of a code, and the basic limitations that the storage scheme can impose.

Arguments such as we have made about the $0(n^2)$ versus $0(n)$ execution time are undoubtedly valid in an asymptotic sense. However, we are dealing with finite problems. The constants involved are important! It may very well be the case that for all practical problem sizes the $0(n^2)$ component in the execution time of our column-linked implementation may be vanishing small, and the algorithm in practice may appear to execute linearly with n. Alternately, the reduction in storage requirements due to reduced numbers of pointers may be far more important than the $0(n^2)$ execution time.

This latter observation is a constantly recurring issue in the design of sparse elimination codes. There is almost always a trade-off to be made between processor overhead and storage overhead, and one can very often exchange increased processor time for a reduction in total storage requirements [12]. Whether or not such an exchange is beneficial in dollar terms is very much dependent on the charging algorithm used for the various computing resources, which in turn is often keyed to the cost and/or availability and utilization of those resources.

We now give an example to illustrate these remarks. In order to reduce overhead storage over the column list representation discussed earlier (which would in general require three storage locations per nonzero component), the nonzero components of each row may be stored in contiguous locations, along with their column subscripts. Then only two storage locations per nonzero component are required.

Note, however, that the storage management is now more complicated; as the elimination progresses, the number of nonzeros in a row will normally increase or decrease, so that some form of allocation and deallocation of variable length records from a storage pool is necessary [20, pp.435-455; 4].

Another idea, employed by Gustavson [13], in connection with efficient storage utilization, is based on the observation that for many sparse matrices the triangular factors also remain sparse until very nearly the end of the computation. His idea is to monitor the density of the part of the matrix remaining to be factored, and to switch to the standard dense matrix format when it is beneficial to do so. Again, if a priori ordering is numerically stable, such a storage scheme can be used in a static manner.

§4 SOME USEFUL TOOLS FOR SPARSE MATRIX SOFTWARE DEVELOPMENT

As we stated in the introduction, a sparse matrix package is typically a large and complicated piece of software whose development involves (or should involve) a great deal of analysis, performance measurement, field testing and independent certification. In the course of software development, we have found a number of tools and techniques to be of significant help. Some are not specific to sparse matrix software, but are nevertheless highly important. (It seems strange to have to refer to some of them in 1975, but we are amazed by the primitive environments within which many numerical analysts work.)

General tools

First and foremost, it is of immeasurable help to work in an environment that has a reasonable on-line file system and interactive editor; the advantages of not having to physically manipulate card decks are immense. Second, the number of interrelated modules is typically very large, so that mechanical aids for editing and managing the library are desirable. Typical tasks are identifying modules affected by a change, automatically generating closely related versions of a module, and similar global manipulations. Some of these functions can be performed by any reasonably powerful interactive, programmable text editor [6,9]. Third, it is helpful when assembling a code from modules in a library to be able to mechanically confirm the correctness of intermodular interfaces [27]. Fourth, preprocessors or other aids for automatically instrumenting a program for performance measurement are particularly useful in assessing and improving a program. An example of such an aid is a profiler [19], which provides counts of the number of times each source statement in a program has been executed. Fifth, interactive execution, combined with a source level interactive debugging system [1], facilitates debugging and studying the dynamic behaviour of the program in selected circumstances. Finally, to achieve portability, it is very helpful to have a metaprogram to detect machine dependencies [27], and another one to produce machine specific codes from a basic core version, thus making portability practical.

Specific tools

There are six specific tools we have found useful in developing and studying the behaviour of sparse matrix codes. They are described below.

1. A picture generator

This is a utility program used to display the zero-nonzero structure of a matrix. Several versions are usually required as it is necessarily keyed to a particular storage scheme. It is often convenient to call it at various stages in the computation so that fill can be examined.

The construction of a useful display takes some imagination; the usual picture with X's for nonzeros and blanks for zeros is not very helpful except for small problems because it is difficult to tell which items are in the same row or column. Test problems for sparse matrix codes typically have to be at least a few hundred equations, so the 132 columns by 66 rows of the typical line printer page is inadequate. The use of an electrostatic dot printer [30] is a great improvement in this regard, since such devices have a much higher resolution. Including some horizontal and vertical lines which partition the matrix into small submatrices provide a significantly more helpful picture than the usual ones, at far less expense than that incurred by using standard plotting hardware.

2. Selective print routines

Since most data structures used in sparse matrix computations are fairly complicated, it is helpful to have "utility" routines for each data structure which will print rows, columns or selected portions of matrices or more general lists, in an intelligible form. Thus, we can avoid manually tracing pointers and unpacking coded data.

3. Bit matrix manipulation routines

Although the use of bit matrices in storage schemes and ordering algorithms are seldom the most economical, it is still sometimes very helpful in development work to be able to model real codes using bit matrix codes. By using logical machine instructions, these bit manipulation routines can be made competitive with alternate schemes as long as the problems are of moderate size.

4. Symbolic elimination routines

These are routines which accept the sparsity structure of a matrix and simulate various decompositions of the matrix, ultimately providing the sparsity structure of the factors. These routines are mostly useful when a priori numbering is numerically stable. Combined with appropriate picture generation, they can provide snapshots of the matrix as elimination proceeds. Such routines may also be used in connection with initializing storage schemes and setting up "customised" code.

5. Audit routines

These are storage structure-specific subroutines whose job is to test as much as possible whether the data contained in a storage structure is valid. When a sparse matrix code aborts, it very often does so some considerable time after the actual error producing instructions were executed. Typically, the program has been manipulating an invalid structure for some time. Evidence indicating the problem is usually hopelessly lost in the general debris. To track down the problem, an audit routine is inserted into the program, along with calls to it from strategic points in the program, and this expanded program is then re-run. The audit subroutine does nothing when it is called as long as it finds all pointers etc. in acceptable ranges; if they are not, it prints out appropriate error messages and calls the appropriate print routines described in 2) above. Another way to use audit routines is in combination with an interactive debugging facility - the audit routine is periodically called by the debugger. At first report of invalid structure, control variables and pointers can be examined to see why the invalid structure was created.

6. Test problems and test problem generators

It is a great help in developing and evaluating new and modified packages to have readily available sets of problems for which answers and benchmark solution times are known. In our opinion, it is important that these test sets can be chosen or generated from real applications rather than randomly generated, and also that they be chosen from widely different applications areas. This is important not only to exercise different parts of the package in different ways, but also because a significant aspect of real sparse matrix problems is that they invariably exhibit some kind of structure. The coding of the package may or may not attempt to exploit that structure, but in any case the performance of the package is usually affected by it.

§5 CONCLUDING REMARKS

We have described how the characteristics of a given sparse system influences the choice of method for storing the matrix during Gaussian elimination. The advantages of being able to choose an ordering before the computation begins, and the subsequent use of a static storage scheme, appear to be considerable, particularly with regard to program efficiency, software modularity, and general flexibility. We saw how the selection of a particular storage scheme can impose some inherent limitations on the speed with which a sparse Gaussian elimination code can execute, irrespective of the amount of storage used or arithmetic performed. We also observed that there is usually a trade-off between execution time and storage requirements. This points up what we regard as a weakness of some analyses of sparse matrix computations; they too often ignore the computer implementation of the algorithm under study.

In section 4 we described some software tools which we have found valuable in connection with developing sparse matrix software. It would be beneficial to the development and comparison of sparse matrix packages to make available a comprehensive, representative set of matrix test problems and comparisons to serve as a benchmark, just as has been done for ordinary differential equation solvers [16].

§6 REFERENCES

[1] G. Ashby, L. Salmonson, and R. Heilman, *Design of an interactive debugger for FORTRAN: MANTIS*, Software Practice and Experience, 3 (1973), pp.65-74.

[2] A.V. Aho, J. Hopcroft, and J.D. Ullman, *The design and analysis of computer algorithms*, Addison Wesley, 1974.

[3] D.M. Brandon, Jr., *The implementation and use of sparse matrix techniques in general simulation programs*, Computer J., 17 (1974), pp.165-171.

[4] W.S. Brown, *An operating environment for dynamic-recursive computer programming systems*, CACM 8 (1965), pp.371-377.

[5] A.R. Curtis and J.K. Reid, *Solution of large sparse sets of unsymmetric linear equations*, JIMA 8 (1971), pp.344-353.

[6] L.P. Deutch and B.W. Lampson, *An online editor*, CACM 10 (1967), pp.793-799, 803.

[7] I.S. Duff and J.K. Reid, *A comparison of sparsity orderings for obtaining a pivotal sequence in Gaussian elimination*, AERE Technical Report 73, March 1973.

[8] R.R. Fenichel and J.C. Yochelson, *A LISP garbage-collector for virtual memory computer systems*, CACM 12 (1967), pp.611-612.

[9] R. Fajman and John Borgelt, *WYLBUR: An interactive text editing and remote job entry system*, CACM 16 (1973), pp.314-323.

[10] G. Von Fuchs, J.R. Roy, and E. Schrem, *Hypermatrix solution of large sets of symmetric positive-definite linear equations*, Comp. Meth. in Appl. Mech. and Engrg., 1 (1972), pp.197-216.

[11] W.M. Gentleman, *Row elimination for solving sparse linear systems and least squares problems*, Dundee Biennial Conference on Numerical Analysis, July 1975. (Proceedings to appear in the Springer Verlag "Lecture Notes in Mathematics" Series.)

[12] Alan George, *Numerical experiments using dissection methods to solve n by n grid problems*, SIAM J. Numer. Anal., to appear.

[13] F.G. Gustavson, private communication.

[14] F.G. Gustavson, W. Liniger, and R. Willoughby, *Symbolic generation of an optimal Crout algorithm for sparse systems of linear equations*, JACM 17 (1970), pp.87-109.

[15] F.G. Gustavson, *Some basic techniques for solving sparse systems*, in Sparse Matrices and their Applications, D.J. Rose and R.A. Willoughby, eds., Plenum Press, New York, 1972.

[16] T.E. Hull, W.H. Enright, B.M. Fellen, and A.E. Sedgwick, *Comparing numerical methods for ordinary differential equations*, SIAM J. Numer. Anal. 9 (1972), pp.603-637.

[17] A. Jennings, *A compact storage scheme for the solution of simultaneous equations*, Computer J., 9 (1966), pp.281-285.

[18] J.E. Kalan, *Aspects of large-scale in-core linear programming*, Proc. ACM Annual Conference, Chicago, Illinois, 1971.

[19] D.E. Knuth, *Empirical study of Fortran programs*, Software Practice and Experience, 1 (1971), pp.105-133.

[20] D.E. Knuth, *The art of computer programming*, vol.1, Fundamental Algorithms, Addison Wesley, 1969.

[21] R.S. Martin and J.H. Wilkinson, *Solution of symmetric and unsymmetric band equations and calculation of eigenvectors of band matrices*, Numer. Math., 9 (1967), pp.279-301.

[22] J.E. Morrison, *User program performance in virtual storage systems*, IBM Systems Journal, 12 (1973), pp.216-237.

[23] U.W. Pooch and A. Nieder, *A survey of indexing techniques for sparse matrices*, ACM Computing Surveys, 5 (1973), pp.109-133.

[24] W.C. Rheinboldt and C.K. Mesztenyi, *Programs for the solution of large sparse matrix problems based on the arc-graph structure*, Univ. of Maryland Technical Report TR-262, Sept. 1973.

[25] L.D. Rogers, *Optimal paging strategies and stability considerations for solving large linear systems*, Doctoral thesis, Computer Science Dept., Univ. of Waterloo, 1973.

[26] D.J. Rose, *A graph-theoretic study of the numerical solution of sparse positive definite systems of linear equations*, in Graph Theory and Computing, edited by R.C. Read, Academic Press, New York, 1972.

[27] B.G. Ryder, *The PFORT verifier*, Software Practice and Experience, 4 (1974), pp.359-377.

[28] M.A. Saunders, *The complexity of LU updating in the simplex method*, in Complexity of Computational Problem Solving, R.S. Anderssen and R.P. Brent (eds.), Queensland Univ. Press, Queensland, 1975.

[29] J.H. Wilkinson, The Algebraic Eigenvalue Problem, Clarendon Press, Oxford, 1965.

[30] R. Zaphiropoulous, *Nonimpact printers*, Datamation 19 (1973), pp.71-76.

CONSIDERATIONS IN THE DESIGN OF SOFTWARE FOR SPARSE GAUSSIAN ELIMINATION*

S. C. Eisenstat[1], M. H. Schultz[1], and A. H. Sherman[2]

1. Introduction

Consider the large sparse system of linear equations

$$Ax = b \tag{1.1}$$

where A is an N x N sparse nonsymmetric matrix and x and b are vectors of length N. Assume that A can be factored in the form

$$A = LDU \tag{1.2}$$

where L is a unit lower triangular matrix, D a nonsingular diagonal matrix, and U a unit upper triangular matrix. Then an important method for solving (1.1) is sparse Gaussian elimination, or equivalently, first factoring A as in (1.2) and then successively solving the systems

$$Ly = b, \ Dz = y, \ Ux = z \tag{1.3}$$

Recently, several implementations of sparse Gaussian elimination have been developed to solve systems like (1.1) (cf. Curtis and Reid [2], Eisenstat, Schultz, and Sherman [5], Gustavson [6], Rheinboldt and Mesztenyi [7]). The basic idea of all of these is to factor A and compute x without storing or operating on zeroes in A, L, or U. Doing this requires a certain amount of storage and operational overhead, i.e., extra storage for pointers in addition to that needed for nonzeroes, and extra nonnumeric "book-keeping" operations in addition to the required arithmetic operations. All these implementations of sparse Gaussian elimination generate the same factorization of A and avoid storing and operating on zeroes. Thus, they all have the same costs as measured in terms of the number of nonzeroes in L and U or the number of arithmetic operations performed. However, the implementations do have different overhead requirements and

(1) Department of Computer Science, Yale University, New Haven, Connecticut 06520

(2) Department of Computer Science, University of Illinois, Urbana, Illinois 61801

* This research was supported in part by NSF Grant GJ-43157 and ONR Grant N0014-67-A-0097-0016.

thus their total storage and time requirements vary a great deal.

In this paper we discuss the design of sparse Gaussian elimination codes. We are particularly interested in the effects of certain flexibility and cost constraints on the design, and we examine possible tradeoffs among the design goals of flexibility, speed, and small size.

In Section 2 we describe a basic design due to Chang [1] which has been used effectively in the implementations referenced above. Next, in Section 3 we discuss the storage of sparse matrices and present two storage schemes for use with sparse Gaussian elimination. In Section 4 we describe three specific implementations which illustrate the range of possible tradeoffs among design goals. Finally, in Section 5 we give some quantitative comparisons of the three implementations.

2. A Basic Implementation Design

Several years ago Chang [1] suggested a design for the implementation of sparse Gaussian elimination which has proved to be particularly robust. He proposed breaking the computation up into three distinct steps: symbolic factorization (SYMFAC), numeric factorization (NUMFAC), and forward- and back-solution (SOLVE). The SYMFAC step computes the zero structures of L and U (i.e., the positions of the nonzeroes in L and U) from that of A, disregarding the actual numerical entries of A. The NUMFAC step then uses the structure information generated by SYMFAC to compute the numerical entries of L, D, and U. Finally, the SOLVE step uses the numerical factorization generated by NUMFAC to solve the system (1.3).

The main advantage of splitting up the computation as described here is flexibility. If several linear systems have identical coefficient matrices but different righthand sides, then only one SYMFAC and one NUMFAC are needed; the different righthand sides require only separate SOLVE steps. (This situation arises in the use of the chord method to solve a system of nonlinear equations.) Similarly, a sequence of linear systems, all of whose coefficient matrices have identical zero structure but different numerical entries, can be solved using just one SYMFAC combined with separate NUMFAC and SOLVE steps for each system. (This situation arises when Newton's method is used to solve a system of nonlinear equations.)

A drawback to the three-step design is that it is necessary to store the descriptions of the zero structures and the actual numerical entries of both L and U. In essence, the great flexibility is paid for with a large amount of extra storage. By giving up some flexibility, it is possible to substantially reduce the storage requirements. For example, combining NUMFAC and SOLVE into a single NUMSLV step eliminates the need for storing the numerical entries of L. However, it is no longer possible to handle multiple righthand sides as efficiently. Alternatively, if all three steps are combined into one TRKSLV step, it is unnecessary to store even a description of the zero structure of L. But by combining steps in this way, we lose the ability to efficiently solve sequences of systems all of whose coefficient matrices have identical zero structure.

3. Storage of Sparse Matrices

In this section we describe two storage schemes which can be used to store A, L, and U. The schemes are designed specifically for use with sparse Gaussian elimination and they exploit the fact that random access of sparse matrices is not required.

We call the first storage scheme the _uncompressed storage scheme_. It has been used previously in various forms by Gustavson [6] and Curtis and Reid [2]. The version given here is a row-oriented scheme in which nonzero matrix entries are stored row-by-row, although a column-oriented version would work as well. Within each row, nonzero entries are stored in order of increasing column index. To identify the entries of any row, it is necessary to know where the row starts, how many nonzero entries it contains, and in what columns the nonzero entries lie. This extra information describes the zero structure of the matrix and is the storage overhead mentioned earlier.

Storing the matrix A with the uncompressed scheme requires three arrays (IA, JA, and A), as shown in Figure 3.1. The array A contains the nonzero entries of A stored row-by-row. IA contains N+1 pointers which delimit the rows of nonzeroes in the array A -- A(IA(I)) is the first stored entry of the I-th row. Since the rows are stored consecutively, the number of entries stored for the I-th row is given by IA(I+1) - IA(I). (IA(N+1) is defined so that this holds for the N-th row.) The array JA contains the column indices which correspond to the nonzero entries in the array A -- if A(K) contains a_{IJ}, then JA(K) = J. The storage overhead incurred by using the uncompressed storage scheme for A is the storage for IA and JA. Since IA has N+1 entries and JA has one entry per entry of the array A, the storage overhead is approximately equal to the number of nonzeroes in the matrix A.

Previous implementations of sparse Gaussian elimination have also used variants of the uncompressed storage scheme for storing L and U, as shown in Figure 3.2. Again the storage overhead is approximately equal to the number of nonzero entries in the two matrices. Storing L and U in this way has the advantage that the operational overhead in implementation is quite small, since the data structures are simple and the matrix entries can be accessed quickly.

In certain situations where storage is at a premium, however, it is essential to reduce the storage overhead, even if the operational overhead is increased. This can be done by storing L and U with a more complex compressed storage scheme (cf. Eisenstat, Schultz, and Sherman [4], Sherman [8]). The compressed storage scheme will incur more operational overhead than the uncompressed scheme but the storage requirement will be substantially reduced. In the compressed storage scheme L is stored by columns (or, equivalently, L^T is stored by rows) and U is stored by rows. Figure 3.3 illustrates the derivation of the compressed storage scheme for U. (The derivation for L is similar, using a column-oriented scheme.)

Figure 3.3a shows the data structures required to store U in the uncompressed storage scheme. It is immediately evident that the diagonal entries do not need to be stored, since they are always equal to 1 and occur as the first stored entry of each row.

Figure 3.3b shows the data structures required when the diagonal entries are omitted. We now note that the indices in JU for certain rows of U are actually final subsequences of the indices for previous rows. For example, the indices for row 3 are 4,5, while those for row 2 are 3,4,5. Instead of storing the indices for row 3 separately, we can simply make use of the last two indices stored for row 2. All that is required is a pointer to locate the indices for row 3.

In general, the indices in JU can be compressed by deleting the indices for any row if they already appear as a final subsequence of some previous row (see Figure 3.4). It is possible to compress the indices in certain other cases as well, but tests have shown that very little is gained by so

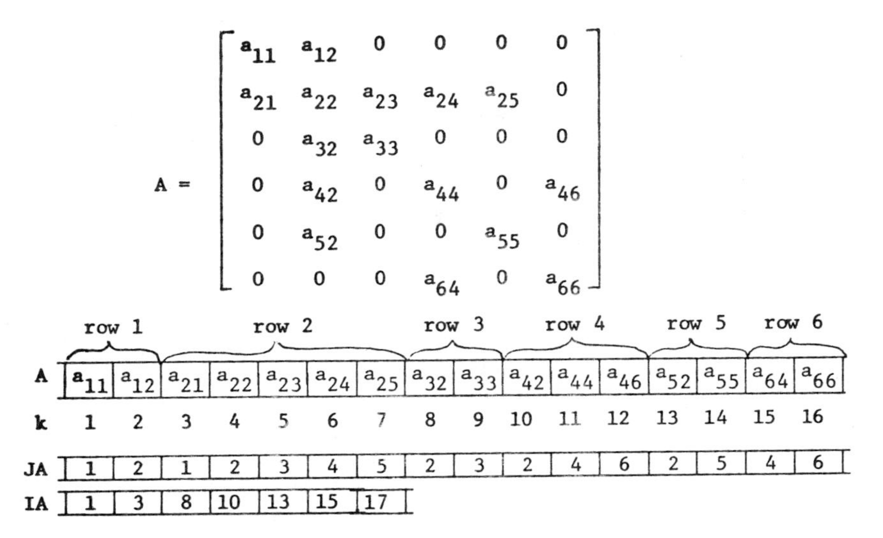

Figure 3.1.

$$
L = \begin{bmatrix}
1 & & & & & \\
\ell_{21} & 1 & & & & \\
0 & \ell_{32} & 1 & & & \\
0 & \ell_{42} & \ell_{43} & 1 & & \\
0 & \ell_{52} & \ell_{53} & \ell_{54} & 1 & \\
0 & 0 & 0 & \ell_{64} & \ell_{65} & 1
\end{bmatrix}
\qquad
U = \begin{bmatrix}
1 & u_{12} & 0 & 0 & 0 & 0 \\
 & 1 & u_{23} & u_{24} & u_{25} & 0 \\
 & & 1 & u_{34} & u_{35} & 0 \\
 & & & 1 & u_{45} & u_{46} \\
 & & & & 1 & u_{56} \\
 & & & & & 1
\end{bmatrix}
$$

L	1	ℓ_{21}	1	ℓ_{32}	1	ℓ_{42}	ℓ_{43}	1	ℓ_{52}	ℓ_{53}	ℓ_{54}	1	ℓ_{64}	ℓ_{65}	1
k	1	2	3	4	5	6	7	8	9	10	11	12	13	14	15
JL	1	1	2	2	3	2	3	4	2	3	4	5	4	5	6
IL	1	2	4	6	9	13	16								

U	1	u_{12}	1	u_{23}	u_{24}	u_{25}	1	u_{34}	u_{35}	1	u_{45}	u_{46}	1	u_{56}	1
k	1	2	3	4	5	6	7	8	9	10	11	12	13	14	15
JU	1	2	2	3	4	5	3	4	5	4	5	6	5	6	6
IU	1	3	7	10	13	15	16								

Figure 3.2.

$$U = \begin{bmatrix} 1 & u_{12} & 0 & 0 & 0 & 0 \\ 0 & 1 & u_{23} & u_{24} & u_{25} & 0 \\ 0 & 0 & 1 & u_{34} & u_{35} & 0 \\ 0 & 0 & 0 & 1 & u_{45} & u_{46} \\ 0 & 0 & 0 & 0 & 1 & u_{56} \\ 0 & 0 & 0 & 0 & 0 & 1 \end{bmatrix}$$

(a)

U:	1	u_{12}	1	u_{23}	u_{24}	u_{25}	1	u_{34}	u_{35}	1	u_{45}	u_{46}	1	u_{56}	1
k	1	2	3	4	5	6	7	8	9	10	11	12	13	14	15
JU:	1	2	2	3	4	5	3	4	5	4	5	6	5	6	6
IU:	1	3	7	10	13	15	16								

(b)

U:	u_{12}	u_{23}	u_{24}	u_{25}	u_{34}	u_{35}	u_{45}	u_{46}	u_{56}
k	1	2	3	4	5	6	7	8	9
JU:	2	3	4	5	4	5	5	6	6
IU:	1	2	5	7	9	10	10		

(c)

U:	u_{12}	u_{23}	u_{24}	u_{25}	u_{34}	u_{35}	u_{45}	u_{46}	u_{56}
k	1	2	3	4	5	6	7	8	9
JU:	2	3	4	5	5	6			
IU:	1	2	5	7	9	10	10		
ISU:	1	2	3	5	6	6			

Figure 3.3.

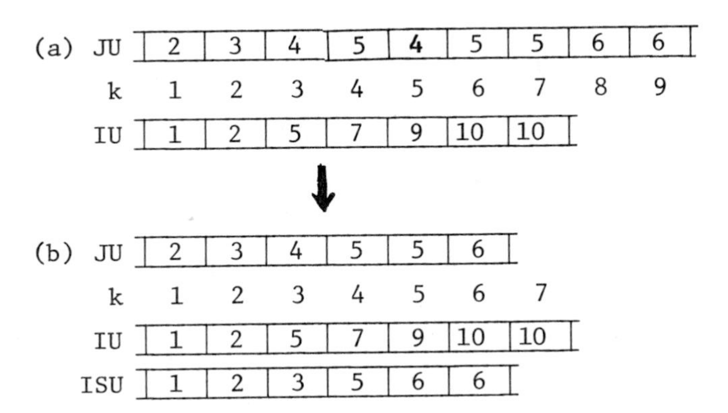

Locations of Column Indices

Row	Before Compaction (a)	After Compaction (b)
1	JU(1)	JU(1)
2	JU(2) - JU(4)	JU(2) - JU(4)
3	JU(5) - JU(6)	JU(3) - JU(4)
4	JU(7) - JU(8)	JU(5) - JU(6)
5	JU(9)	JU(6)
6	-	-

Figure 3.4.

doing.

Since the compressed indices in JU do not correspond directly to the nonzeroes stored in the array U, an extra array of pointers (IJU) is required to locate the indices for each row (see Figure 3.3c). Thus the storage overhead for the compressed storage of U is the number of locations required for IU, JU, and IJU. Although this overhead can be larger than that with the uncompressed scheme, there are important cases in which it is substantially smaller (cf. Section 6; Eisenstat, Schultz, and Sherman [3,4]; Sherman [8]).

4. Three Implementation Designs

In this section we describe three specific implementation designs, which illustrate some of the tradeoffs mentioned earlier. Designs other than these three can also be derived, but these indicate the broad spectrum of implementations that are possible.

The first implementation (SGE1) is designed for speed. It uses the uncompressed storage scheme for A, L, and U because of the smaller operational overhead associated with it. Furthermore, we combine the NUMFAC and SOLVE steps to avoid saving the numeric entries of L, so that the computation consists of a SYMFAC step followed by the NUMSLV step.

The second implementation (SGE2) is designed to reduce the storage requirements. The entire computation is performed in a TRKSLV step to avoid storing either the description or the numerical entries of L. Moreover, U is stored with the compressed storage scheme to reduce the storage overhead. This design incurs more operational overhead than SGE1; the total storage requirements, however, are much smaller.

Finally, the third implementation (SGE3) attempts to balance the design goals of speed and small size. It splits the computation as in SGE1 to avoid storing the numerical entries of L and it uses the compressed storage scheme as in SGE2 to reduce storage overhead.

5. Quantitative Comparisons

In order to compare the designs of Section 4, we tested them on five-point model problems, i.e. linear systems of the form

$$A_n x = b, \qquad (5.1)$$

where A_n is a permutation $P_n \tilde{A}_n P_n^T$ of the n×n block tridiagonal matrix $\tilde{A}_n$ given by

$$\tilde{A}_n = \begin{bmatrix} B & C & & \\ D & \ddots & \ddots & \\ & \ddots & \ddots & C \\ & & D & B \end{bmatrix}.$$

Here the blocks B, C, and D are n×n with B tridiagonal and C and D diagonal, and P_n is an $n^2 \times n^2$ permutation matrix chosen to reduce the number of nonzeroes in the factors of A_n and the amount of arithmetic required to solve (5.1). Systems like (5.1) arise frequently in the numerical solution of partial differential equations in rectangular domains (cf. Woo, Eisenstat, Schultz, and Sherman [9] for a specific example).

Our experiments were designed to answer two questions. First, what are the storage and CPU-time costs of solving the model systems with each of the three implementations? Second, how large a model system can each of the implementations solve in a fixed amount of core storage (262,144 words)?

The results shown in Tables 5.1 and 5.2 were obtained on an IBM System/370 Model 158 using the FORTRAN IV Level H Extended compiler.* For n = 60, SGE1 is the fastest implementation, requiring 40-45% less time than SGE2 and 10-15% less time than SGE3. On the other hand, SGE2 requires the least storage, using 35-40% less than SGE1 and 15-20% less than SGE3. Furthermore, we see that SGE2 can solve larger problems (n = 80) than either SGE1 (n = 65) or SGE3 (n = 70). Evidently, then, the qualitative comparisons suggested in Section 4 are borne out in practice.

Table 5.1. Storage Requirements for the Model Problem

n	SGE1	SGE2	SGE3	n	SGE1	SGE2	SGE3
20	17,324	12,789	15,508	60	214,788	137,621	164,364
25	29,038	20,971	25,519	65	257,510	--	194,547
30	44,226	30,909	37,334	70	>262,144	--	227,740
40	85,696	57,653	69,352	75	>262,144	--	>262,144
50	142,382	94,021	113,346	80	>262,144	256,155	>262,144

* We are indebted to Dr. P. T. Woo of the Chevron Oil Field Research Company for running these experiments for us.

Table 5.2. Timings for the Model Problem (in seconds)

n	SGE1 (NUMSLV)	SGE2 (TRKSLV)	SGE3 (NUMSLV)	n	SGE1 (NUMSLV)	SGE2 (TRKSLV)	SGE3 (NUMSLV)
20	.58	1.15	.71	60	18.04	31.55	20.31
25	1.14	2.23	1.40	65	--	--	25.73
30	2.05	3.86	2.45	70	--	--	31.86
40	5.00	9.25	5.91	80	--	78.15	--
50	10.22	18.07	11.52				

6. Conclusion

In this paper we have considered the design of implementations of sparse Gaussian elimination in terms of the competing goals of flexibility, speed, and small size. We have seen that by varying certain aspects of the design, it is possibly to vary the degree to which each of these goals is attained. Indeed, there seems to be almost a continuous spectrum of possible designs -- SGE1 and SGE2 are its endpoints, while SGE3 is just one of many intermediate points. There is no single implementation that is always best; the particular implementation that should be used in a given situation depends on the problems to be solved and the computational environment in which the calculations are to be performed.

References

[1] A. Chang. Application of sparse matrix methods in electric power system analysis. In Willoughby, editor, *Sparse Matrix Proceedings*, 113-122. IBM Research Report RA1, Yorktown Heights, New York, 1968.

[2] A. R. Curtis and J. K. Reid. The solution of large sparse unsymmetric systems of linear equations. *JIMA* 8:344-353, 1971.

[3] S. C. Eisenstat, M. H. Schultz, and A. H. Sherman. Application of sparse matrix methods to partial differential equations. In *Proceedings of the AICA International Symposium on Computer Methods for PDE's*, 40-45. Bethlehem, Pennsylvania, 1975.

[4] S. C. Eisenstat, M. H. Schultz, and A. H. Sherman. Efficient implementation of sparse symmetric Gaussian elimination. In *Proceedings of the AICA International Symposium on Computer Methods for PDE's*, 33-39. Bethlehem, Pennsylvania, 1975.

[5] S. C. Eisenstat, M. H. Schultz, and A. H. Sherman. Subroutines for the efficient implementation of sparse Gaussian elimination.

[6] F. G. Gustavson. Basic techniques for solving sparse systems of linear equations. In Rose and Willoughby, editors, *Sparse Matrices and Their Applications*, 41-52. Plenum Press, New York, 1972.

[7] W. C. Rheinboldt and C. K. Mesztenyi. Programs for the solution of large sparse matrix problems based on the arc-graph structure. Technical Report TR-262, Computer Science Center, University of Maryland, 1973.

[8] A. H. Sherman. On the efficient solution of sparse systems of linear and nonlinear equations. PhD dissertation, Yale University, 1975.

[9] P. T. Woo, S. C. Eisenstat, M. H. Schultz, and A. H. Sherman. Application of sparse matrix techniques to reservoir simulation. This proceedings.

FINDING THE BLOCK LOWER TRIANGULAR FORM
OF A SPARSE MATRIX

Fred Gustavson
IBM Thomas J. Watson Research Center
Yorktown Heights, New York 10598

1.0 INTRODUCTION

This paper is concerned with efficient algorithms that find the block lower triangular form of a sparse matrix. We assume that the best algorithm has two stages; i.e., consists of finding a maximal assignment for A followed by finding the strong components of the associated directed graph. The recent recursive algorithm of Tarjan [1] finds the strong components of a directed graph in O(V,E) time. We present a non recursive version of Tarjan's algorithm, called STCO, along with a detailed analysis and a sparse matrix interpretation. The analysis reveals the constants of the linear bounds in V and E. Based on the low values of these constants we can argue convincingly that STCO is near optimal.

Algorithms for finding a maximal assignment abound in the literature. The recent thesis of I.Duff [2; chapter 2] has an extensive bibliography; he also reviews the history of research on this problem. We agree with Duff that some form of Hall algorithm [3] is best if one wishes to stretch the assignment by a small amount, say length one. In section 2.0 we present a Depth First Search (DFS) version of Hall's algorithm, called ASSIGN ROW. We have not seen ASSIGN ROW in the literature; however A. Hoffman and I. Duff recognized it immediately. We believe that ASSIGN ROW combined with the enhancements of Section 2.1 is new.

The recent work of Hopcroft and Karp [4] is an important contribution to the maximal assignment problem. They give an algorithm that allows one to simultaneously stretch an assignment with several paths of minimal length. We have not included their work because we lacked the space and because we want to emphasize the use of heuristics. In section 2.2 and 2.3 we describe some new heuristics (see Duff [2; chapter 2]) and supplement his results with a further statistical study. The results show that usually one does not have to stretch the assignment; if one has to it is by a small amount. This says that the more complicated parts of ASSIGN ROW and the Hopcroft-Karp algorithm are hardly ever used.

In section 2.4 we describe how a covering in the sense of König can be cheaply added to ASSIGN ROW. We then exhibit a block diagram of sparse singular matrix and indicate in what sense it might be called canonical.

2.0 A DEPTH FIRST VERSION OF HALL'S ALGORITHM

The algorithm ASSIGN ROW (Figure 2.1) assigns row k of a sparse matrix A to some unassigned column j. (p(j)=k) It attempts to make a cheap assignment (step I) by scanning row k for an unassigned column. If step I fails a DFS is made of all assigned rows accessible from row k to find an unassigned column. If this search succeeds an alternating chain will have been established between $\nu+1$ new assignments and ν old. Step V assign $\nu+1$ values to p; it erases ν old values. If the search fails then $\nu+1$ rows will have been visited whose union possess only ν distinct column indices.

ASSIGN ROW is a difficult algorithm to analyze. It is not easy to predict how many times step I will fail. Each time step I fails it may be necessary to search all assigned rows of A to get a new assignment. Assuming the worst, ASSIGN ROW has a operation count O(M) where

$M = \sum_{\nu=1}^{n} N(\nu)$; $N(\nu)$ is the number of nonzeros in the first ν rows of A.

ASSIGN ROW k

Step I	Is there an unassigned column in row k? If yes, set j= to such a column index, turn switch SW on, and go to step V
Step II	(There is no unassigned column in row k) Assign RV(k)=1 stating that row k has been visited
Step III	For every nonzero (column index j) in row k DO; Has row p(j) been visited (RV(p(j))=1)? If yes, get next column index If no, execute ASSIGN ROW p(j); END;
Step IV	If switch SW is OFF RETURN
Step V	(Switch SW is ON here) Assign column j to row k by setting p(j)=k

Notes: Arrays p and RV and switch SW are global variables. Switch SW must set OFF and RV=0 before entering ASSIGN ROW. If SW=OFF on exiting from ASSIGN ROW then the union of r rows for which RV=1 contains a set of only r-1 column indicies.

Figure 2.1

2.1 IMPROVEMENTS TO ASSIGN ROW k

Suppose one tries to find an assignment for A by executing ASSIGN ROW ν, for $\nu=1,\ldots,N$. A first improvement relates to Step I. It is only necessary to scan the nonzero of A once to determine all cheap assignments. In this regard we can introduce an Auxiliary Row Pointer array ARP and a boolean Row Done array RD. If step I fails for row ν we set RD(ν)=1; we will skip step I whenever RD(ν)=1. When step I succeeds we set ARP(ν) equal to the next nonzero in row ν; on possible re-entry to step I with k=ν we continue scanning from the nonzero pointed at by ARP. We can see why this enhancement works by examining a generic extension chain of length 2(ℓ+1 produced at step III: $(k,j_1),(k_1,j_1),(k_1,j_2),\ldots,(k_\ell,j_\ell)$, (k_ℓ,j) where $k_\nu=p(j_\nu)$. The old (new) assignment is given by the even (odd) number terms in this chain. Hence the new column set is augmented by column index j; this says that previously assigned column indices are still assigned.

A second improvement concerns the row visited array RV. In searching for a new assignment via recursive entry at step III we must mark the rows visited so that we do not return to them. Since RV should be empty on entry to ASSIGN ROW we face the problem of zeroing RV. When step I succeeds RV remains unchanged; this means RV need zeroing only when ASSIGN ROW is recursively entered. Suppose this happens at stage ν=k. Unless we remember which rows are visited zeroing RV requires k operations. We can, however, completely avoid the zeroing problem: Note that ASSIGN ROW k is executed via a normal entry only once. Hence during execution of ASSIGN ROW k we may set RV(ν)=k to indicate that row ν has been visited. This assignment scheme effects an introduction of N separate switches where off (at step k) means RV(ν)$\neq$k. All we have to do to maintain the integrity of testing is to make sure that initially RV$\leq$0 or RV>N.

Table 2.1 gives an indication on how good ASSIGN ROW and its improvements are. It contains statistics of four matrices taken from the literature; Duff-10×10 [2], Yaspan-10×10 [5], Ford & Fulkerson-9×9 [6], and Willoughly 199-199 [7]

TABLE 2.1

MATRIX	SIZE	NONZEROS	NONZERO TESTED FOR CHEAP ASSIGN	RV SET TO ZERO	ROWS CHEAPLY ASSIGNED	# TIME STEP 3 EXECUTED	# of NONZEROS TESTED DURING STEP 3
DUFF	10	31	22(20)	9(2)	9	2	3
YASPAN	10	22	50(22)	25(16)	7	16	30
FORD & FULKERSON	9	34	41(30)	17(7)	8	6	18
WILLOUGHBY	199	701	1706(661)	8000(413)	144	413	766

The second entry of column four in Table 2.1 gives the improvement obtained by scanning A once for cheap assignment. The second entry of column five shows the number times RV is set ON during DFS; the second improvement obviates resetting RV completely.

2.2 THE USE OF HEURISTICS WITH HALL'S ALGORITHM

If we are clairvoyant in choosing the rows and columns in the cheap assign part (Step 1) of ASSIGN ROW we could have an optimal algorithm. The use of heuristics is an attempt to reduce the need to enter Step III by being clever with Step I. Following the lead of Duff [2;Chapter 2] we propose three new heuristics and provide statistical evidence on how well they work.

Heuristic 1: Choose the rows of A in the order they are given. In executing step I pick an unassigned column of minimal column count.

Heuristic 2: Order the rows of A according to density (ascending row count). Apply Heuristic 1.

Heuristic 3: Set $A^{(1)}=A$. For each ν, $\nu=1,\ldots,N$ do the following:

1. Find the rows (columns) of $A^{(\nu)}$ that possess minimum row (column) count.
2. If minimum column count is greater than minimum row count go to 3b.

3a. Select an element of minimum column count from the set of minimum rows. Go to 4.

3b. Select an element of minimum row count from the set of minimum columns.

4. Let $a_{ij}^{(\nu)}$ be the element selected in step 3. Assign the corresponding element a_{IJ} of A via P(J)=I. Delete row i and j of $A^{(\nu)}$ and call the resulting matrix $A^{(\nu+1)}$. Stop if $A^{(\nu+1)} = 0$.

In addition to Heuristics 1,2, and 3 (H1,H2, and H3) we shall include methods 1 and 4 of Duff, [2; chapter 2] (M1, and M4). M1 is the straight application of ASSIGN ROW k, k=1,...,N; some results appear in Table 2.1. M4 is similar to H3. In M4 singleton rows and columns are taken from $A^{(\nu)}$ as assignments. If none exist an $a_{ij}^{(\nu)}$ is found that minimizes the sum of the row and column count.

2.3 STATISTICAL STUDY OF THE HEURISTICS

We generate a random sparse matrix of order N with NT elements as follows: Shuffle a deck of N^2 cards and deal out the first NT cards; i.e., select the first NT element of a random permutation of order N^2. Next form a boolean vector v where $v_\nu=1$ iff ν belongs to set of NT numbers. Finally reshape the N^2 bits of v into an N by N matrix. We shall be interested in random matrices with $NT=\alpha N$ where $2\le\alpha\le6$. It turns out that many of these matrices are symbolically singular*; we must, therefore, call ASSIGN ROW to find out if the heuristic finds a maximal transversal. Table 2.2 below contain the results of our study. Each row entry in Table 2.2 represents a sample of 100 matrices. The FAILURE column contains two entries α/β; α is the number of matrices for which the heuristic failed to get a maximal assignment and β is the total number of assignment extensions found for the subset of α matrices. Our results show M4 and H3 best followed by H2,H1, and M1. The distribution column shows the symbolic nullity** of the set of matrices; e.g. in the second row of Table 2.2 the sample of 100 matrices partitioned into 11, 54, 26, 7, and 2 matrices which had transversal lengths of 50, 49, 48, 47, and 46 respectively. This column points out that randomly generated sparse matrices are mostly singular.

TABLE 2.2
Statistical Study of Heuristic Methods

N	NT	Method	Failure	Distribution
50	100	M4	0/0	7 8 9 10 11 12 13 14 15 16
50	100	H3	0/0	1 14 16 21 28 10 9 0 0 1
50	200	M4	7/7	0 1 2 3 4
50	200	H3	1/1	11 54 26 7 2
50	200	M1	100/≈800	
50	125	M4	2/2	4 5 6 7 8 9 10 11
50	125	H3	2/2	5 13 23 27 16 12 3 1
				0 1
50	300	H3	0/0	87 13
10	20	M4	0/0	0 1 2 3 4
10	20	H3	0/0	3 31 52 13 1
				0 1 2 3
10	25	M4,H3	0/0	19 66 14 1
10	25	M4,H3	0/0	19 61 20
100	500	H3	0/0	0 1 2 3 4
100	500	M1	100/1221	23 45 29 2 1
				0 1 2 3 4 5 6 7 8
25	50	M1	95/226	0 0 1 9 30 30 16 10 4
50	200	H2	75/124	0 1 2 3 4
50	200	H1	98/291	10 55 26 7 2
50	200	M1	100/690	
100	400	H2	96/287	0 1 2 3 4 5 6 7
100	400	H1	100/594	2 15 26 32 21 2 1 1
100	400	M1	100/1371	

---

* A matrix is symbolically singular if there is no transversal of length N.

** Symbolic Nullity of $A\equiv N-\rho$; Symbolic Rank of $A\equiv\rho=$ length of of a maximal transversal.

2.4 CANONICAL FORM FOR A SPARSE MATRIX

The result of König [8] states that nonzeros of a sparse matrix can be covered by ρ lines* where ρ is the symbolic rank of A. By using ASSIGN ROW we can get a cover very easily. Suppose CR and CC (standing for cover row and cover column) are a set of $r+c = \nu$ lines that cover $\mu \geq \nu$ rows of A. Consider what happens when ASSIGN ROW attempt to assign a $\mu+1$-st row, call it k, of A. If ASSIGN ROW succeeds then CR $\cup \{k\}$ and CC cover the $\mu+1$ rows. If ASSIGN ROW fails one finds that η rows +row k are covered by η columns. Hence subtract the η rows from set CR and add η columns to set CC and the cover remains in tact.

Suppose a cover of A has c and r elements in the sets CC and CR where $r+c=\rho$ and n, the symbolic nullity of A, equals $N-\rho$. Figure 2.2 depicts the canonical form. Matrices X and Y can be assumed to have full assignments of length c and r down the diagonal. Matrices U,V, and W are in general non empty. Let u_{ij} be a nonzero of U; we may interchange row i of U with row j of X and maintain the assignment of X. Hence there are up to $\binom{c+n}{c}$ possible

	c	n	r
c	X		
n	U	0	
r	V	W	Y

Figure 2.2

matrices that could represent X. Each of these matrices has a block lower triangular form (BTF); we have a set of BTF's to choose from. Similarly there are up to $\binom{r+n}{r}$ matrices, each possessing a BTF, that could represent Y. Perhaps a BTF that gives rise to minimal operation count for finding a singular solution could serve as the definition of a canonical form.

---

* A line is either a row or column of A.

3.0 A PROGRAM IMPLEMENTATION OF STCO - INTRODUCTION:

Our hope in Section 3 is to present a clear, yet detailed, sparse matrix interpretation of Tarjan's recursive strong component algorithm. We also present an efficient program called STCO (STrong COmponents), along with a detailed analysis of STCO. It is our opinion that STCO is near optimal in both storage and operation count; at the very least we present data for one to make a valid comparison. We feel the best way to describe STCO is by an example (Section 3.1). The example hopefully will impart enough of what STCO does so that its statements (Figure 3.4) become understandable (see Sections 3.2 and 3.3). The Figures 3.2, 3.3, and OUTPUT from STCO contain ample information to then allow the reader to carefully trace the program flow. We conclude Section 3 by giving detailed operation counts in Section 3.4.

3.1 A SPARSE MATRIX INTERPRETATION OF PROGRAM STCO-AN EXAMPLE

Consider A = BR where B is the matrix of Figure 1 in [9,p.68] and R is the permutation 10 5 9 7 1 12 3 16 4 8 2 11 14 15 6 13. The matrix A is given below (Figure 3.1).

	1	2	3	4	5	6	7	8	9	10	11	12	13	14	15	16
1	x						x									x
2	x	x	x							x			x			x
3			x													
4	x		x	x	x	x						x	x			
5	x				x		x									
6	x					x		x						x		
7							x	x								
8							x	x								
9	x								x		x	x				x
10	x				x					x			x			
11									x	x	x					
12						x			x		x	x				
13										x			x			
14			x					x					x			
15			x	x				x		x				x	x	
16		x			x					x						x

Figure 3.1

We assume the sparse matrix A is stored rowwise via Row Pointers, RP and Column Indices, CI. [10]. These arrays contain N+1 and NT elements where N is the matrix order and NT is the number of its nonzero elements. A permutation array Q of size N is needed so that $A_{Q(i)i} \neq 0$ for i=1,...,N; otherwise STCO will not work.

Input data to STCO describing A is

Q =	1	2	3	4	5	6	7	8	9	10	11	12	13	14	15	16				
RP =	1	4	10	11	18	21	25	27	29	34	38	41	45	47	50	56	60			
CI =	1	7	16	1	2	3	10	13	16	3	1	3	4	5	6	12	13	1	5	7
=	1	6	8	14	7	8	7	8	1	9	11	12	16	1	5	10	13	9	10	11
=	6	9	11	12	10	13	3	8	14	3	4	8	10	14	15	2	5	10	16	

We associate a directed graph G with A by identifying a nonzero a_{ij} with edge (i,j). A forest tree, generated by a DFS of vertex 1 of G, is depicted in Figure 3.2.

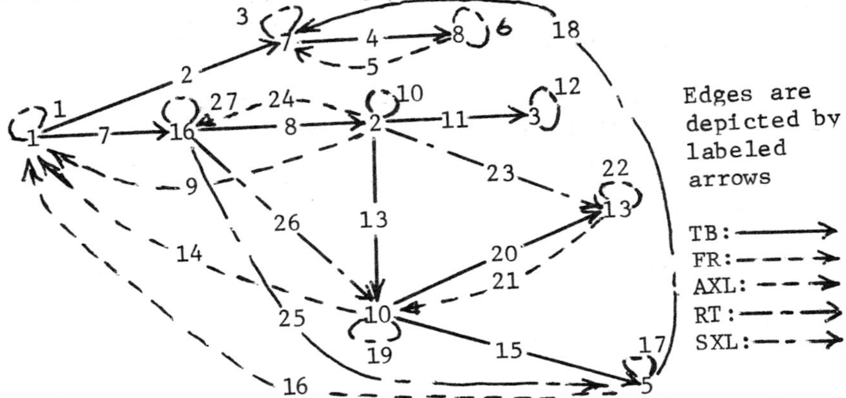

Figure 3.2 - First Tree generated by DFS of graph associated with A

The edges are labeled according to the order they are processed in DFS; e.g., the edge from vertex 13 to vertex 10 is the 21st edge processed by STCO. Figures 3.2 and 3.3 (next page) give a picture of the processing STCO does on matrix A; the relationship of the nonzeros of A to the forest of trees generated by STCO will be made using these figures.

We now describe Figure 3.3. Dotted lines encompass the Trees in the Forest; there are 3(=TF) areas of sizes 9,6 and 1. The tree rooted at vertex 1 generates blocks of size 2, 1, and 6; the information in area 1 is the same as the information in Figure 3.2. The tree rooted at vertex 4 generates blocks of size 1,1,3, and 1; vertex 15 generates a block of size 1. We have used a positive integer I (I is the Ith edge encountered in the DFS) to represent nonzero instead of the x of Figure 3.1; e.g., nonzero (4,6) (nonzero 13, 10 of Matrix A) is processed on the 21st time statements 11* of STCO are executed. (See corresponding edge 21 in Figure 3.2.)

*See Figure 3.4 ahead.

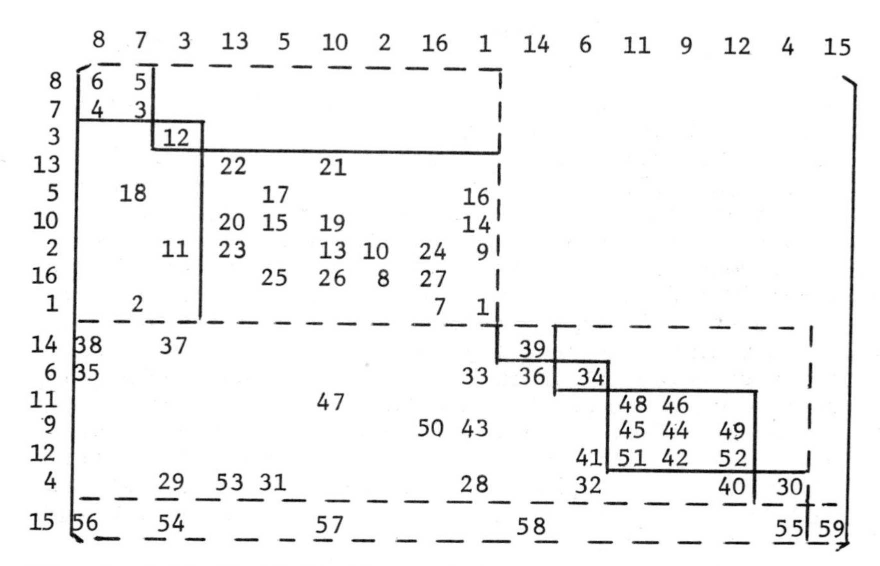

RT = 5 3 12 22 17 19 23 10 25 26 27 1 39 34 48 44 51 52 30 59
TB = 4 20 15 11 13 8 2 7 36 45 42 32 40
FR = 6 21 16 14 24 9 46 49
AXL= ∅
SXL=18 38 37 35 33 47 50 43 41 29 53 31 28 56 54 57 58 55

Figure 3.3

There are 20 Redundant Trees(RT), 13 Tree Branches(TB); 8 FRonds(FR), no Active cross Link(AXL), and 18 Sterile cross Links(SXL). Tarjan associated with graph G a forest of palm-treelike graphs whose vertices (see definition of NUMBER ahead) are numbered. Figure 3.2 is an example of one such tree. The edges of the associated graphs are classified according to their numbering via a DFS:

An edge (V,W) is
- RT: if it runs from ancestor to descendant: i.e., (NUMBER(V) ≥ NUMBER (W))
- FR: if it runs from descendant to ancestor: i.e., (NUMBER(V) < NUMBER (W))
- XL: if it runs from one subtree to another subtree: i.e., (NUMBER(V) < NUMBER (W))
- TB: if vertex W is not yet visited in DFS

RT are neglected by STCO; we have, however, included them in Figures 3.2 and 3.3. We have further classified a cross link (XL) as active (sterile) if it enters (does not enter) into the calculation of LOWLINK. (See definition of LOWLINK ahead.) The sizes of the arrays RT,TB,FR,AXL, and SXL in Figure 3.3 are 20,13,8,0,and 18; these arrays list the 59 edges of G according to their designation. For example, nonzero 13,10 of A is a frond. Element (13,5) of A, if it existed, would have been an active cross link; it would have been processed between edges 20 and 21. The fronds and active cross links <u>always</u> occur in the strictly upper triangular part of the diagonal blocks. The redundant trees <u>always</u> occur in the lower triangular part of the diagonal blocks. The sterile cross links <u>always</u> occur outside the diagonal blocks. One tree branch occurs in each column that is <u>not</u> a root of a tree in the forest. For each such column the tree branch is the edge encountered first in the DFS; e.g., the edges of column 3 were processed at the 12,11,37,29, and 54th time statements 11* were executed-edge (2,3) is the tree branch. Output from STCO is permutation array P and pointer (into P) array BLK. It follows that 4N+NT+1 integers are required to describe the input and output of STCO. Tarjan determines the roots of strong components by introducing two bookkeeping functions NUMBER and LOWLINK; vertex V is a root iff NUMBER(V)=LOWLINK(V). (See [1,p.156].) The important thing about these functions is the low overhead required for their computation (See Section 3.4.) The definition of NUMBER and LOWLINK now follow.

NUMBER (V) = K if V is K^{th} visited during a DFS; e.g.,row 16 has NUMBER value 4 since it is encountered fourth-see Figure 3.2. LOWLINK(V) is the smallest vertex belonging to the same block as V that is reachable by traversing zero or more tree branches followed by at most one frond or active cross link. For example,LOWLINK(16)=1 since edges 8 and 9 constitute a path consisting of a tree branch and frond to vertex 1 having NUMBER value 1. For matrix A, the OUTPUT from STCO is:

	1	2	3	4	5	6	7	8	9	10	11	12	13	14	15	16
P=	8	7	3	13	5	10	2	16	1	14	6	11	9	12	4	15
BLK=	1	3	4	10	11	12	15	16								
NUMBER=	1	5	6	10	8	11	2	3	14	7	15	13	9	12	16	4
LOWLINK=	$\binom{1}{0}$	$\binom{5}{8}$	$\binom{6}{11}$	$\binom{10}{27}$	$\binom{8}{15}$	$\binom{11}{32}$	$\binom{2}{2}$	$\binom{3}{4}$	$\binom{14}{42}$	$\binom{7}{13}$	$\binom{15}{45}$	$\binom{13}{40}$	$\binom{9}{20}$	$\binom{12}{36}$	$\binom{16}{53}$	$\binom{4}{7}$
		$\binom{1}{9}$			$\binom{1}{16}$			$\binom{2}{5}$	$\binom{13}{49}$	$\binom{1}{14}$	$\binom{14}{46}$		$\binom{7}{21}$			$\binom{1}{27}$

*See Figure 3.4 ahead.

The matrix $M=PQAP^T$ is in block lower triangular form; M is depicted in Figure 3.3. Note that diagonal blocks begin in rows BLK(I),I=1,...,8. The entries in LOWLINK are given as a number pair $\binom{\alpha}{\beta}$,α stands for LOWLINK value and β for edge number when value α was set. For example, LOWLINK(5) was set to 8 during processing of tree branch 15; during processing of frond 16 LOWLINK(5) was set to 1.

3.2 STORAGE OVERLAP AND OTHER VARIABLES OF STCO:

In the example we defined and discussed the seven arrays Q, RP,CI,P,BLK,NUMBER,and LOWLINK. Below we define four more arrays.

LSTACK(N): Array to see if vertex is on stack. Vertex V on stack if 0<NUMBER(V)<N+1 (LSTACK and NUMBER share the same space)

$$\text{LSTACK(V)} = \begin{cases} \text{0 if vertex V is not yet visited} \\ \text{N+1 after vertex V leaves stack} \end{cases}$$

VSTACK(N): Vertex stack(P and VSTACK share the same space)

$$\text{VSTACK( )} = \begin{cases} \text{Positions 1,2,... contain output permutation} \\ \text{Positions N,N-1,... contain vertices currently on stack} \end{cases}$$

SUBTREE(N): Array to save vertices due to recursive calling of STCO(SUBTREE and BLK share the same space)

$$\text{SUBTREE( )} = \begin{cases} \text{Positions 1,2,... contain pointers to the irreducible blocks} \\ \text{Positions N,N-1,... . contain the vertices of the current SUBTREE} \end{cases}$$

EDGEPTR(N): Array of row pointers(one for each node)pointing at column indices(current edges)under consideration

The number in parenthesis is the size of the array; e.g.,array LSTACK requires space for N elements. Altogether we have eleven arrays requiring space of 10N+NT+1. However, arrays LSTACK,VSTACK, and SUBTREE share the same space as NUMBER,P, and BLK respectively. The total storage is 7N+NT+1. Note that because of sharing the NUMBER array contains N+1 on output. For matrix A we have displayed its values prior to the 17 being stored. Next we define the scalar variable of STCO.

N: Order of sparse matrix (input)

NP1: N+1

NB: Latest NUMBER value.Next available NUMBER value is NB+1

NBLK: Counts the Number of irreducible BLocKs (output)

CNT: Vertex CouNTer,CNT counts the vertices in output permutation P.

LFP1: Last Found unvisited vertex Plus 1.

SP: Stack Pointer. VSTACK(SP) is top stack element

VP: Vertex Pointer. Associated with vertex V

V: Current vertex V

WP: Vertex pointer associated with vertex W

W: Current vertex W. Edge(V,W) is under consideration

QVP1: Q(V)+1. Needed to find end of row Q(V)

I: Dummy index

3.3 PROGRAM STCO AND COMMENTS:

Figure 3.4 contains a block diagram of STCO. We now expand on the remarks we made about STCO for the example matrix A. In what follows it will be helpful to refer to Figure 3.4. The introduction of variable LFP1 which points to the first possible unvisited node reduces the operation count X of statements 4 and 5 to no more than N. The value of X is the number value given to the root of the last tree in the forest. If we always started our search from vertex one this count could be as high 1/2N(N+1). Note that 15 is the value of X for the example matrix A. LFP1 takes on values 1,2,5,and 16. Execution of statement 2 fills EDGEPTR(I) with a pointer to first nonzero of row I of matrix QA. We decrement the pointers VP and SP (Statements 8a and 10a) because we begin the stacks at the top (position N) of arrays BLK and P. Similarly to remove stack elements we increment the pointers (Statements 18c and 21c). The act of setting NUMBER(V) at statement 9 automatically implies V is on the stack. When statement 12 is executed 0<NUMBER(V)<N+1. If NUMBER(W)=N+1 then edge (V,W) is a sterile cross link because vertex W is already fully processed and off the stack. Otherwise edge (V,W) is a redundant tree branch and can be ignored. Statement 13 determines whether edge (V,W) is a tree branch; the edges processed by statement 14 are the fronds and active cross-links. At statement 19 we can check V vs. W instead of NUMBER(V) vs. NUMBER(W) because vertices W above V on the stack satisfy LOWLINK(W)<NUMBER(W) (statement 16, executed previously, gave true answers for these vertices).

PROGRAM DESCRIPTION OF STCO

(1): Initialization

```
1a,b          LFP1←CNT←1+NB←NBLK←0; SP←VP←NP1←N+1
2             EDGEPTR(I)←RP(Q(I))+LSTACK(I)←0 for I=1,...,N
```

(2): Exit when output permutation is full; otherwise begin search at vertex LFP1 for another tree in forest

```
3           α If (CNT=NP1)  RETURN
4             Do β   I=LFP1, N
5           β If (LSTACK(I)=0)    GO TO γ
6             ERROR STOP
7a,b        γ LFP1←1+V←I;  GO TO δ
```

(3): Prepare for recursive entry into STCO

```
8a,b,c      ε  VP←VP-1; SUBTREE(VP)←V; W←V
```

(4): Add tree branch to current tree

```
9           δ LOWLINK(V)←NUMBER(V)←NB←NB+1
10a,b,c       SP←SP-1; VSTACK(SP)←V; QVP1←Q(V)+1
```

(5): Examine all edges leading out of V

```
11a,b,c     ζ WP←EDGEPTR(V); W←CI(WP); EDGEPTR(V)←WP+1
12            IF(NUMBER(W)≥NUMBER(V)) GO TO η
13            IF(LSTACK(W)=0) GO TO ε
14            LOWLINK(V)←MIN(LOWLINK(V), NUMBER(W))
15          η IF(EDGEPTR(V)<RP(QVP1)) GO TO ζ
```

(6): Check if V is strong component root. Gather component if it is.

```
16            IF (LOWLINK(V)<NUMBER(V)) GO TO θ
17a,b         NBLK←NBLK+1; BLK(NBLK)←CNT
18a,b,c     ι W←VSTACK(SP); LSTACK(W)←NP1; SP←SP+1
18d,e         P(CNT)←W; CNT←CNT+1
19            IF(V≠W) GO TO ι
```

(7): GO TO label α when present tree is finished; otherwise recursively RETURN from STCO

```
20            IF (SP=NP1) GO TO α
21a,...,d   θ W←V; V←SUBTREE(VP); VP←VP+1; QVP1←Q(V)+1
22a           LOWLINK(V)←MIN(LOWLINK(V), LOWLINK(W))
22b,23        GO TO η;  END
```

Figure 3.4

3.4 OPERATION COUNTS FOR PROGRAM STCO

We may apply a direct analog of Kirchhoff's Law: "the number of times an instruction is executed must equal the number of times we transfer to that instruction". (See [11, pp.166-169].) For program STCO the solution of the Kirchhoff equations are simple; the results are given in Table 3.1 below.

STMT #	# times executed	STMT #	# times executed	STMT #	# times executed
1	once	8	TB	16	N
2	N	9,10	N	17	NBLK
3	TF+1	11,12	NT	18,19	N
4,5	X	13	TB+FR+AXL	20	NBLK
6	0	14	FR+AXL	21,22	TB
7	TF	15	NT		

Table 3.1

We have defined the symbols N,NT,TF,RT,TB,FR,AXL in Section 3.1 and the symbols X and NBLK were defined in Sections 3.3 and 3.2 Note that TB+TF=N and NT=TB+FR+AXL+RT+SXL. Program STCO uses fast and simple computer instructions; i.e., assignments, adds, compares, and indexing. We could carry this analysis further; however, it would then be appropriate to write STCO in assembly language and then to give exact estimates in terms of machine instructions. We shall not do this. Statements 14 and 22a did not succumb to complete analysis; i.e., we could not express the number of times LOWLINK(V) changed in terms of the other variables above. This value depends also on the location of the edges of G. Referring back to the example matrix A we can see that eight edges, (7 FR and 1TB), caused LOWLINK to change at statements 14 and 22a, they are listed in the LOWLINK array there. For this example LOWLINK is set 24 times; the factor α of Table 3.2 below has the value 8/21. We end this section by listing in Table 3.2 how many times each array of STCO is set during execution.

ARRAY	# times set	ARRAY	# times set
NUMBER	N	SUBTREE	2TB
LOWLINK	N+α(TB+FR+AXL) where $0 \le \alpha \le 1$	EDGEPTR	N+NT
		P	N
LSTACK	2N	BLK	NBLK
VSTACK	2N		

Table 3.2

References

[1] Tarjan, R. J. "Depth First Search and Linear Graph Algorithms" SIAM J. Comput. Vol. 1, No. 2, June 1972, pp. 146-160.

[2] Duff, I. S. "Selecting a Maximal Transversal", Chapter 2- Thesis. "Analysis of Sparse Systems", University of Oxford, October 1972.

[3] Hall, M. "An Algorithm for Distinct Representatives" Am. Math. Monthly 63, pp. 716-717.

[4] Hopcroft, J. E., Karp, R. M. "An $n^{5/2}$ Algorithm for Maximum Matchings in Bipartite Graphs" SIAM J. Comput. Vol. 2, No. 4, December 1973, pp. 225-231.

[5] Yaspan, A. "On Finding a Maximal Assignment" Operations Research, 14, pp. 646-651.

[6] Ford, L. R., Fulkerson, D.R. "Flows in Networks" Book, Princeton University Press, N. J., 1962.

[7] Willoughby, R. A. "Sparse Matrix Algorithms and their Relation to Problem Classes and Computer Architecture" Book "Large Sparse Sets of Linear Equations" (Reid, J., Editor) Academic Press, London and New York, pp. 255-277.

[8] König, D. "Theorie der Endlichen and Unendlichen Graphen" Book, Chelsea, New York, 1950.

[9] Hellerman E., Rarick, D. "The Partitioned Preassigned Pivot Procedure", Book "Sparse Matrices and their Applications" (Rose, D. and Willoughly, R., Editors) Plenum Press, N. Y., January 1973, pp. 67-76.

[10] Gustavson, F. G. "Some Basic Techniques for Solving Sparse Systems of Linear Equations", ibid., pp. 41-52.

[11] Knuth, D. E. "The Art of Computer Programming", Vol. 1, "Fundamental Algorithms," Book, Addison Wesley, Mass. 1968.

V

Matrix Methods for Partial Difference Equations

Marching Algorithms and Block Gaussian Elimination

Randolph E. Bank

Department of Mathematics
The University of Chicago

Abstract: Fast direct methods incorporating marching techniques are considered from the viewpoint of sparse Gaussian elimination. It is shown that the algorithms correspond to the nonstandard block backsolution of particular *LU* decompositions.

1. Introduction

Consider the n×n block tridiagonal linear system Mx = b given by

$$(1.1)\qquad \begin{bmatrix} T & -I & & & \\ -I & T & -I & & \\ & \ddots & \ddots & \ddots & \\ & & -I & T & -I \\ & & & -I & T \end{bmatrix} \begin{bmatrix} x_1 \\ x_2 \\ \vdots \\ x_{n-1} \\ x_n \end{bmatrix} = \begin{bmatrix} b_1 \\ b_2 \\ \vdots \\ b_{n-1} \\ b_n \end{bmatrix}$$

Here M is a block tridiagonal $n^2 \times n^2$ matrix with n×n blocks, which we shall denote by the triple M = [-I T -I], T is the n×n tridiagonal matrix T=[-1 4 -1], and I is the identity matrix of order n. The x_i and b_i, $1 \le i \le n$ are n-vectors. This linear system arises in the solution of Poisson's equation in a square domain, using the usual 5-point approximation of the Laplacian and a uniform n×n mesh [12].

In recent years, a number methods for solving the system (1.1) have been proposed (see the survey papers by Dorr [10] and Birkhoff and George [6], and their references). Among the most successful of these have been the <u>fast direct methods</u>, employing either the fast Fourier transform, as in Hockney [14], or a variant of the cyclic reduction algorithm as in Buzbee, Golub and Nielson [9]. Solving (1.1) using either of these methods requires $O(n^2 \log n)$ arithmetic operations (additions, subtractions, multiplications and divisions) [10]. Our concern here is with several new fast direct methods, which employ <u>marching</u> techniques [1-5].

The <u>marching algorithm</u> is an $O(n^2)$ method for solving (1.1) [1,2,5]. This is optimal from the viewpoint of computational complexity, in the sense that the amount of computation grows linearly with the size of the input vector b. Unfortunately, this algorithm suffers numerically from exponential growth in roundoff error [2,5], and thus may be impractical for many calculations. However, the marching algorithm can be successfully employed as a component of more stable fast direct methods.

The <u>generalized marching algorithm</u> is an $O(n^2 \log(n/k))$ method for solving (1.1) [2,5]. This algorithm employs the marching algorithm to solve several small problems, whose solutions are then combined to form the solution of a large problem. The parameter k is a measure of the size of the smaller problems.

The <u>k-reduction-matrix decomposition</u> algorithm combines the marching algorithm with fast Fourier transforms techniques to produce a second $O(n^2)$ algorithm for solving (1.1) [2,5]. This algorithm can be shown to be numerically stable, essentially up to the condition number of M [2,5].

Although detailed derivations of these algorithms can be tedious, their essential features can be revealed by exploring their relationship to block Gaussian elimination. In §3, we describe the marching and generalized marching algorithms in terms of the non-standard block backsolution of LU decompositions of PM or PMP^T, where P is an appropriate permutation matrix. Our presentation here is similar to that given in [2]. The LU decompositions are computed analytically by exploiting the properties of the Chebyshev polynomials $S_n(x)$ and $C_n(x)$, with zeroes in (-2,2) [5,13,16]. Relevant properties of these polynomials are summarized in the Appendix. The non-standard techniques in the block backsolution are described in §2.

In §4, we describe the solution of Poisson's equation with mixed boundary conditions, as a simple illustration of the applicability of the generalized marching algorithm to more general block tridiagonal systems.

2. Preliminaries

In this section, we describe two techniques which play important roles in our derivations. The first is the method of <u>deferred back solution</u>, described more generally in Bunch and Rose [7]. Suppose we are given the linear system $Ax = b$, which we write as

$$(2.1) \qquad Ax = \begin{bmatrix} G & C \\ R & B \end{bmatrix} \begin{bmatrix} x_1 \\ x_2 \end{bmatrix} = \begin{bmatrix} b_1 \\ b_2 \end{bmatrix} = b \ ,$$

where A is a sparse nonsingular matrix. The submatrices G

and B are square, with G nonsingular. The matrices R and C are in general rectangular, and the vectors x and b are partitioned conformably with A. We assume the existence of a fast method for solving linear systems of the form $Gy = f$.

The block LU decomposition of A can be formally written as

$$(2.2) \qquad A = \begin{bmatrix} I & 0 \\ RG^{-1} & I \end{bmatrix} \begin{bmatrix} G & C \\ 0 & \Delta \end{bmatrix} = LU \ ,$$

where $\Delta = B - RG^{-1}C$. The solution of (2.1) using (2.2) is carried out as follows. First, we solve $Ly = b$, $y^T = [y_1^T \; y_2^T]$, giving

$$(2.3) \qquad \begin{aligned} y_1 &= b_1 \\ y_2 &= b_2 - (RG^{-1})y_1 = b_2 - (RG^{-1})b_1 \ . \end{aligned}$$

The implied matrix multiplication $(RG^{-1})b_1$ is avoided by solving

$$(2.4) \qquad Gz = b_1$$

and then setting

$$(2.5) \qquad y_2 = b_2 - Rz \ .$$

The use of (2.4)-(2.5) requires the solution of a linear system involving G and a multiplication by the matrix R, but avoids the calculation (and storage) of the potentially full matrix (RG^{-1}), yielding substantial (order of magnitude) savings in arithmetic operations in the applications we shall study.

We next solve the system $Ux = y$, giving

$$(2.6) \qquad \Delta x_2 = y_2 \ ;$$

$$(2.7) \qquad Gx_1 = y_1 - Cx_2 = b_1 - Cx_2 \ .$$

Equation (2.7) requires a second solution of a linear system involving G, and a multiplication by the matrix C. The calculation of Δ, and the solution of (2.6) may be accomplished numerically in several ways [7,8]. In our applications, however, we shall determine Δ analytically and employ some fast method for solving the linear system.

The second technique deals with the solution of the linear system

$$(2.8) \qquad P_\ell^{-1}(T)P_k(T)x = y \ , \ (\text{or } P_k(T)x = P_\ell(T)y) \ ,$$

where x and y are n-vectors, T is a sparse $n \times n$ matrix and

$P_k(T)$ and $P_\ell(T)$ are nonsingular polynomials in T, of degrees k and ℓ respectively, with $k > \ell \geq 0$. Linear systems of this form often arise in connection with fast Poisson solvers [1-5,9,10,17]. The polynomials $P_k(t)$ and $P_\ell(t)$ may be written in factored form as

$$(2.9)\qquad \begin{aligned} P_k(t) &= c_k \prod_{i=1}^{k} (t-r_k(i)) \ ; \\ P_\ell(t) &= c_\ell \prod_{i=1}^{\ell} (t-r_\ell(i)) \ . \end{aligned}$$

We assume the roots and leading coefficients of both polynomials are known. Then (2.8) may be solved by using $P_k(T)$ and $P_\ell(T)$ in factored form, and solving

$$(2.10)\qquad \begin{aligned} &\text{a.} \quad v_0 = y \\ &\text{b.} \quad (T-r_k(i)I)v_i = (T-r_\ell(i)I)v_{i-1} \ , \qquad 1 \leq i \leq \ell; \\ &\text{c.} \quad (T-r_k(i)I)v_i = v_{i-1} \ , \qquad \ell+1 \leq i \leq k; \\ &\text{d.} \quad x = (c_\ell/c_k)v_k \ . \end{aligned}$$

The implied matrix multiplication in (2.10b) is avoided by using

$$(2.11)\qquad \begin{aligned} (T-r_k(i)I)w_{i-1} &= v_{i-1} \ ; \\ v_i &= v_{i-1} + (r_k(i)-r_\ell(i))w_{i-1} \ . \end{aligned}$$

The nonsingularity of $P_k(T)$ implies the nonsingularity of each of its linear factors, so that the algorithm (2.10)-(2.11) is well defined. Using (2.10)-(2.11), we avoid the explicit computation of $P_k(T)$, $P_\ell(T)$ and the *LU* decomposition of $P_k(T)$, which result in order of magnitude savings in arithmetic operations in our applications.

3. Marching Techniques

We now derive the marching algorithm for solving (1.1), and its generalizations, using the techniques developed in §2. The marching algorithm itself may be formally interpreted in terms of solving an initial value problem once appropriate initial conditions have been determined, e.g., as a shooting method [2,5,15]. In terms of block Gaussian elimination, we consider the linear system PMx = Pb where P is the permutation matrix which yields

$$(3.1)\qquad \left[\begin{array}{ccccccc|c} -I & T & -I & & & & & \\ & -I & T & -I & & & & \\ & & \ddots & \ddots & \ddots & & & \\ & & & -I & T & -I & & \\ & & & & -I & T & & -I \\ & & & & & -I & & T \\ \hline T & -I & & & & & & 0 \end{array}\right] \left[\begin{array}{c} x_1 \\ x_2 \\ \vdots \\ x_{n-2} \\ x_{n-1} \\ \hline x_n \end{array}\right] = \left[\begin{array}{c} b_2 \\ b_3 \\ \vdots \\ b_{n-1} \\ b_n \\ \hline b_1 \end{array}\right]$$

For convenience, we write this as

$$(3.2)\qquad \left[\begin{array}{c|c} U & C \\ \hline R & 0 \end{array}\right] \left[\begin{array}{c} \bar{x} \\ \hline x_n \end{array}\right] = \left[\begin{array}{c} \bar{b} \\ \hline b_1 \end{array}\right]$$

where the symbols U, R, C, $\bar{x}$, and $\bar{b}$ are used to denote their corresponding terms in (3.1). Observing that the "marching matrix" U is already upper triangular, in applying Gaussian elimination, we may focus attention on the nth block row of PM. Using straightforward algebraic manipulations, and (A1.1), the block *LU* decomposition of PM is

$$(3.3)\qquad \left[\begin{array}{c|c} U & C \\ \hline R & 0 \end{array}\right] = \left[\begin{array}{c|c} I & 0 \\ \hline RU^{-1} & I \end{array}\right] \left[\begin{array}{c|c} U & C \\ \hline 0 & \Delta \end{array}\right]$$

where

$$(3.4)\qquad \Delta = -RU^{-1}C = S_n(T)$$

and $S_n(T)$ is a Chebyshev polynomial in T. We shall hereafter let $S_n(T) = S_n$, where its use is unambiguous.

The block backsolution using (3.3)-(3.4) proceeds using deferred backsolution to perform the matrix multiplication by RU^{-1}. Since the zeroes of $S_n(t)$ are well known (see (A1.3)), the polynomial backsolution algorithm can be applied to the linear system involving Δ.

Each backsolution involving U requires $O(n^2)$ operations and solving $S_n y = d$ also requires $O(n^2)$ operations (n tri-diagonal systems), that the overall operation count is $O(n^2)$ [1,2,5,11]. If one neglects either the deferred backsolution or the polynomial backsolution, then the marching algorithm requires at least $O(n^3)$ operations [1,10].

Viewed as a shooting method, each backsolution involving U corresponds to solving an initial value problem, and solving the linear system involving S_n corresponds to calculating the appropriate initial conditions [2,5,15]. Although quite satisfactory from the viewpoint of computational complexity, like many shooting methods, the marching algorithm suffers from the exponential growth in roundoff error. Under suitable hypotheses on T, the error grows as $(||S_n||_2 + n^2)$ [2,5]. For example, for $T = [-1\ 4\ -1]$, $||S_n||_2 \doteq 1.05(5.83)^n$.

The generalized marching algorithm is an attempt to overcome this numerical instability, by breaking a large problem into several smaller ones, in analogy with multiple shooting methods [15]. Here we consider the *LU* decomposition of PMP^T, where now P is the permutation matrix which yields

$$PMP^T = \left[\begin{array}{cc|c} M_{i-1} & 0 & v_{i-1} \\ 0 & M_{n-i} & w_{n-i} \\ \hline v_{i-1}^T & w_{n-i}^T & T \end{array}\right] \tag{3.5}$$

where

$M_p = [-I\ T\ -I]$, of block dimension $p \times p$, $1 \le p \le n$;

$v_p^T = [0\ 0\ \ldots\ 0\ -I]$ of block dimension $1 \times p$, $2 \le p \le n$;

$w_p^T = [-I\ 0\ 0\ \ldots\ 0]$ of block dimension $1 \times p$, $2 \le p \le n$.

This permutation orders the ith line of unknowns last, breaking a single large problem into two smaller ones. The block *LU* decomposition of PMP^T is given by

$$\left[\begin{array}{cc|c} M_{i-1} & 0 & v_{i-1} \\ 0 & M_{n-i} & w_{n-i} \\ \hline v_{i-1}^T & w_{n-i}^T & T \end{array}\right] = \left[\begin{array}{cc|c} I & 0 & 0 \\ 0 & I & 0 \\ \hline v_{i-1}^T M_{i-1}^{-1} & w_{n-i}^T M_{n-i}^{-1} & I \end{array}\right] \left[\begin{array}{cc|c} M_{i-1} & 0 & v_{i-1} \\ 0 & M_{n-i} & w_{n-i} \\ \hline 0 & 0 & \Delta \end{array}\right] \tag{3.6}$$

where, using (A1.13) and (A1.15)

$$\begin{aligned} \Delta &= T - v_{i-1}^T M_{i-1}^{-1} v_{i-1} - w_{n-i}^T M_{n-i}^{-1} w_{n-i} \\ &= T - S_{i-1}^{-1} S_{i-2} - S_{n-i}^{-1} S_{n-i-1} \\ &= S_{i-1}^{-1} S_{n-i}^{-1} S_n . \end{aligned} \tag{3.7}$$

As in the marching algorithm, the backsolution using (3.6)-(3.7) proceeds using deferred backsolution on the lower triangular system, and the polynomial backsolution algorithm on the system involving Δ. This procedure requires partially solving two linear systems involving M_{i-1} (and two involving M_{n-i}). Two complete backsolutions are not required since most of the computations done in the first backsolution are repeated in the second. If $||S_{i-1}||_2$ is sufficiently small, then the marching algorithm can be used; otherwise, the factorization (3.6)-(3.7) is applied recursively to M_{i-1}, and is terminated when the M_p matrices become small enough for the marching algorithm to be successfully applied.

An alternative matrix interpretation for the generalized marching algorithm can be given. In this interpretation, we view the algorithm as a "k-reduction" process. For convenience, suppose $n = kd-1$, and $||S_{k-1}||_2$ is sufficiently small that the marching algorithm can be applied to linear systems involving M_{k-1}. We begin by partitioning M_n as

$$(3.8)\quad M_n = \begin{bmatrix} M_{k-1} & v_{k-1} & & & & \\ v_{k-1}^T & T & w_{k-1}^T & & & \\ & w_{k-1} & M_{k-1} & v_{k-1} & & \\ & & \ddots & \ddots & \ddots & \\ & & & v_{k-1}^T & T & w_{k-1}^T \\ & & & & w_{k-1} & M_{k-1} \end{bmatrix}$$

For the remainder of this section, we will uniformly suppress the subscript k-1.

We now form the block *LU* decomposition of PM_nP^T, where P is the permutation matrix which yields

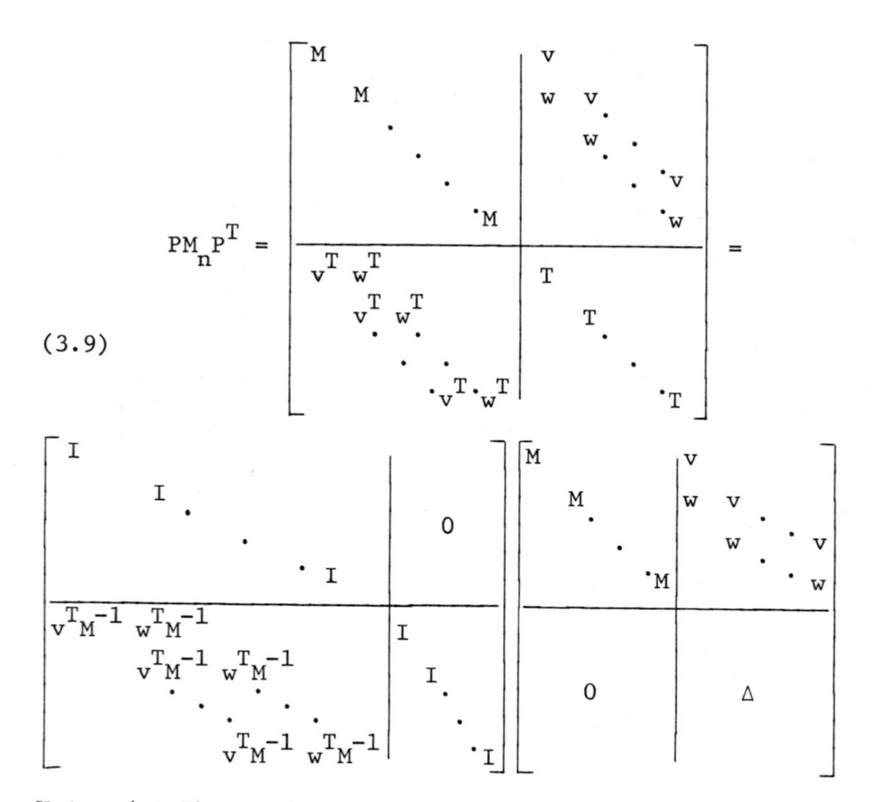

(3.9)

Using (A1.7) and (A1.15), we find Δ to be the block $(d-1)\times(d-1)$ tridiagonal matrix given by

$$\begin{aligned}\Delta &= [\{-v^TM^{-1}w\}\{T-v^TM^{-1}v-w^TM^{-1}w\}\{-w^TM^{-1}v\}] \\ &= [\{-S_{k-1}^{-1}\}\{S_{k-1}^{-1}C_k\}\{-S_{k-1}^{-1}\}] \\ &= \text{Diag }[S_{k-1}^{-1}][-I\ C_k\ -I]\ ,\end{aligned} \tag{3.10}$$

where $\text{Diag}[S_{k-1}^{-1}]$ is a block diagonal matrix.

The block backsolution using (3.9)-(3.10) proceeds using deferred backsolution, and the marching algorithm to solve the linear systems involving M_{k-1}. To solve the linear system involving Δ, we could proceed recursively; letting $d = k_1d_1$, Δ has block dimension $(k_1d_1-1)\times(k_1d_1-1)$, and we could apply a k_1-reduction step, using (A1.11)-(A1.12) to write polynomials in $C_k(T)$, which would appear, in terms of polynomials in T itself.

In the generalized marching algorithm, by selecting $n = k2^{\ell}-1$, we can solve (1.1) using one step of k-reduction and $(\ell-1)$ steps of 2-reduction. Here the two recursive matrix factorizations coincide, since, in view of the analysis following (3.7), we could implement the recursion (3.5)-(3.7), at each step partitioning M_p such that $M_{i-1} = M_{p-i}$, until $M_{i-1} = M_{p-i} = M_{k-1}$.

If n is arbitrary, and M_n is partitioned with arbitrarily sized submatrices M_p, the two factorizations still coincide, although we must then deal with more complicated rational functions involving the S_j, as in (3.7), and the C_k polynomials will not appear, in general.

The k-reduction factorization (3.8)-(3.10) also forms the basis of the k-reduction-matrix decomposition algorithm [2,5], a numerically stable $O(n^2)$ extension of the marching algorithm. After one step of k-reduction has been carried out, the matrix decomposition algorithm, employing the fast Fourier transform, is applied to the "reduced" set of equations involving the matrix $[-I \; C_k \; -I]$. The operation count for this algorithm is $O(n^2/k \log_2 n)$. Selecting $k \doteqdot c\log_2(n)$, for a fixed constant c, this becomes $O(n^2)$. By properly selecting c, the roundoff error growth due to the k-reduction step can be made proportional to the condition number of M. For the model problem (1.1), $c \doteqdot .79$ [5].

4. Poisson's Equation with Mixed Boundary Conditions

As a final example of the application of the methods of §2, consider the block n×n system $\overline{M}x = b$, given by

$$(4.1)\quad \begin{bmatrix} T+\rho I & -2I & & & \\ -I & T & -I & & \\ & \ddots & \ddots & \ddots & \\ & & -I & T & -I \\ & & & -2I & T+\rho I \end{bmatrix} \begin{bmatrix} x_1 \\ x_2 \\ \vdots \\ x_{n-1} \\ x_n \end{bmatrix} = \begin{bmatrix} b_1 \\ b_2 \\ \vdots \\ b_{n-1} \\ b_n \end{bmatrix}$$

where T is the n×n tridiagonal matrix

$$(4.2)\quad T = \begin{bmatrix} 4+\rho & -2 & & & \\ -1 & 4 & -1 & & \\ & \ddots & \ddots & \ddots & \\ & & -1 & 4 & -1 \\ & & & -2 & 4+\rho \end{bmatrix}$$

and $\rho > 0$.

This block tridiagonal system can arise, for example, from 5 point discretizations of Poisson's equation with mixed boundary conditions [12].

We partition $\overline{M}$ as

$$(4.3) \qquad \overline{M} = \begin{bmatrix} T+\rho I & 2v_{n-2}^T & 0 \\ v_{n-2} & M_{n-2} & w_{n-2} \\ 0 & 2w_{n-2}^T & T+\rho I \end{bmatrix}$$

where M_{n-2}, v_{n-2} and w_{n-2} are defined as in §3. (We now suppress the subscript n-2.) We form the block *LU* decomposition of $P\overline{M}P^T$, given by

$$P\overline{M}P^T = \left[\begin{array}{c|cc} M & v & w \\ \hline 2v^T & T+\rho I & 0 \\ 2w^T & 0 & T+\rho I \end{array}\right] =$$

$$(4.4) \qquad \left[\begin{array}{c|cc} I & 0 & \\ \hline 2v^T M^{-1} & I & 0 \\ 2w^T M^{-1} & 0 & I \end{array}\right] \left[\begin{array}{c|c} M & v \quad w \\ \hline 0 & \Delta \end{array}\right] .$$

The solution of (4.1) using (4.3)-(4.4) is carried out using deferred backsolution, employing the marching or generalized marching algorithm to solve the linear systems involving M.

The matrix Δ is the block 2×2 matrix

$$(4.5) \qquad \Delta = \begin{bmatrix} S_{n-2}^{-1} & 0 \\ 0 & S_{n-2}^{-1} \end{bmatrix} \begin{bmatrix} P_{n-1} & -2I \\ -2I & P_{n-1} \end{bmatrix}$$

$$= \begin{bmatrix} I & 0 \\ -2P_{n-1}^{-1} & I \end{bmatrix} \begin{bmatrix} S_{n-2}^{-1}P_{n-1} & -2S_{n-2}^{-1} \\ 0 & P_{n-1}^{-1}P_n \end{bmatrix} .$$

where, using (A1.7)-(A1.10) and (A1.15)

$$(4.6) \qquad P_{n-1} = C_{n-1} + \rho S_{n-2} \; ;$$
$$P_n = S_{n-2}(T^2-4I+\rho^2 I) + 2\rho C_{n-1} .$$

The linear system involving Δ may be solved using the block *LU* factorization given in (4.5), employing deferred backsolution and the polynomial backsolution algorithms in the usual fashion.

To successfully carry out this procedure, we must first determine the zeroes $r_{n-1}(i)$, $1 \le i \le n-1$, and $r_n(i)$, $1 \le i \le n$, of the polynomials P_{n-1} and P_n, respectively. The zeroes $r_{n-1}(i) = -\lambda_i$, where the λ_i are the eigenvalues of the $(n-1)\times(n-1)$ tridiagonal matrix

$$(4.7) \qquad A_{n-1} = \begin{bmatrix} \rho & -2 & & & & \\ -1 & 0 & -1 & & & \\ & \ddots & \ddots & \ddots & & \\ & & \ddots & \ddots & \ddots & \\ & & & -1 & 0 & -1 \\ & & & & -1 & 0 \end{bmatrix}$$

and $r_n(i) = -\overline{\lambda}_i$, where the $\overline{\lambda}_i$ are the eigenvalues of the $n\times n$ tridiagonal matrix

$$(4.8) \qquad A_n = \begin{bmatrix} \rho & -2 & & & & \\ -1 & 0 & -1 & & & \\ & \ddots & \ddots & \ddots & & \\ & & \ddots & \ddots & \ddots & \\ & & & -1 & 0 & -1 \\ & & & & -2 & \rho \end{bmatrix}.$$

This can be verified by finding $\mathrm{Det}(A_{n-1}-\lambda I)$ and $\mathrm{Det}(A_n-\lambda I)$, using the well known three term recurrence relation for determinants [18]. Very fast methods, such as the QR or QL algorithms [18], can be used to find these eigenvalues.

The case $\rho = 0$, corresponding to the Neumann boundary conditions, follows as a special case of (4.4)-(4.6). (Note that in this case, the zeroes of P_{n-1} and P_n can be analytically determined, and the factor $(T-2I)$ of P_n is singular.) The case of periodic boundary conditions can be easily handled using essentially the same techniques [5].

5. Concluding Remarks

Consider the block n×n linear system Mx = b given by

$$(5.1)\quad \begin{bmatrix} T+\alpha_1 I & -\beta_1 I & & & \\ -\beta_1 I & T+\alpha_2 I & -\beta_2 I & & \\ & \ddots & \ddots & \ddots & \\ & & -\beta_{n-2} I & T+\alpha_{n-1} I & -\beta_{n-1} I \\ & & & -\beta_{n-1} I & T+\alpha_n I \end{bmatrix} \begin{bmatrix} x_1 \\ x_2 \\ \vdots \\ x_{n-1} \\ x_n \end{bmatrix} = \begin{bmatrix} b_1 \\ b_2 \\ \vdots \\ b_{n-1} \\ b_n \end{bmatrix}$$

where T is the n×n tridiagonal matrix

$$(5.2)\qquad T = \begin{bmatrix} \gamma_1 & -\sigma_1 & & & \\ -\sigma_1 & \gamma_2 & -\sigma_2 & & \\ & \ddots & \ddots & \ddots & \\ & & -\sigma_{n-2} & \gamma_{n-1} & -\sigma_{n-1} \\ & & & -\sigma_{n-1} & \gamma_n \end{bmatrix} .$$

Each of the linear systems we have discussed in §§3-4 can be viewed as special cases of (5.1)-(5.2). (The system 4.1 requires a simple diagonal similarity transformation to be put into this form.) Swarztrauber [17] has studied the solution of this system by cyclic reduction methods.

Marching and generalized marching algorithms can be applied to systems of the form (5.1)-(5.2), requiring $O(n^2)$ and $O(n^2 \log_2(n/k))$ operations, respectively. The analysis of these algorithms is similar to that for the special cases we have described here. The main difference is that the polynomials involved in the more general problem are not necessarily Chebyshev polynomials, and one must routinely solve tridiagonal eigenproblems, as in (4.7)-(4.8), to determine their zeroes. A complete analysis of these algorithms is given in [3,5].

Acknowledgment: The author acknowledges and thanks Professor Donald J. Rose of Harvard University for many helpful conversations before and during the preparation of this manuscript, and the staff of the Applied Mathematics Division, Argonne National Laboratory, for the use of their facilities.

Appendix: Properties of Chebyshev Polynomials

Here we state the properties of the modified Chebyshev polynomials $S_n(x)$ and $C_n(x)$, which are relevant to this work. The interested reader is referred to the literature [5,13,16] for a more complete analysis.

The modified Chebyshev polynomials of the first kind, $S_n(x)$, satisfy the recurrence

$$\text{(A1.1)} \qquad \begin{aligned} S_0(x) &= 1 \ ; \\ S_1(x) &= x \ ; \\ S_{i+1}(x) &= xS_i(x) - S_{i-1}(x), \quad i=1,2,\ldots \end{aligned}$$

other representations of $S_n(x)$ are:

$$\text{(A1.2)} \qquad S_n(x) = \begin{cases} \dfrac{\sin[(n+1)\theta]}{\sin[\theta]}, & \cos\theta = \dfrac{x}{2}, \quad 0 \le x < 2 \\ n+1, & x = 2 \\ \dfrac{\sinh[(n+1)\psi]}{\sinh[\psi]}, & \cosh\psi = \dfrac{x}{2}, \quad 2 < x \end{cases} \ ;$$

$$\text{(A1.3)} \qquad S_n(x) = \prod_{i=1}^{n} (x - r_n(i)), \quad r_n(i) = 2\cos\left[\frac{i\pi}{(n+1)}\right].$$

The modified Chebyshev polynomials of the second kind, $C_n(x)$, satisfy the recurrence

$$\text{(A1.4)} \qquad \begin{aligned} C_0(x) &= 2 \ ; \\ C_1(x) &= x \ ; \\ C_{i+1}(x) &= xC_i(x) - C_{i-1}(x), \quad i=1,2,\ldots \end{aligned}$$

Other representations of $C_n(x)$ are:

$$\text{(A1.5)} \qquad C_n(x) = \begin{cases} 2\cos[n\theta], & \cos\theta = \dfrac{x}{2}, \quad 0 \le x < 2 \\ 2, & x = 2 \\ 2\cosh[n\psi], & \cosh\psi = \dfrac{x}{2}, \quad 2 < x \end{cases}$$

$$\text{(A1.6)} \qquad C_n(x) = \prod_{i=1}^{n} (x - r'_n(i)), \quad r'_n(i) = 2\cos\left[\frac{(2i-1)\pi}{2n}\right]$$

Some relevant identities involving the Chebyshev polynomials are given below:

(A1.7) $C_k(x) = S_k(x) - S_{k-2}(x)$, $k \geq 2$;

(A1.8) $S_{k-1}(x)C_k(x) = S_{2k-1}(x)$;

(A1.9) $C_k^2(x) - 2 = C_{2k}(x)$;

(A1.10) $C_{k+1}^2(x) - 4 = S_k^2(x)(x-2)(x+2)$;

(A1.11) $S_{\ell-1}(C_k(x)) = S_{k\ell-1}(x)/S_{k-1}(x)$, $S_{k-1}(x) \neq 0$;

(A1.12) $C_\ell(C_k(x)) = C_{\ell k}(x)$;

(A1.13) $S_\ell(x)S_{i-k}(x) = S_i(x)S_{\ell-k}(x) - S_{k-1}(x)S_{\ell-i-1}(x)$.

The inverse of the block tridiagonal matrix $M = [-I\ T\ -I]$ of (1.1) can be expressed in terms of rational functions of the matrices $S_k(T)$. In particular, $\mathrm{Det}(S_n(T)) \neq 0$, and M^{-1} is given blockwise by

(A1.14) $M^{-1} = [M_{ij}^{-1}]$, $1 \leq i,j \leq n$

where

$$\text{(A1.15)} \quad M_{ij}^{-1} = \begin{cases} S_n^{-1}(T)S_{i-1}(T)S_{n-j}(T) , & j \geq i \\ S_n^{-1}(T)S_{j-1}(T)S_{n-i}(T) , & i \geq j \end{cases} .$$

This can be verified by forming the product MM^{-1} blockwise, using (A1.1) and (A1.13) when necessary.

References

1. R. E. Bank and D. J. Rose. An $O(n^2)$ method for solving constant coefficient boundary value problems in two dimensions. *SIAM J. Numer. Anal.*, 12, (1975), p. 529-540.

2. R. E. Bank and D. J. Rose. Marching Algorithms for Elliptic Boundary Value Problems I: The Constant Coefficient Case. Submitted *SIAM J. Numer. Anal.*

3. R. E. Bank. Marching Algorithms for Elliptic Boundary Value Problems II: The Nonconstant Coefficient Case. Submitted *SIAM J. Numer. Anal.*

4. R. E. Bank. *Fortran Implementations of Marching Algorithms*. Harvard University Technical Report TR 17-75, May, 1975.

5. R. E. Bank. *Marching Algorithms for Elliptic Boundary Value Problems.* Thesis, Harvard University, April, 1975.

6. G. Birkhoff and J. A. George. Elimination by nested dissection *Complexity of Sequential and Parallel Numerical Algorithms*, Academic Press, New York, 1973.

7. J. R. Bunch and D. J. Rose. Partitioning, tearing and modification of sparse linear systems. *J. Math. Anal. Appl.*, 48, (1974), p. 574-593.

8. B. L. Buzbee, F. W. Dorr, J. A. George, and G. H. Golub. The direct solution of the discrete Poisson equation on irregular regions. *SIAM, J. Numer. Anal.*, 4, (1971), p. 722-736.

9. B. L. Buzbee, G. H. Golub, and C. W. Nielson. On direct methods for solving Poisson's equation. *SIAM J. Numer. Anal.*, 7, (1970), p. 627-656.

10. F. W. Dorr. The direct solution of the discrete Poisson equation on a rectangle. *SIAM Rev.*, 12, (1970), p. 248-263.

11. G. E. Forsythe and C. B. Moler. *Computer Solution of Linear Algebraic Systems*, Prentice Hall, Englewood Cliffs, New Jersey, 1967.

12. G. E. Forsythe and W. R. Wasow. *Finite Difference Methods for Partial Differential Equations*, Wiley, New York, 1960.

13. L. Fox and I. B. Parker. *Chebyshev Polynomials in Numerical Analysis*, Oxford Univ. Press, 1968.

14. R. W. Hockney. The potential calculation and some applications. *Methods in Computational Physics*, 9, (B. Adler, S. Fernbach, S. Rotenberg, eds.), Academic Press, New York, 1969, p. 131-211.

15. H. B. Keller. *Numerical Methods for 2-Point Boundary Value Problems*, Blaisdell, Waltham, Mass., 1968.

16. C. Lanczos. *Tables of the Chebyshev Polynomials* $S_n(x)$ *and* $C_n(x)$, AMS 9, National Bureau of Standards, 1952.

17. P. N. Swarztrauber. A direct method for the discrete solution of separable elliptic equations. *SIAM J. Numer. Anal.*, 11, (1974), p. 1136-1150.

18. J. H. Wilkinson. *The Algebraic Eigenvalue Problem.* The Clarendon Press, Oxford, 1965.

A more extensive bibliography on the subject of marching algorithms may be found in Bank [5].

A GENERALIZED CONJUGATE GRADIENT METHOD FOR THE NUMERICAL SOLUTION OF ELLIPTIC PARTIAL DIFFERENTIAL EQUATIONS

Paul Concus
Lawrence Berkeley Laboratory
University of California
Berkeley, CA 94720

Gene H. Golub
Computer Science Dept.
Stanford University
Stanford, CA 94305

Dianne P. O'Leary
Dept. of Mathematics
University of Michigan
Ann Arbor, MI 48104

0. INTRODUCTION

In 1952, Hestenes and Stiefel [0] proposed the conjugate gradient method (CG) for solving the system of linear algebraic equations

$$A\underset{\sim}{x} = \underset{\sim}{b} ,$$

where A is an $n \times n$, symmetric, positive-definite matrix. This elegant method has as one of its important properties that in the absence of round-off error the solution is obtained in at most n iteration steps. Furthermore, the entire matrix A need not be stored as an array in memory; at each stage of the iteration it is necessary to compute only the product $A\underset{\sim}{z}$ for a given vector $\underset{\sim}{z}$.

Unfortunately the initial interest and excitement in CG was dissipated, because in practice the numerical properties of the algorithm differed from the theoretical ones; *viz.* even for small systems of equations ($n \leq 100$) the algorithm did not necessarily terminate in n iterations. In addition, for large systems of equations arising from the discretization of two-dimensional elliptic partial differential equations, competing methods such as successive overrelaxation (SOR) required only $O(\sqrt{n})$ iterations to achieve a prescribed accuracy [1]. It is interesting to note that in the proceedings of the Conference on Sparse Matrices and Their Applications held in 1971 [2] there is hardly any mention of the CG method.

In 1970, Reid [3] renewed interest in CG by giving evidence that the method could be used in a highly effective manner as an iterative procedure for solving large sparse systems of linear equations. Since then a number of authors have described the use of CG for solving a variety of problems (cf. [4], [5], [6], [7], [8]). Curiously enough, although CG was generally discarded during the sixties as a useful method for solving linear equations, except in conjunction with other methods [9], there was considerable interest in it for solving nonlinear equations (cf. [10]).

The conjugate gradient method has a number of attractive properties when used as an iterative method:

(i) It does not require an estimation of parameters.

(ii) It takes advantage of the distribution of the eigenvalues of the iteration operator.

(iii) It requires fewer restrictions on the matrix A for optimal behavior than do such methods as SOR.

Our basic view is that CG is most effective when used as an iteration acceleration technique.

In this paper, we derive and show how to apply a generalization of the CG method and illustrate it with numerical examples. Based on our investigations, we feel that the generalized CG method has the potential for widespread application in the numerical solution of boundary value problems for elliptic partial differential equations. Additional experience should further indicate how best to take full advantage of the method's inherent possibilities.

1. DERIVATION OF THE METHOD

Consider the system of equations

$$A\underset{\sim}{x} = \underset{\sim}{b} \; , \tag{1.1}$$

where A is an $n \times n$, symmetric, positive-definite matrix and $\underset{\sim}{b}$ is a given vector. It is frequently desirable to rewrite (1.1) as

$$M\underset{\sim}{x} = N\underset{\sim}{x} + \underset{\sim}{c} \; , \tag{1.2}$$

where M is positive-definite and symmetric and N is symmetric. In § 4 we describe several decompositions of the form (1.2). We are interested in those situations for which it is a much simpler computational task to solve the system

$$M\underset{\sim}{z} = \underset{\sim}{d} \tag{1.3}$$

than it is to solve (1.1).

We consider an iteration of the form

$$\underset{\sim}{x}^{(k+1)} = \underset{\sim}{x}^{(k-1)} + \omega_{k+1}(\alpha_k \underset{\sim}{z}^{(k)} + \underset{\sim}{x}^{(k)} - \underset{\sim}{x}^{(k-1)}) \; , \tag{1.4}$$

where

$$M\underset{\sim}{z}^{(k)} = \underset{\sim}{c} - (M-N)\underset{\sim}{x}^{(k)} \; . \tag{1.5}$$

Many iterative methods can be described by (1.4); e.g. the Chebyshev semi-iterative method and the Richardson second order method (cf. [11]). The generalized CG method is also of this form.

For the Richardson or Chebyshev methods, the optimal parameters (ω_{k+1}, α_k) are given as simple, easy-to-compute functions of the smallest and largest eigenvalues of the iteration matrix $M^{-1}N$ [11]; thus good estimates of these eigenvalues are required for the methods to be efficient. The methods do not take into account the values of any of the interior eigenvalues of the iteration matrix.

The CG method, on the other hand, needs no a priori information on the extremal eigenvalues and does take into account the interior ones, but at a cost of increased computational requirements for evaluating ω_{k+1} and α_k. In § 3, we describe a technique to provide directly from the CG method good estimates for the extreme eigenvalues of the iteration matrix.

From equations (1.4) and (1.5), we obtain the relation

$$M\underset{\sim}{z}^{(k+1)} = M\underset{\sim}{z}^{(k-1)} - \omega_{k+1}(\alpha_k(M-N)\underset{\sim}{z}^{(k)} + M(\underset{\sim}{z}^{(k-1)} - \underset{\sim}{z}^{(k)})). \quad (1.6)$$

For the generalized CG method the parameters $\{\alpha_k, \omega_{k+1}\}$ are computed so that

$$\underset{\sim}{z}^{(p)^T} M\underset{\sim}{z}^{(q)} = 0 \quad (1.7)$$

for $p \neq q$ and $p, q = 0, 1, \ldots, n-1$.

Since M is $n \times n$ positive-definite, (1.7) implies that for some $k \leq n$

$$\underset{\sim}{z}^{(k)} = \underset{\sim}{0}$$

and hence

$$\underset{\sim}{x}^{(k)} = \underset{\sim}{x} \quad . \quad (1.8)$$

That is, the iteration converges in no more than n steps.

We derive the above result by induction. Assume

$$\underset{\sim}{z}^{(p)^T} M\underset{\sim}{z}^{(q)} = 0 \tag{1.9}$$

for $p \neq q$ and $p,q = 0,1,\ldots, k$.

Then if

$$\alpha_k = \underset{\sim}{z}^{(k)^T} M\underset{\sim}{z}^{(k)} / \underset{\sim}{z}^{(k)^T} (M-N)\underset{\sim}{z}^{(k)} , \tag{1.10}$$

there holds

$$\underset{\sim}{z}^{(k)^T} M\underset{\sim}{z}^{(k+1)} = 0 ,$$

and if

$$\omega_{k+1} = \left(1 - \alpha_k \frac{\underset{\sim}{z}^{(k-1)^T} N\underset{\sim}{z}^{(k)}}{\underset{\sim}{z}^{(k-1)^T} M\underset{\sim}{z}^{(k-1)}} \right)^{-1} \tag{1.11}$$

then

$$\underset{\sim}{z}^{(k-1)^T} M\underset{\sim}{z}^{(k+1)} = 0 .$$

We can simplify the above expression for ω_{k+1} as follows. From (1.6) we obtain

$$M\underset{\sim}{z}^{(k)} = M\underset{\sim}{z}^{(k-2)} - \omega_k (\alpha_{k-1} (M-N)\underset{\sim}{z}^{(k-1)} + M(\underset{\sim}{z}^{(k-2)} - \underset{\sim}{z}^{(k-1)})),$$

and then from (1.9)

$$\underset{\sim}{z}^{(k)^T} N\underset{\sim}{z}^{(k-1)} = \underset{\sim}{z}^{(k)^T} M\underset{\sim}{z}^{(k)} / (\omega_k \alpha_{k-1}) .$$

Since

$$\underset{\sim}{z}^{(k-1)^T} N\underset{\sim}{z}^{(k)} = \underset{\sim}{z}^{(k)^T} N\underset{\sim}{z}^{(k-1)} ,$$

it follows

$$\omega_{k+1} = \left(1 - \frac{\alpha_k}{\alpha_{k-1}} \frac{\underset{\sim}{z}^{(k)^T} M\underset{\sim}{z}^{(k)}}{\underset{\sim}{z}^{(k-1)^T} M\underset{\sim}{z}^{(k-1)}} \frac{1}{\omega_k}\right)^{-1} .$$

From (1.6), for $j < k-1$

$$\underset{\sim}{z}^{(j)^T} M\underset{\sim}{z}^{(k+1)} = \alpha_k \omega_{k+1} \underset{\sim}{z}^{(j)^T} N\underset{\sim}{z}^{(k)} .$$

But

$$M\underset{\sim}{z}^{(j+1)} = M\underset{\sim}{z}^{(j-1)} - \omega_{j+1}(\alpha_j (M-N)\underset{\sim}{z}^{(j)} + M(\underset{\sim}{z}^{(j-1)} - \underset{\sim}{z}^{(j)})) ,$$

so that

$$\underset{\sim}{z}^{(k)^T} N\underset{\sim}{z}^{(j)} = 0 .$$

Thus, since $N = N^T$,

$$\underset{\sim}{z}^{(j)^T} M\underset{\sim}{z}^{(k+1)} = 0 \quad \text{for} \quad j < k-1 .$$

Hence by induction we obtain (1.7) and (1.8).

The _generalized CG method_ is summarized as follows.

Algorithm

Let $\underset{\sim}{x}^{(0)}$ be a given vector and arbitarily define $\underset{\sim}{x}^{(-1)}$. For $k = 0, 1, \ldots$

(1) Solve $M\underset{\sim}{z}^{(k)} = \underset{\sim}{c} - (M-N)\underset{\sim}{x}^{(k)}$.

(2) Compute

$$\alpha_k = \frac{\underset{\sim}{z}^{(k)^T} M\underset{\sim}{z}^{(k)}}{\underset{\sim}{z}^{(k)^T} (M-N)\underset{\sim}{z}^{(k)}} ,$$

$$\omega_{k+1} = \left(1 - \frac{\alpha_k}{\alpha_{k-1}} \frac{z^{(k)^T} M z^{(k)}}{z^{(k-1)^T} M z^{(k-1)}} \cdot \frac{1}{\omega_k} \right)^{-1}, \quad (k \geq 1),$$

$$\omega_1 = 1 .$$

(3) Compute

$$x^{(k+1)} = x^{(k-1)} + \omega_{k+1}(\alpha_k z^{(k)} + x^{(k)} - x^{(k-1)}) .$$

Note that the algorithm can be viewed as an acceleration of the underlying first order iteration $(\omega_{k+1} \equiv 1)$, $x^{(k+1)} = x^{(k)} + \alpha_k z^{(k)}$. As with other higher order methods, the storage requirements of the algorithm are greater than those of the underlying first order iteration being accelerated.

The algorithm presented above is given primarily for expository purposes. For actual computation, the following equivalent form can be more efficient in terms of storage [3].

Algorithm (alternative form)

Let $x^{(0)}$ be a given vector and arbitrarily define $p^{(-1)}$. For $k = 0, 1, \ldots$

(1) Solve $Mz^{(k)} = c - (M-N)x^{(k)}$.

(2) Compute

$$b_k = \frac{z^{(k)^T} M z^{(k)}}{z^{(k-1)^T} M z^{(k-1)}}, \quad k \geq 1 ,$$

$$b_0 = 0,$$

$$p^{(k)} = z^{(k)} + b_k p^{(k-1)} .$$

(3) Compute

$$a_k = \frac{\underset{\sim}{z}^{(k)^T} M\underset{\sim}{z}^{(k)}}{\underset{\sim}{p}^{(k)^T}(M-N)\underset{\sim}{p}^{(k)}} ,$$

$$\underset{\sim}{x}^{(k+1)} = \underset{\sim}{x}^{(k)} + a_k \underset{\sim}{p}^{(k)} .$$

In the computation of the numerators of a_k and b_k one need not recompute $M\underset{\sim}{z}^{(k)}$, since it can be saved from step (1). Also, instead of computing the right hand side of step (1) explicitly at each iteration, it is often advantageous to compute it recursively from

$$[\underset{\sim}{c} - (M-N)\underset{\sim}{x}^{(k+1)}] = [\underset{\sim}{c} - (M-N)\underset{\sim}{x}^{(k)}] - a_k (M-N)\underset{\sim}{p}^{(k)} , \qquad (1.12)$$

which equation is obtained from step (3). The quantity $(M-N)\underset{\sim}{p}^{(k)}$ appearing in (1.12) may be saved from the computation of a_k. Similar remarks hold for the algorithm in its first form as well. There is evidence that the use of (1.12) is no less accurate than use of the explicit computation (see [18], [3] for particular examples).

The calculated vectors $\{\underset{\sim}{z}^{(k)}\}_{k=0}^{n}$ will not generally be M-orthogonal in practice because of rounding errors. One might consider forcing the newly calculated vectors to be M-orthogonal by a procedure such as Gram-Schmidt. However, this would require the storage of all the previously obtained vectors.

Our basic approach is to permit the gradual loss of orthogonality and with it the finite termination property of CG. We consider primarily the iterative aspects of the algorithm. In fact, for solving large sparse systems arising from the discretization of elliptic partial differential equations, the application of principal interest for us and for which the generalized CG method seems particularly effective, convergence to desired accuracy often occurs within a number of iterations small compared with n.

2. OPTIMALITY PROPERTIES

From (1.6), we obtain

$$\underset{\sim}{z}^{(k+1)} = \underset{\sim}{z}^{(k-1)} - \omega_{k+1}(\alpha_k(I-M^{-1}N)\underset{\sim}{z}^{(k)} + \underset{\sim}{z}^{(k-1)} - \underset{\sim}{z}^{(k)}). \quad (2.1)$$

Define

$$K = I - M^{-1}N . \quad (2.2)$$

We have $\underset{\sim}{z}^{(1)} = (I - \alpha_0 K)\underset{\sim}{z}^{(0)}$, and there follows by induction that

$$\underset{\sim}{z}^{(\ell+1)} = [I - KP_\ell(K)]\underset{\sim}{z}^{(0)} \quad (2.3)$$

where

$$P_\ell(K) = \sum_{j=0}^{\ell} \beta_j^{(\ell)} K^j . \quad (2.4)$$

We denote

$$p_\ell(\lambda) = \sum_{j=0}^{\ell} \beta_j^{(\ell)} \lambda^j \quad (2.5)$$

and from (2.1) we have for $k = 2, 3, \ldots , \ell$

$$p_k(\lambda) = \omega_{k+1}(1-\alpha_k\lambda)p_{k-1}(\lambda) - (\omega_{k+1}-1)p_{k-2}(\lambda) + \alpha_k\omega_{k+1} ,$$

and

$$p_0(\lambda) = \alpha_0, \qquad p_1(\lambda) = \omega_2(\alpha_0 + \alpha_1 - \alpha_0\alpha_1\lambda) .$$

The coefficients $\{\beta_j^{(\ell)}\}_{j=0}^{\ell}$ can be generated directly. From (2.3) and the relation $\underset{\sim}{z}^{(\ell+1)} = \underset{\sim}{z}^{(0)} + K(\underset{\sim}{x}^{(\ell+1)} - \underset{\sim}{x}^{(0)})$, there follows

$$\underset{\sim}{x}^{(\ell+1)} = \underset{\sim}{x}^{(0)} + P_\ell(K)\underset{\sim}{z}^{(0)} .$$

Then if

$$Z = [\underset{\sim}{z}^{(0)}, K\underset{\sim}{z}^{(0)}, \ldots , K^\ell \underset{\sim}{z}^{(0)}] , \tag{2.6}$$

$$\underset{\sim}{x}^{(\ell+1)} = \underset{\sim}{x}^{(0)} + Z\underset{\sim}{\beta}^{(\ell)} . \tag{2.7}$$

Consider the <u>weighted</u> <u>error</u> <u>function</u>:

$$E(\underset{\sim}{x}^{(\ell+1)}) = \frac{1}{2} (\underset{\sim}{x} - \underset{\sim}{x}^{(\ell+1)})^T (M-N)(\underset{\sim}{x} - \underset{\sim}{x}^{(\ell+1)}) . \tag{2.8}$$

Assuming that $(M-N)$ is nonsingular, we obtain, using

$$\underset{\sim}{z}^{(0)} = K(\underset{\sim}{x} - \underset{\sim}{x}^{(0)}) ,$$

the relations

$$E(\underset{\sim}{x}^{(\ell+1)}) = \frac{1}{2} \underset{\sim}{z}^{(0)^T} (I-KP_\ell(K))^T M(M-N)^{-1} M(I-KP_\ell(K))\underset{\sim}{z}^{(0)}$$

$$= \frac{1}{2} \underset{\sim}{e}^{(0)^T} (I-KP_\ell(K))^T (M-N)(I-KP_\ell(K))\underset{\sim}{e}^{(0)}, \tag{2.9}$$

where

$$\underset{\sim}{e}^{(0)} = \underset{\sim}{x} - \underset{\sim}{x}^{(0)} .$$

Equivalently, we can use (2.7) and re-write (2.8) as

$$E(\underset{\sim}{x}^{(\ell+1)}) = \frac{1}{2} (K^{-1}\underset{\sim}{z}^{(0)} - Z\underset{\sim}{\beta}^{(\ell)})^T (M-N)(K^{-1}\underset{\sim}{z}^{(0)} - Z\underset{\sim}{\beta}^{(\ell)}) . \tag{2.10}$$

The quantity $E(\underset{\sim}{x}^{(\ell+1)})$ is minimized when we choose $\underset{\sim}{\beta}^{(\ell)}$ so that

$$G\underset{\sim}{\beta}^{(\ell)} = \underset{\sim}{h} ,$$

where

$$G = Z^T(M-N)Z, \quad \underset{\sim}{h} = Z^T M\underset{\sim}{z}^{(0)} .$$

Let

$$\kappa = \lambda_{max}(K)/\lambda_{min}(K) .$$

Then using arguments similar to those given in [12], the following can be shown:

(A)
$$\frac{E(\underset{\sim}{x}^{(\ell+1)})}{E(\underset{\sim}{x}^{(0)})} \leq 4\left(\frac{\sqrt{\kappa}-1}{\sqrt{\kappa}+1}\right)^{2(\ell+1)} . \qquad (2.11)$$

(B) The generalized CG method is optimal in the class of all algorithms for which

$$\underset{\sim}{x}^{(\ell+1)} = \underset{\sim}{x}^{(0)} + P_\ell(K)\underset{\sim}{z}^{(0)}.$$

That is, the approximation $\underset{\sim}{x}^{(\ell+1)}$ generated by the generalized CG method satisfies

$$E(\underset{\sim}{x}^{(\ell+1)}) = \min_{P_\ell} \frac{1}{2}\, \underset{\sim}{e}^{(0)^T}(I-KP_\ell(K))^T(M-N)(I-KP_\ell(K))\underset{\sim}{e}^{(0)} ,$$

where the minimum is taken with respect to all polynomials P_ℓ of degree ℓ.

Recall that we have assumed that M and $(M-N)$ are positive definite and symmetric. Thus the eigenvalues of $K = (I-M^{-1}N)$ are all real and K is similar to a diagonal matrix. Hence, if K has only $p < n$ distinct eigenvalues, there exists a matrix polynomial $Q_p(K)$ so that

$$Q_p(K) = 0 .$$

In this case, $E(\underset{\sim}{x}^{(p)}) = 0$ and hence

$$\underset{\sim}{x}^{(p)} = \underset{\sim}{x} ,$$

so that the iteration converges in only p steps. The same result also holds if K has a larger number of distinct eigenvalues but $\underset{\sim}{e}^{(0)}$ lies in a subspace generated by the eigenvectors associated with only p of these eigenvalues.

We remark also that Statement (B) implies CG is optimal for the particular eigenvector mix of the initial error $\underset{\sim}{e}^{(0)}$, taking into account interior as well as extremal eigenvalues. As will be discussed in the next section, the extremal eigenvalues are approximated especially well as CG proceeds, the iteration then behaving as if the corresponding vectors are not present. Thus the error estimate (2.11), which is based on the extremal eigenvalues, tends to be pessimistic asymptotically. One often observes, in practice (see § 5), a superlinear rate of convergence for the CG method.

3. EIGENVALUE COMPUTATIONS

The CG method can be used in a very effective manner for computing the extreme eigenvalues of the matrix $K = I - M^{-1}N$. We write (see (2.1))

$$\underset{\sim}{z}^{(k+1)} = \underset{\sim}{z}^{(k-1)} - \omega_{k+1}(\alpha_k K\underset{\sim}{z}^{(k)} + \underset{\sim}{z}^{(k-1)} - \underset{\sim}{z}^{(k)}), \tag{3.1}$$

as

$$K\underset{\sim}{z}^{(k)} = c_{k-1}\underset{\sim}{z}^{(k-1)} + a_k\underset{\sim}{z}^{(k)} + b_{k+1}\underset{\sim}{z}^{(k+1)},$$

or

$$K[\underset{\sim}{z}^{(0)},\underset{\sim}{z}^{(1)},\ldots,\underset{\sim}{z}^{(n-1)}]$$

$$= [\underset{\sim}{z}^{(0)},\underset{\sim}{z}^{(1)},\ldots,\underset{\sim}{z}^{(n-1)}] \begin{bmatrix} a_0 & c_0 & & & \bigcirc \\ b_1 & a_1 & c_1 & & \\ & \ddots & \ddots & \ddots & \\ & & & & c_{n-2} \\ \bigcirc & & & b_{n-1} & a_{n-1} \end{bmatrix}$$

thus defining a_k, b_k, and c_k. In matrix notation, the above equation can be written as

$$KZ = ZJ \ . \tag{3.2}$$

Assuming that the columns of Z are linearly independent, there follows from (3.2) that

$$K = ZJZ^{-1} \ ,$$

hence the eigenvalues of K are equal to those of J. As pointed out in § 2, if K has repeated eigenvalues or if the vector $\underset{\sim}{z}^{(0)}$ is deficient in the direction of some eigenvectors of K, iteration (3.1) will terminate in $k < n$ steps.

The process described by (3.1) is essentially the Lanczos algorithm [13]. It has been shown by Kaniel [14] and by Paige [15] that good estimates of the extreme eigenvalues of K often can be obtained from the truncated matrix

$$J_k = \begin{bmatrix} a_0 & c_0 & & & & \bigcirc \\ b_1 & a_1 & c_1 & & & \\ & \cdot & \cdot & \cdot & & \\ & & \cdot & \cdot & \cdot & \\ & & & \cdot & \cdot & c_{k-2} \\ \bigcirc & & & & b_{k-1} & a_{k-1} \end{bmatrix}$$

where k is considerably less than n. This result holds even in the presence of round-off error [16].

It was pointed out in § 1 that the equation describing the CG method is of the same form as that describing the Chebyshev semi-iterative method and Richardson second order method, but that a knowledge of the extreme eigenvalues of K

is required for obtaining parameters for the latter two methods. Thus one could construct a polyalgorithm in which the CG method is used initially to obtain good approximations to the solution and to the extreme eigenvalues of K, after which the Chebyshev semi-iterative method (say) is used, thereby avoiding the additional work of repeatedly calculating CG parameters. This technique has been used in an effective manner by O'Leary [17].

4. CHOICE OF M

For the splitting $M = I$, $N = I-A$ one obtains the basic, unmodified CG algorithm, for which

$$\underset{\sim}{z}^{(k)} = \underset{\sim}{r}^{(k)} = \underset{\sim}{b} - A\underset{\sim}{x}^{(k)}$$

is simply the residual at the kth step. Since the rate of convergence of the generalized CG method, as given by the estimate (2.11), decreases with increasing

$$\kappa = \lambda_{max}(K)/\lambda_{min}(K),$$

it is desirable to choose a splitting for which κ is as small as possible. If $A = L + D + U$, where D consists of the diagonal elements of A and $L(U)$ is a strictly lower (upper) triangular matrix, then it is reasonable to consider the choice

$$M = D, \qquad N = -(L + U) .$$

This M, which is equivalent to a rescaling of the problem, is one for which (1.3) can be solved very simply for $\underset{\sim}{z}$. It has been shown by Forsythe and Straus [19] that if A is two-cyclic then among all diagonal matrices this choice of M will minimize κ.

In many cases, the matrix A can be written in the form

$$A = \left(\begin{array}{c|c} M_1 & F \\ \hline F^T & M_2 \end{array} \right), \tag{4.3}$$

where the systems

$$M_1 \underset{\sim}{z}_1 = \underset{\sim}{d}_1 \qquad \text{and} \qquad M_2 \underset{\sim}{z}_2 = \underset{\sim}{d}_2$$

are easy to solve, and for such matrices, it is convenient to choose

$$M = \left(\begin{array}{c|c} M_1 & 0 \\ \hline 0 & M_2 \end{array} \right) \text{and} \qquad N = \left(\begin{array}{c|c} 0 & -F \\ \hline -F^T & 0 \end{array} \right).$$

Using (4.3), we can write the system (1.1) in the form

$$M_1 \underset{\sim}{x}_1 + F \underset{\sim}{x}_2 = \underset{\sim}{b}_1 \tag{4.4a}$$

$$F^T \underset{\sim}{x}_1 + M_2 \underset{\sim}{x}_2 = \underset{\sim}{b}_2 . \tag{4.4b}$$

Let the initial approximation for $\underset{\sim}{x}_1$ be $\underset{\sim}{x}_1^{(0)}$, and obtain $\underset{\sim}{x}_2^{(0)}$ as the solution to (4.4b) so that

$$M_2 \underset{\sim}{x}_2^{(0)} = \underset{\sim}{b}_2 - F^T \underset{\sim}{x}_1^{(0)} .$$

This implies that

$$\underset{\sim}{z}_2^{(0)} = \underset{\sim}{0} ,$$

and hence by (1.10)

$$\alpha_0 = 1 ,$$

and thus

$$\underset{\sim}{x}^{(1)} = \begin{pmatrix} \underset{\sim}{x}^{(0)} + \underset{\sim}{z}_1^{(0)} \\ \underset{\sim}{x}_2^{(0)} \end{pmatrix}.$$

A short calculation shows that $\underset{\sim}{z}_1^{(1)} = \underset{\sim}{0}$ and hence $\alpha_1 = 1$. Using (1.6), a simple inductive argument then yields that for $j = 0, 1, 2, \ldots$

$$\alpha_j \equiv 1, \quad \underset{\sim}{z}_1^{(2j+1)} = \underset{\sim}{0}, \quad \underset{\sim}{z}_2^{(2j)} = \underset{\sim}{0}. \qquad (4.5)$$

This result was first observed by Reid [8] for the case in which M_1 and M_2 are diagonal, i.e., in which the matrix A has "Property A" and is suitably ordered. Other cases for elliptic boundary value problems in which matrices of the form (4.3) arise will be discussed in § 5. For these cases convergence can be rapid because K has only a few distinct eigenvalues, even though κ is not especially small.

Various other splittings of the matrix A can occur quite naturally in the solution of elliptic partial differential equations. For example, if one wishes to solve

$$-\Delta u + \sigma(x,y)u = f \qquad (x,y) \in R$$

$$u = g \qquad (x,y) \in \partial R ,$$

where R is a rectangular region, it is convenient to choose M as the finite difference approximation to a separable operator, such as the Helmholtz operator $-\Delta + C$, for which fast direct methods can be used [23]. A numerical example for this case is discussed in § 5. If one wishes to solve a separable equation, but on a nonrectangular region S, then by extending the problem to one on a rectangle R in which S is embedded, M can be chosen as the discrete approximation to the separable operator on R, for which fast direct methods can be used. Such a technique provides an alternative to the related capacitance matrix method [25] for handling such problems. Forms of this method utilizing CG, but in a different manner than here, are described in [26] and [27].

Several authors [4], [20], [21] have used CG in combination with symmetric successive overrelaxation (SSOR). For this method the solution of the equation $M\underset{\sim}{z}^{(k)} = \underset{\sim}{c} - (M-N)\underset{\sim}{x}^{(k)}$ reduces to the solution of

$$(D+\omega L)\, D^{-1}(D+\omega U)\underset{\sim}{z}^{(k)} = \omega(2-\omega)\underset{\sim}{r}^{(k)}$$

where D, L, and U are as described previously in this section (although D may be block diagonal), $\underset{\sim}{r}^{(k)} = \underset{\sim}{b} - A\underset{\sim}{x}^{(k)}$, and ω is a parameter in the open interval (0,2). SSOR is particularly effective in combination with CG because of the distribution of the eigenvalues of K (cf. [22]).

Meijerink and van der Vorst [7] have proposed that the following factorization of A be used:

$$A = FF^T + E,$$

so that

$$M = FF^T, \qquad N = -E .$$

The matrix F is chosen with a sparsity pattern resembling that of A. This splitting appears to yield a matrix K with eigenvalues that also are favorably distributed for CG. A block form of this technique recently developed by Underwood [24] achieves a more accurate approximate factorization of A with less computer storage and about the same number of arithmetic operations per iteration.

Generally, in addition to the requirement that (1.5) be "easy" to solve, M should have the following features if the generalized CG algorithm is to be computationally efficient. For rapid convergence one seeks a splitting so that

(i) $M^{-1}N$ has small or nearly equal eigenvalues

or (ii) $M^{-1}N$ has small rank.

Often a choice for M satisfying these restrictions comes about naturally from the inherent features of a given problem.

5. NUMERICAL EXAMPLES

For the first example, we consider the test problem discussed in [23]

$$-\text{div}(a(x,y)\nabla u) = f \qquad (x,y) \in R$$

$$u = g \qquad (x,y) \in \partial R ,$$

where $a(x,y) = [1 + \frac{1}{2}(x^4 + y^4)]^2$ and R is the unit square $0 < x,y < 1$. After a transformation the problem becomes

$$\begin{aligned} -\Delta w + \sigma(x,y)w &= a^{-1/2}f \qquad (x,y) \in R \\ w &= a^{1/2}g \qquad (x,y) \in \partial R, \end{aligned} \tag{5.1}$$

where $\sigma(x,y) = 6(x^2 + y^2)/a^{1/2}$. As in [23] we discretize (5.1) on a uniform mesh of width h, using for Δ the standard five-point approximation Δ_h, and we choose the splitting

$$M = A + N = -\Delta_h + CI$$

with $C = 0 = \sigma_{min}$ or $C = 3 = \frac{1}{2}(\sigma_{max} + \sigma_{min})$.

In [23] Chebyshev acceleration was used, which requires an estimate of the ratio of the extremal eigenvalues of the iteration matrix. Here we use the modified CG algorithm of

§ 1. For an initial guess $\underset{\sim}{W}^{(0)} \equiv \underset{\sim}{0}$ and choice of f and g corresponding to the solution $w = 2[(x-1/2)^2 + (y-1/2)^2]$, the results are given in Table 1 for $h = 1/64$. The results obtained for $h = 1/32$ were essentially identical, as the iteration is basically independent of h for this problem (see [23]).

Note that the Chebyshev method is sensitive both to the value of C and to the accuracy of the eigenvalues from which the parameters are calculated. The parameters used for the middle column were based on Gerschgorin estimates from the Rayleigh quotient, which gave a ratio of largest to smallest eigenvalue about three times too large. The CG method appears to be less sensitive to the value of C. After several iterations CG begins to converge more rapidly than does the optimal Chebyshev method, which behavior is typical of the CG superlinear convergence property discussed in § 2. This example is one for which rapid convergence results because the eigenvalues of $M^{-1}N$ are small.

	Chebyshev (from [23])			CG	
	C = 0	C = 3	C = 3		
iteration	exact eigenvalues	approximate eigenvalues	exact eigenvalues	C = 0	C = 3
1			1.6(-2)	4.5(-2)	1.6(-2)
2			7.4(-4)	2.6(-3)	6.7(-4)
3			1.1(-5)	3.0(-5)	1.0(-5)
4			2.7(-7)	5.7(-7)	1.1(-7)
5	2.4(-6)	1.1(-6)	4.3(-9)	5.1(-9)	8.2(-10)
6			1.2(-10)	4.4(-11)	5.7(-12)

TABLE 1
Maximum error vs. iteration number for first example

We give as the second example

$$-\Delta u = f \qquad (x,y) \in T$$

$$u = g \qquad (x,y) \in \partial T$$

where T is the domain shown in Fig. 1. For a uniform square mesh of width h, and $0 < \ell < (2h)^{-1}$ a whole number, so that all boundary segments are mesh lines, the coefficient matrix A for the standard five-point discretization and natural ordering has the form (4.3).

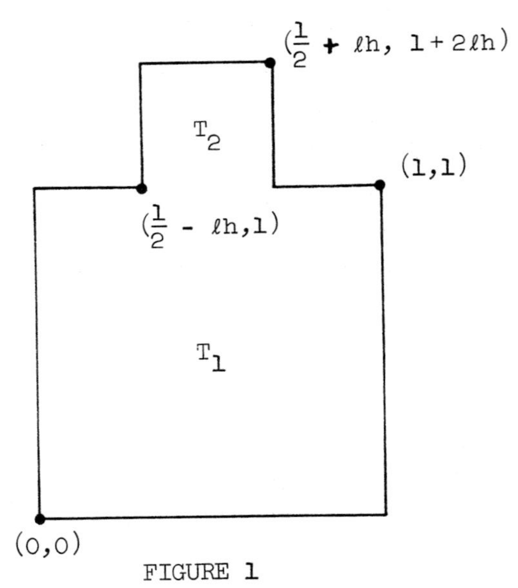

FIGURE 1

T-shaped domain

M_1 and M_2 correspond to the mesh points in each of the two squares, T_1 and T_2, and F to the coupling between them. F has non-zero entries in only $p = 2\ell - 1$ of its rows.

According to the discussion following (4.3) we choose

$$M = \begin{pmatrix} M_1 & 0 \\ 0 & M_2 \end{pmatrix}$$

and for initial approximation

$$\underset{\sim}{U}^{(0)} = \begin{pmatrix} \underset{\sim}{U}_1^{(0)} \\ M_2^{-1}(\underset{\sim}{b}_2 - F^T \underset{\sim}{U}_1^{(0)}) \end{pmatrix} .$$

Then for the generalized CG algorithm, there holds $\alpha_k \equiv 1$ and that $\underset{\sim}{z}_1$ and $\underset{\sim}{z}_2$ are alternately zero, thereby reducing computational and storage requirements. We use a fast direct Poisson solver for the systems involving M_1 and M_2.

The results for $\underset{\sim}{U}_1^{(0)}$ uniformly distributed random numbers in $(0,2)$ and $f(x,y)$ and $g(x,y)$ such that $u = x^2 + y^2$ is the solution are given in Table 2. Here the average error per point, the two norm of the error divided by the square root of the number of interior mesh points, is given for each of the test problems.

For this example, the eigenvalues of $M^{-1}N$ are not especially small in magnitude, however since $M^{-1}N$ has rank of only $2p$, convergence is obtained in only a moderate number of iterations. For Case I and Case II the last row represents full convergence to machine accuracy subject to rounding errors, as would be expected since $2p = 14$ for these cases.

	Case I	Case II	Case III
h	1/32	1/64	1/64
ℓ	4	4	8
p	7	7	15
iteration	ave. error/pt	ave. error/pt	ave. error/pt
1	8.58(-2)	3.70(-2)	1.08(-1)
2	7.05(-2)	3.13(-2)	9.82(-2)
3	1.30(-2)	6.66(-3)	4.94(-2)
4	3.35(-3)	2.53(-3)	1.80(-2)
5	2.71(-4)	6.03(-4)	4.28(-3)
10	2.65(-7)	5.13(-8)	7.35(-5)
15	1.14(-13)	5.60(-13)	4.71(-8)

TABLE 2

Average error per point vs. iteration number

We wish to thank Myron Stein of the Los Alamos Scientific Laboratory for his careful computer programming of the second test problem. This work was supported in part by the Energy Research and Development Administration, by the Hertz Foundation, and by the National Science Foundation.

REFERENCES

[0] M. Hestenes and E. Stiefel, "Methods of conjugate gradients for solving linear systems," J. Research NBS 49 (1952), 409-436.

[1] D. Young, Iterative Solution of Large Linear Systems, Academic Press, New York, 1972.

[2] D. J. Rose and R. A. Willoughby (ed.), Sparse Matrices and Their Applications, Plenum Press; New York-London, 1972.

[3] J. K. Reid, "On the method of conjugate gradients for the solution of large sparse systems of linear equations," Proc. Conference on "Large sparse sets of linear equations," Academic Press, New York, 1971.

[4] O. Axelsson, "On preconditioning and convergence acceleration in sparse matrix problems," Report CERN 74-10 of the CERN European Organization for Nuclear Research, Data Handling Division, Laboratory I, 8 May 1974.

[5] R. Bartels and J.W. Daniel, "A conjugate gradient approach to nonlinear elliptic boundary value problems in irregular regions," Proc. Conf. on "Numerical solution of differential equations," Springer-Verlag, Berlin, 1974.

[6] R. Chandra, S.C. Eisenstat, and M.H. Schultz, "Conjugate gradient methods for partial differential equations," in Advances in Computer Methods for Partial Differential Equations, R.Vichnevtsky (ed), Publ. A.I. C.A.-1975.

[7] J. A. Meijerink and H.A. van der Vorst, "Iterative solution of linear systems arising from discrete approximations to partial differential equations," Academisch Computer Centrum, Utrecht, The Netherlands, 1974.

[8] J. Reid, "The use of conjugate gradients for systems of linear equations possessing 'Property A'," SIAM J. Numer. Anal. 9 (1972), 325-332.

[9] E.L. Wachspress, "Extended applications of alternating direction implicit iteration model problem theory," SIAM J. 11 (1963), 994-1016.

[10] R. Fletcher and C.M. Reeves, "Function minimization by conjugate gradients," Comput. J., 7 (1964), 145-154.

[11] G. H. Golub and R.S. Varga, "Chebyshev semi-iterative methods, successive over-relaxation iterative methods, and second order Richardson iterative methods," Numer. Math. 3 (1961), 147-68.

[12] D.K. Faddeev and V.N. Faddeeva, Computational Methods of Linear Algebra, W.H. Freeman and Co., San Francisco, and London, 1963.

[13] C. Lanczos, "An iteration method for the solution of the eigenvalue problem of linear differential and integral operators," J. Research NBS 45 (1950), 255-282.

[14] S. Kaniel, "Estimates for some computational techniques in linear algebra," Math. Comp. 20 (1966), 369-378.

[15] C.C. Paige, "The computation of eigenvalues and eigenvectors of very large sparse matrices," Ph.D. Thesis, London Univ., Institute of Computer Science, 1971.

[16] C.C. Paige, "Computational variants of the Lanczos method for the eigenproblem," J. Inst. Math. Appl. 10 (1972), 373-381.

[17] D. O'Leary, "Hybrid conjugate gradient algorithms," Ph.D. Thesis, Computer Science Dept., Stanford Univ., 1975.

[18] M. Engeli, T. Ginsburg, H. Rutishauser and E. Stiefel, Refined Iterative Methods for Computation of the Solution and the Eigenvalues of Self-Adjoint Boundary Value Problems, Birkhäuser Verlag, Basel/Stuttgart, 1959.

[19] G. Forsythe and E.G. Straus, "On best conditioned matrices," Proc. Amer. Math. Soc. 6 (1955), 340-345.

[20] D.M. Young, L. Hayes, and E. Schleicher, "The use of the accelerated SSOR method to solve large linear systems," Abstract, 1975 SIAM Fall Meeting, San Francisco.

[21] L.W. Ehrlich, "On some experience using matrix splitting and conjugate gradient," Abstract, 1975 SIAM Fall Meeting, San Francisco.

[22] L.W. Ehrlich, "The block symmetric successive overrelaxation method," J. SIAM 12 (1964), 807-826.

[23] P. Concus and G.H. Golub, "Use of fast direct methods for the efficient numerical solution of nonseparable elliptic equations," SIAM J. Numer. Anal. 10 (1973), 1103-1120.

[24] R.R. Underwood, "An approximate factorization procedure based on the block Cholesky factorization and its use with the conjugate gradient method," Tech. Rept., General Electric Nuclear Energy Division, San Jose, CA (to appear).

[25] B.L. Buzbee, F.W. Dorr, J.A. George, and G.H. Golub, "The direct solution of the discrete Poisson equation in irregular regions," SIAM J. Num. Anal. 8 (1971), 722-736.

[26] J.A. George, "The use of direct methods for the solution of the discrete Poisson equation on non-rectangular regions," Report CS-70-159, Computer Science Dept., Stanford Univ., Stanford, CA. (1970).

[27] W. Proskurowski and O. Widlund, "On the numerical solution of Laplace's and Helmholtz's equations by the capacitance matrix method," Tech. Rept., Courant Institute, NYU (to appear).

Preconditioned Conjugate Gradient Iteration Applied to Galerkin Methods for a Mildly Nonlinear Dirichlet Problem

Jim Douglas, Jr., and Todd Dupont
University of Chicago

Abstract. A conjugate gradient iteration procedure based on the use of a "fast Poisson solver" as a preconditioner is applied to solve the nonlinear algebraic equations arising from a Galerkin method for approximating the solution of a mildly nonlinear Dirichlet problem.

Introduction. The primary concern of this article is the efficient solution of systems of nonlinear algebraic equations arising from the approximate solution of mildly nonlinear Dirichlet problems by Galerkin methods. The underlying iterative technique will be a modified successive substitution procedure combined with a conjugate gradient inner iteration preconditioned by the use of a "fast Poisson solver" associated with some fixed elliptic operator. Inherent in such a procedure is a setup calculation, such as an LU-factorization of a matrix associated with the Laplace operator, that will usually require more calculation than the remaining work to produce the solution of a single nonlinear algebraic system of the type to be considered; thus, the results presented here are most profitably applied to situations in which many problems of the same form, but with different coefficients in the differential equation or different data, are solved on the same domain (which does not need to be a rectangle).

Let Ω be a bounded domain in the plane and let $a = a(x)$ be a bounded, measurable function on Ω such that

$$(1.1) \qquad 0 < a_0 \leq a(x) \leq a_1 < \infty , \quad x \in \Omega.$$

Let $f = f(v) = f(x, v)$ be bounded and measurable on $\overline{\Omega} \times \mathbb{R}$ and satisfy

$$(1.2) \qquad \begin{aligned} &|f(x, v)| \leq K , \quad x \in \overline{\Omega} , \; v \in \mathbb{R} \\ &0 \leq \frac{\partial f}{\partial v}(x, v) \leq M , \; x \in \overline{\Omega} , \; v \in \mathbb{R}; \end{aligned}$$

while (1.2) may seem at first glance to be rather restrictive, it corresponds from a practical point of view to the requirements that f be smooth in a neighborhood of the solution of the differential problem (1.3) and that the Galerkin method be selected in such a way that uniform convergence (at any rate whatsoever) of the approximate solution u_h of (1.8) to the exact solution be assured as the discretization refines to zero.

The Dirichlet problem whose solution is to be approximated is given by

$$(1.3) \qquad \begin{aligned} -\nabla \cdot (a \nabla u) + f(u) &= 0 , \quad x \in \Omega, \\ u &= g , \; x \in \partial\Omega . \end{aligned}$$

Existence and uniqueness of a solution of (1.3) follows easily from the theory of monotone operators [12].

Some notation is needed to introduce the Galerkin method. It will be kept as near consistent with that of [7] as feasible. Let $W_2^k(\Omega)$ denote the Sobolev space of functions having distributional derivatives in Ω through order k lying in $L^2(\Omega)$. Let $\partial\Omega$ be Lipschitzian, and let

$$\|v\| = \|v\|_{L^2(\Omega)}, \quad \|v\|_k = \|v\|_{W_2^k(\Omega)} = [\sum_{|\alpha| \leq k} \|D^\alpha v\|^2]^{1/2},$$

$$|v| = \|v\|_{L^2(\partial\Omega)}, \quad (p, q) = \int_\Omega pq\,dx, \quad \langle p,q \rangle = \int_{\partial\Omega} pq\,ds .$$

Set

$\mathcal{H} = \{v \in W_2^1(\Omega)$: i) for almost every $x \in \partial\Omega$, there exists an open ball $B(x)$ about x such that $v \in W_2^2(\Omega \cap B(x))$, ii) the outward normal derivative $\partial v/\partial n \in L^2(\partial\Omega)\}$.

For $h \in (0,1)$, define a norm on $\mathcal{H}$ by

$$|||\varphi|||^2 = |||\varphi|||_h^2 = \|\varphi\|_1^2 + h^{-1}|\varphi|^2 + h\left|\frac{\partial\varphi}{\partial n}\right|^2 .$$

Consider a family of finite-dimensional subspaces $\mathcal{M}_h$, $0 < h < 1$, of $\mathcal{H}$ satisfying the following hypotheses:

(1.4.i) There exists a positive integer r and a constant c_1 such that, if $2 \leq s \leq r+1$ and $v \in W_2^s(\Omega)$,

$$\inf_{\chi \in \mathcal{M}_h} |||v - \chi|||_h \leq c_1 h^{s-1} \|v\|_s .$$

(1.4.ii) For all $\chi \in \mathcal{M}_h$

$$\left|\frac{\partial\chi}{\partial n}\right| \leq c_1 h^{-1/2} \|\chi\|_1 .$$

Next, introduce Nitsche's symmetric, bilinear form [13] associated with the operator $-\nabla \cdot (a \nabla u)$:

$$B(\varphi,\psi) = B_a(\varphi,\psi) = (a\nabla\varphi,\nabla\psi) - \langle a\frac{\partial\varphi}{\partial n},\psi\rangle - \langle\varphi, a\frac{\partial\psi}{\partial n}\rangle \tag{1.5}$$
$$+ \gamma h^{-1}\langle\varphi,\psi\rangle,$$

where γ is a positive constant. Nitsche [13] has shown that γ can be chosen so that

$$B(v,v) \ge \rho\,|||v|||_h^2, \quad v \in \mathcal{M}_h, \quad 0 < h < 1,$$
$$|B(\varphi,\psi)| \le C\,|||\varphi|||_h \cdot |||\psi|||_h, \quad \varphi,\psi \in \mathcal{H}, \tag{1.6}$$

and it will be assumed that γ has been so chosen. It is easily seen from integration by parts that the solution u of (1.3) satisfies the relation

$$B(u,v) + (f(u),v) + \langle g, a\frac{\partial v}{\partial n} - \gamma h^{-1}v\rangle = 0, \quad v \in \mathcal{H}. \tag{1.7}$$

The Galerkin method employed here will be based on (1.7); i.e., seek $u_h \in \mathcal{M}_h$ such that

$$B(u_h,v) + (f(u_h),v) + \langle g, a\frac{\partial v}{\partial n} - \gamma h^{-1}v\rangle = 0, \quad v \in \mathcal{M}_h. \tag{1.8}$$

Again, standard monotone operator theory implies the existence and uniqueness of a solution of (1.8). A standard argument shows that

$$\|u-u_h\|_1 \le |||u-u_h|||_h \le C\|u\|_s h^{s-1}, \quad 2 \le s \le r+1. \tag{1.9}$$

In addition, if $\partial\Omega$, $a(x)$, and $f(x,u)$ are sufficiently smooth, the usual duality argument using the auxiliary problem

$$-\nabla\cdot(a\nabla\varphi) + F(x)\varphi = u - u_h, \quad x \in \Omega,$$
$$\varphi = 0, \quad x \in \partial\Omega,$$

where

$$F(x) = \begin{cases} \dfrac{f(x,u) - f(x,u_h)}{u - u_h} , & u(x) \neq u_h(x) , \\ \dfrac{\partial f}{\partial u}(x,u) , & u(x) = u_h(x) , \end{cases}$$

shows that

$$\|u - u_h\| \leq C \|u\|_s h^s , \quad 2 \leq s \leq r+1 . \tag{1.10}$$

Consequently, the solution u_h provides an optimal order approximation to u in the usual norms, and the practicality of the method reduces to the development of an efficient method of evaluating u_h.

The next section will treat a modified successive substitution iteration [5]. Then a two-level iteration using a conveniently chosen preconditioned conjugate gradient inner iteration will be presented. Afterwards, counts will be given for the arithmetic operations needed in these iterative methods, and finally some comparisons to other methods will be given.

2. A modified successive substitution method. Let $\lambda > 0$ and let $v^0 \in \mathcal{M}_h$. Define v^m, $m \geq 1$, recursively as follows:

$$\begin{aligned} & B(v^m - v^{m-1}, w) + \lambda(v^m - v^{m-1}, w) \\ & + \{B(v^{m-1}, w) + (f(v^{m-1}), w) + \langle g, a\frac{\partial w}{\partial n} - \gamma h^{-1} w\rangle\} = 0 , \\ & \qquad w \in \mathcal{M}_h . \end{aligned} \tag{2.1}$$

It is an obvious consequence of (1.6) and the Lax-Milgram lemma that (2.1) has a unique solution; moreover, each step of (2.1) can consist of applying a fast Poisson solver associated with the Nitsche method for the space $\mathcal{M}_h$ and the differential operator $-\nabla\cdot(a\nabla u)+\lambda u$ on Ω.

Let $\zeta^m = v^m - u_h$. Then, for $w \in \mathcal{M}_h$,

$$B(\zeta^m, w) + \lambda(\zeta^m - \zeta^{m-1}, w) + \left(\frac{\partial f^m}{\partial u}\zeta^{m-1}, w\right) = 0 , \tag{2.2}$$

and the choice $w = \zeta^m$ leads to the relation

$$\begin{aligned} B(\zeta^m, \zeta^m) + \lambda\|\zeta^m\|^2 &= \left(\left(\lambda - \frac{\partial f^m}{\partial u}\right)\zeta^m, \zeta^{m-1}\right) \\ &\le \sup_{\Omega\times\mathbb{R}} \left|\lambda - \frac{\partial f^m}{\partial u}\right| \cdot \|\zeta^m\| \cdot \|\zeta^{m-1}\| . \end{aligned} \tag{2.3}$$

Let

$$\theta = \min_{z\in\mathcal{M}_h} \frac{B(z,z)}{\|z\|^2} . \tag{2.4}$$

Clearly, $\theta > \rho > 0$, where ρ appears in (1.6), and it can be shown that $\theta \ge \rho(1 + \nu_{min} - \epsilon(h))$, where ν_{min} is the least eigenvalue of the operator $-\Delta$ on Ω when Dirichlet boundary conditions are imposed and $\epsilon(h) \to 0$ as $h \to 0$. Hence,

$$\|\zeta^m\| \le \frac{\sup_{\Omega\times\mathbb{R}} \left|\lambda - \frac{\partial f}{\partial u}\right|}{\lambda + \theta} \|\zeta^{m-1}\| , \quad m \ge 1 . \tag{2.5}$$

Let $\lambda \ge \frac{1}{2} M$ and set

$$q = \frac{\sup_{\Omega\times\mathbb{R}} \left|\lambda - \frac{\partial f}{\partial u}\right|}{\lambda + \theta} < 1 , \tag{2.6}$$

Note that an upper bound for q can be found that is independent of h. It follows from (2.3), (2.5), and (2.6) that

$$\begin{aligned} \|\zeta^m\| &\le q^m \|\zeta^0\| , \\ |||\zeta^m||| &\le C_1 q^{m-1/2} \|\zeta^0\| , \end{aligned} \tag{2.7}$$

where

$$C_1^2 \leq \rho^{-1} \sup_{\Omega \times \mathbb{R}} |\lambda - \frac{\partial f}{\partial u}|.$$

Thus, we have proved the following theorem.

Theorem 1. Let $\lambda \geq \frac{1}{2} M$, where M satisfies (1.2). Then the iteration defined by (2.1) converges at a rate described by (2.7) for some $q \in (0,1)$, where q is independent of h.

Recall that $|||u - u_h||| \leq C\|u\|_{r+1} h^r$ for sufficiently smooth u. It seems reasonable to iterate until $|||\zeta^m||| = O(h^r)$ or $\|\zeta^m\| = O(h^{r+1})$. If no particularly good first guess v^0 is available, then $\|\zeta^0\|$ can be expected to be $O(1)$ and $O(\log h^{-1})$ iterations would be predicted to be necessary in order to lower the error in the accepted approximate solution to the level of the truncation error.

In the analysis of two-level iteration to be discussed in the next section it is helpful to express the above results in a matricial form. Let

$$\mathcal{M}_h = \text{Span}[w_1, \ldots, w_N]. \tag{2.8}$$

Let $N \times N$ matrices $A = A_a$ and C be defined by

$$A = [B(w_j, w_i)], \quad C = [(w_j, w_i)]. \tag{2.9}$$

Let μ and ν^m be vectors such that

$$u_h = \sum_{j=1}^{N} \mu_j w_j, \quad v^m = \sum_{j=1}^{N} \nu_j^m w_j. \tag{2.10}$$

Let φ^m be the vector having components $(f(v^m), w_i)$ and

δ the vector having components $\langle g, a\frac{\partial w_i}{\partial n} - \gamma h^{-1} w_i \rangle$. Then, (2.1) is equivalent to

$$(2.11) \qquad (A+\lambda C)(\nu^m - \nu^{m-1}) + A\nu^{m-1} + \varphi^{m-1} + \delta = 0 \;.$$

Let $\xi^m = \mu - \nu^m$ and $\widetilde{C}^m = [(\frac{\partial f^m}{\partial u} w_j, w_i)]$. Then, (2.2) becomes

$$(2.12) \qquad (A+\lambda C)(\xi^m - \xi^{m-1}) + (A + \widetilde{C}^m)\xi^{m-1} = 0.$$

Hence

$$(2.13) \qquad \xi^m = (A+\lambda C)^{-1}(\lambda C - \widetilde{C}^m)\xi^{m-1} \;,$$

and

$$(2.14) \qquad \|(A+\lambda C)^{1/2}\xi^m\| \le q\,\|(A+\lambda C)^{1/2}\xi^{m-1}\| \;,$$

a slightly less precise form of (2.7), where the norm symbol now refers to the Euclidean norm of the vector, since in the matrix norm

$$(2.15) \qquad \|(A+\lambda C)^{-1/2}(\lambda C - \widetilde{C}^m)(A+\lambda C)^{-1/2}\| \le q < 1 \;.$$

A slight generalization of (2.1) can be obtained by using $\lambda = \lambda(x) \ge \frac{1}{2} \max_{t \in \mathbb{R}} \frac{\partial f}{\partial u}(x,t)$ in place of the constant λ. The analysis is essentially unchanged, and there can be examples for which variable $\lambda(x)$ would be valuable.

3. Preconditioned conjugate gradient inner iteration. It will usually be the case that the setup calculation (such as an LU factorization of $A+\lambda C$) is much more costly than the solution calculation for (2.11); consequently, if many problems are to be worked using different values of the coefficient $a(x)$, it can be very useful to do the setup for a

single, conveniently chosen $a(x)$ and λ and then carry the solutions of the equations (2.11) for the differing $a(x)$'s and λ's by conjugate gradient iteration. The computational complexity of this algorithm will be discussed in greater detail in the next section.

The algorithm can be made precise as follows. Again, select any initial guess, say $u^0 = \sum_{j=1}^{N} \mu_j^0 w_j \in \mathcal{M}_h$. Let $D = A + \lambda C$, and assume that D_0 is a symmetric, positive-definite matrix such that

$$(3.1) \qquad 0 < \psi_0 \leq \frac{(D\xi, \xi)}{(D_0\xi, \xi)} \leq \psi_1 , \quad 0 \neq \xi \in \mathbb{R}^N, \; N = \dim \mathcal{M}_h,$$

where ψ_0 and ψ_1 are independent of h. Clearly, D_0 can be constructed in the form $A_0 + \lambda_0 C$, where A_0 corresponds to a coefficient $a = a_0(x)$ in (1.3) with $a(x)a_0(x)^{-1} \in [\psi_0, \psi_1]$, $x \in \Omega$, and $\lambda\lambda_0^{-1} \in [\psi_0, \psi_1]$. Define μ^m recursively as follows. Set (with φ^m now corresponding to u^{m-1})

$$(3.2) \qquad \begin{aligned} x_0 &= x_0^m = 0 \in \mathbb{R}^N , \\ r_0 &= D_0 s_0 = D x_0 + A\mu^{m-1} + \varphi^{m-1} + \delta , \end{aligned}$$

and iterate by conjugate gradients [1,9] for $j = 0, 1, \ldots, k-1$:

$$(3.3) \qquad \begin{aligned} x_{j+1} &= x_j + \alpha_j s_j , \\ r_{j+1} &= r_j + \alpha_j D s_j , \\ s_{j+1} &= D_0^{-1} r_{j+1} + \beta_j s_j , \end{aligned}$$

where

$$(3.4) \qquad \begin{aligned} \alpha_j &= -(D_0^{-1} r_j, r_j)/(s_j, D s_j) , \\ \beta_j &= (D_0^{-1} r_{j+1}, r_{j+1})/(D_0^{-1} r_j, r_j) . \end{aligned}$$

Finally, set

$$\mu^m = \mu^{m-1} + x_k \quad . \tag{3.5}$$

The analysis of the convergence of $u^m = \sum \mu_j^m w_j$ to $u_h = \sum \mu_j w_j$ makes strong use of the estimates of the preceding section. Define $\bar{\mu}^m$ to be the vector such that

$$D(\bar{\mu}^m - \mu^{m-1}) + A\mu^{m-1} + \varphi^{m-1} + \delta = 0. \tag{3.6}$$

Let $\xi^m = \mu - \mu^m$ and $\bar{\xi}^m = \mu - \bar{\mu}^m$. Then, (2.14) implies that

$$\| D^{1/2} \bar{\xi}^m \| \leq q \| D^{1/2} \xi^{m-1} \| . \tag{3.7}$$

Furthermore, it is well known [1-3, 9] that there exists $\tilde{q} = \tilde{q}(k, \psi_0, \psi_1) \leq 2\left\{\dfrac{\psi_1^{1/2} - \psi_0^{1/2}}{\psi_1^{1/2} + \psi_0^{1/2}}\right\}^k$ such that

$$\| D^{1/2}(\mu^m - \bar{\mu}^m) \| \leq \tilde{q} \| D^{1/2}(\mu^{m-1} - \bar{\mu}^m) \| . \tag{3.8}$$

Hence,

$$\begin{aligned} \| D^{1/2}(\xi^m - \bar{\xi}^m) \| &\leq \tilde{q} \| D^{1/2}(\xi^{m-1} - \bar{\xi}^m) \| \\ &\leq \tilde{q}(1+q) \| D^{1/2} \xi^{m-1} \| , \end{aligned}$$

and

$$\| D^{1/2} \xi^m \| \leq [q + \tilde{q}(1+q)] \| D^{1/2} \xi^{m-1} \| . \tag{3.9}$$

Set $Q = q + \tilde{q}(1+q)$ and assume that k is chosen so that $\tilde{q}$ is sufficiently small that $Q < 1$. Note that, since q and $\tilde{q}$ can be bounded independently of h, the value of k can be set independently of h. The following theorem has been proved.

Theorem 2. Let k be selected so that $Q = q + \tilde{q}(1+q) < 1$. Then the interation defined by (3.2)-(3.5) converges, and

$$(3.10) \qquad B(u_h - u^m, u_h - u^m) + \lambda \| u_h - u^m \|^2 \leq Q^{2m}[B(u_h - u^0, u_h - u^0) + \lambda \| u_h - u^0 \|^2] .$$

Consider for a moment the special case arising when (1.3) is linear. Then, in (2.11) $\varphi^{m-1} + \delta$ can be written in the form $\hat{C}\mu^m + \epsilon$, $\hat{C}$ and ϵ being independent of m. The use of preconditioned conjugate gradient iteration to solve the resulting linear system is not new. A relatively primitive form of this technique was proposed by Rutishauser [9] in 1959. Since then Wachspress has discussed its use to accelerate alternating-direction iteration, and Axelsson [1-3] has considered several such procedures with most of his emphasis attached to accelerating SSOR iteration. Concus and Golub have applied variants of these procedures to a number of problems, both linear and nonlinear; see their article in these Proceedings.

4. Operation counts. It is of interest to predict the number of arithmetic operations required to carry out an iterative solution of (1.8) by either (2.1) or (3.2)-(3.5). Assume that $N = \dim \mathcal{M}_h$. For any commonly used piecewise-polynomial finite element space or any standard isoparametric finite element space the generation of the matrices A and C arising in either procedure can be accomplished in $O(N)$ operations. Also, a multiplication of

the form $A\nu$ or Ds can be performed in $O(N)$ operations, as can the formation of the vector φ^{m-1}. The vector δ has nonzero components only for unknowns coming from near the boundary; thus, usually only $O(N^{1/2})$ operations are needed to determine δ.

The nested dissection algorithm of A. George [10] has been described in detail in the literature only in a rather special case, but it is clear that, for any reasonable choice of a triangulation of Ω and a corresponding $\mathcal{M}_h$, the general computational complexity estimate he obtained holds at least with respect to order in N. Hence, assume that an LU factorization of a matrix of the form $D = A + \lambda C$ takes $O(N^{3/2})$ operations and the number of nonzero entries in L and U is $O(N \log N)$; i.e., $O(N \log N)$ operations are required to solve equations given in the form $LU\alpha = \beta$, once the factors L and U are known. In some particular cases (e.g., Ω a rectangle, $\mathcal{M}_h$ one of certain tensor product spaces, and D_0 corresponding to $-\rho\Delta + \lambda$), the $O(N^{3/2})$ and $O(N \log N)$ estimates for setup and solution can be reduced to $O(N)$ and $O(N)$, respectively, by use of some modification of the algorithm of Bank and Rose presented in the paper of Bank in these Proceedings; however, such problems are very special and the estimates below will correspond to the general case.

The successive substitution method (2.1) requires the construction and factorization of the matrix $D = A + \lambda C$ with A being determined by the coefficient $a(x)$ in (1.3); this takes $O(N^{3/2})$ operations. Then, it takes

$O(N \log N)$ operations to carry out an iteration. Since $O(\log N)$ iterations can be required to reduce the error to the level of the truncation error, it requires

$$O(N^{3/2}) + M \cdot O(N(\log N)^2) \tag{4.1}$$

operations to find the solution of M problems (1.8) for which $a(x)$ is held fixed and the data $f(x, u)$ and g are varied, assuming that the same λ can be used for all the $f(x, u)$'s.

One advantage of the use of the conjugate gradient inner iteration is that the setup can be done just once even if the coefficient $a(x)$ is varied along with f and g. Clearly the setup is again an $O(N^{3/2})$ procedure. An outer iteration requires $O(N \log N)$ operations per conjugate gradient inner iteration, mostly for the solution of equations of the form $D_0 \alpha = \beta$. Since there are at most $O(1)$ conjugate gradient iterations per outer iteration and $O(\log N)$ outer iterations, the same complexity bound (4.1) holds here; however, the constant in the second term should be larger than in the successive substitution method. The ability to change $a(x)$ without a new setup calculation should override the larger constant in many practical cases.

5. Some comparisons and extensions. It might seem natural to use a Newton iteration in place of (2.1). In that case the equations would be in the form

$$(A + C^m)(\nu^m - \nu^{m-1}) + \epsilon^m = 0 , \tag{5.1}$$

where C^m has entries given by $(\frac{\partial f}{\partial u}(v^{m-1}), w_j, w_i)$ and ϵ^m depends on v^{m-1}, f, and g. It can be shown that quadratic convergence takes place [8, 14]. Practically, if an LU factorization of $A + C^m$ is made at each step, then the calculation swamps that required by (2.1). The procedure (3.2)-(3.5) can be modified to treat (2.1) in a way that preserves the quadratic convergence of the Newton process; O(log log N) steps of the outer iteration are needed and $O(2^m)$ conjugate gradient iterations are indicated to obtain ν^m from ν^{m-1}. It then follows that the estimate of the total number of calculations per problem is the same as for the methods of this paper. Unfortunately, the Newton process suffers from its usual constraint of being only locally convergent. A similar procedure using Richardson iteration in place of conjugate gradient is in [8, 14].

An alternative to preconditioned conjugate gradient iteration would be a preconditioned Richardson iteration. This method was treated by Gunn [11] for finite difference methods for (1.3). Slightly better results were obtained for this procedure than the ones contained herein, but it is our opinion (without adequate computational experience) that the preconditioned conjugate gradient method should be better in practice.

Extensions of the ideas of this paper to provide means of time-stepping Galerkin methods for nonlinear parabolic problems will be given elsewhere. The extension of the Richardson variant was treated by the authors [6] under the name of Laplace-modified methods. See also [4].

References

1. O. Axelsson, On preconditioning and convergence acceleration in sparse matrix problems, CERN European Organization for Nuclear Research, Geneva, 1974.

2. ______, On the computational complexity of some matrix iterative algorithms, Report 74.06, Dept. of Computer Science, Chalmers University of Technology, Göteborg, 1974.

3. ______, A class of iterative methods for finite element equations, Report 75.03R, Dept. of Computer Science, Chalmers University of Technology, Göteborg, 1975.

4. J. E. Dendy, Jr., An analysis of some Galerkin schemes for the solution of nonlinear time-dependent problems, SIAM J. Numer. Anal. 12(1975), pp. 541-565.

5. J. Douglas, Jr., Alternating direction iteration for mildly nonlinear elliptic difference equations, Numer. Math. 3(1961), pp. 92-98 and Numer. Math. 4(1962), pp. 301-302.

6. J. Douglas, Jr., and T. Dupont, Alternating-direction Galerkin methods on rectangles, Numerical Solution of Partial Differential Equations - II, J.H. Bramble (ed.), Academic Press, Inc. New York 1971.

7. _____ and _____, A Galerkin method for a nonlinear Dirichlet problem, Math. of Comp. 29(1975), pp. 689-696.

8. S. C. Eisenstat, M. H. Schultz, and A. H. Sherman, The application of sparse matrix methods to the numerical solution of nonlinear elliptic partial differential equations, Constructive and Computational Methods for Differential and Integral Equations, Lecture Notes in Mathematics 430, Springer-Verlag, Heidelberg 1974.

9. M. Engeli, Th. Ginsburg, H. Rutishauser, and E. Stiefel, Refined iterative methods for the computation of the solution and the eigenvalues of self-adjoint boundary value problems, Mitteilungen aus dem Institut für angewandte Mathematik, nr. 8, ETH, Zurich 1950.

10. J. A. George, Nested dissection of a regular finite element mesh, SIAM J. Numer. Anal. 10(1973), pp. 345-363.

11. J. E. Gunn, The solution of elliptic difference equations by semi-explicit iterative techniques, SIAM J. Numer. Anal. 2(1965), pp. 24-45.

12. J.-L. Lions, Quelques méthodes de résolution des problèmes aux limites non linéaires, Dunod, Paris 1969.

13. J. A. Nitsche, Über ein Variationsprinzip zur Lösung von Dirichlet-Problemen bei Verwendung von Teilräumen, die keinen Randbedingungen unterworfen sind, Abh. Math. Sem. Univ. Hamburg 36(1970/71) pp. 9-15.

14. A. H. Sherman, Thesis, Dept. of Computer Science, Yale University.

THE SPARSE TABLEAU APPROACH TO FINITE ELEMENT ASSEMBLY

GARY HACHTEL
IBM, Office Products Division, Boulder, Colorado

ABSTRACT

It is shown that the Nodal Assembly or "stiffness" matrix commonly associated with finite element equations can be regarded as the result of a prescribed sequence of pivots on a larger sparser matrix. Unconstrained pivoting in this larger matrix may, therefore, lead to smaller overall operation counts and, possibly, to numerical stability. Computed results that indicate the extent of operation count reduction in some practical cases are given.

1. INTRODUCTION

It has been shown in [1]that the advent of sparse matrix technology can make it profitable from an overall operations viewpoint to reexamine the basic formulation of various problem classes. In this paper, we examine a novel sparse matrix-based formulation of the general finite element analysis equations. We shall refer to this new formulation as the STA (Sparse Tableau Assembly) method and compare it to the "standard" finite element analysis method as defined, for example, in [2] or [3]. We will call this method the NA -- Nodal Assembly -- method. The sparse matrix aspects of the finite element equations have already received much attention, notably from George [4] and Speelpenning [5]. These studies, however, have concentrated on the question of efficient Gauss Elimination of the "stiffness matrix", i.e., the coefficient matrix in the standard finite element equations. In contrast, the present work examines the more basic question of whether or not the stiffness matrix is worth forming at all.

We propose and examine an alternative formulation wherein a larger, sparser tableau of equations and unknowns that can, if desired, be eliminated down to the standard form of the finite element equations by a prescribed sequence of pivots, is formed. Thus, the standard finite element formulation can be regarded as a constrained and, therefore, nonoptimal pivoting order in the larger tableau.

We show that if a truly optimum ordering algorithm were used in both approaches, operations counts for the STA method must be better (or at least as good) as for the NA method. Further, we choose a model problem from the two-dimensional analysis of the PDE of semiconductor transport and show that the STA method is, in fact, advantageous even with practical, i.e., suboptimal, ordering algorithms. We use the OPTORD algorithm of F. G. Gustavson [1].

We begin in Section 2 with an example of the "Sparse Tableau Approach to Least Squares" [6] that illustrates how and why the STA method can be superior to NA in both operations count and numerical stability. In Section 3, a Galerkin formulation of the general finite element equations is then given which leads to a sparse tableau of finite element equations that contains both local and global generalized coordinates. For the sake of example, a second Sparse Finite Element Tableau is given for the case of linear finite element equations formed from a variational principle.

In Section 4, we show that NA is equivalent to STA if an appropriate number of diagonal pivots are taken in the Galerkin Sparse Tableau matrix. Also, we discuss our strategy for ordering this matrix so as to minimize the amount of multiplications, addition, and compiled machine code (cf. the GNSO program of F. G. Gustavson, [1]) required for Gauss Elimination and back substitution of this tableau.

In Section 5, we describe our model problem and give a tabluation of our computed results for the comparison of NA vs. STA. We conclude in Sections 6 and 7 with discussion of some implications of our data and limitations of our computer experiments.

2. THE SPARSE TABLEAU APPROACH -- A LEAST SQUARES EXAMPLE

The Sparse Tableau Approach is the result of applying the following heuristic:

> Describe the problem in the largest possible number of equations and unknowns. Any simplification you make (especially for large, complex problems) may introduce a pivot order constraint which increases op counts and/or numerical instability.

The payoff of this approach is substantial in the electrical network analysis problem [1], and we believe similar payoff is available if applied to large, finite element production codes. Before applying the idea to finite elements, we demonstrate how it works with a brief review of its application to the problem of linear least squares analysis, which has been described elsewhere [6].

Given the (sparse) K x K matrix R, the (sparse) K x N matrix A, and the sparse K-vector b, consider solving the overdetermined system*

$$RA\eta = b. \tag{1}$$

One has the option of obtaining a normal form (cf. Stewart, [10])

$$M\eta = A^T RA\eta = A^T b \tag{2}$$

in which M is symmetric positive definite for appropriate R matrices. However, instead of reducing the number of equations, as in forming (2) from (1) by premultiplication by A^T), one can increase the number, viz.

$$\begin{bmatrix} -I_K & O & A \\ R & -I_K & O \\ O & A^T & O \end{bmatrix} \begin{matrix} v \\ r \\ \eta \end{matrix} = \begin{bmatrix} O \\ b \\ O \end{bmatrix} \tag{3a}$$

*See the Sparse Matrix studies by Duff [7] and BJORCK [8] elsewhere in this volume.

Equation (3a) is numerically equivalent to (2) and is operationally equivalent *if* the first $2\bar{K}$ pivots are taken down the diagonal. However, for the case

$$A = \begin{bmatrix} 1 & 0 \\ 0 & 2.1 \\ 1/\varepsilon & -1/\varepsilon \end{bmatrix}, \quad R = I_K, \quad b = \begin{bmatrix} 2.5 \\ 0 \\ 0 \end{bmatrix}, \tag{3b}$$

(3) can be solved in four MULTS and two ADDS, whereas (2) requires four MULTS and three ADDS using the OPTORD-GNSO* code of F. Gustavson [1]. Further, (2) becomes numerically singular for small ε, whereas pivoting in (3) is stable for all ε.

In this example, the pivot flexibility gained by larger, sparse formulations of a given problem can be turned into real gains in operations count and in numerical stability. We now show that a remarkably similar situation exists in the problem of finite element analysis.

3. A SPARSE TABLEAU FOR FINITE ELEMENTS

Suppose the PDE to be solved via finite elements are of the form:

$$\nabla \cdot F(u, \nabla u) = c(u, \dot{u}, x, y), \quad x, y \varepsilon \Omega, \tag{4a}$$

$$\nabla_N u = 0, \quad x, y \varepsilon \partial\Omega_N. \tag{4b}$$

$$u(x,y) = u_D(x,y), \quad x, y \varepsilon \partial\Omega_D. \tag{5}$$

Whether or not the functions F and c are linear or are related to a variational principle, the finite element equations may be written in the Galerkin form:

$$r_n = \int_\Omega \Psi_n \left\{ \nabla \cdot F - c \right\} d\Omega = \sum_{\ell=1}^{L} \int_\Omega \Psi_{n_\ell} \left\{ \nabla \cdot F - c \right\} d\Omega = 0, \tag{6}$$

---

*These codes regard multiplication by ± 1 as a no-op.

for $n = 1,2...(N = |\Psi|)$, where

Ψ = vector of global basis functions,

$u = \eta^T \Psi$ (global approximation),

η = vector of global generalized coordinates.

For each finite element Ω^ℓ, $\ell = 1,2,...L$ of the domain Ω, we may write:

$$r^\ell = \rho^\ell(v^\ell) = \int_{\Omega^\ell} \phi^\ell \{\nabla \cdot F - c\} d\Omega. \tag{7}$$

where

ϕ_m^ℓ = local basis functions $m = 1,2,...(|\phi|^\ell = M^\ell)$

$u^\ell = (v^\ell)^T \phi^\ell$ (local approximation)

The local basis ϕ_m^ℓ is just the restriction of the global basis Ψ_η to the finite element Ω^ℓ, so we may write:

$$\phi^\ell = A^\ell \Psi \quad , \quad v^\ell = A^\ell \eta \quad , \tag{8}$$

where

$$A_{mn}^\ell = \begin{cases} 1 \text{ if } \phi_m^\ell \text{ is restriction of } \Psi_n \text{ to } \Omega^\ell \\ 0 \text{ otherwise} \end{cases} \tag{9}$$

If we collect the local arrays r^ℓ, v^ℓ, and A^ℓ into the concatenations

$$r = \begin{bmatrix} r^1 \\ r^2 \\ \vdots \\ r^L \end{bmatrix}, \quad v = \begin{bmatrix} v^1 \\ v^2 \\ \vdots \\ v^L \end{bmatrix}, \quad A = \begin{bmatrix} A^1 \\ A^2 \\ \vdots \\ A^L \end{bmatrix} \tag{10}$$

Then it may be seen that the Galerkin equations (6) are equivalent to

$$-v + A\eta = 0 \tag{11a}$$

$$\rho(v) - r = 0 \tag{11b}$$

$$A^T r = 0 \tag{11c}$$

The Newton Iteration for (11) may be written

$$\begin{bmatrix} -I_K & 0 & A \\ \frac{\partial\rho}{\partial v} & -I_K & 0 \\ 0 & A^T & 0 \end{bmatrix} \begin{matrix} \nabla v \\ \nabla r \\ \nabla\eta \end{matrix} = \begin{bmatrix} 0 \\ -\rho(v) \\ 0 \end{bmatrix} \tag{12}$$

where

$$K = |v^1| + |v^2| + \ldots |v^L| = |r^1| + |r^2| + \ldots |r^L| \tag{13}$$

We shall refer to (11) as the "Sparse Tableau" and to the coefficient matrix in (12) as the "Sparse Tableau Matrix" of finite element analysis.

Note that by taking $2\bar{K}$ diagonal pivots in (12), we obtain

$$A^T \frac{\partial\rho}{\partial v} A \nabla\eta = A^T(-\rho(v)) \tag{14}$$

where $\frac{\partial\rho}{\partial v}$ is a block diagonal K x K matrix with full $M^\ell \times M^\ell$ submatrices $\frac{\partial\rho}{\partial v}^\ell$ on the ℓth diagonal position.

It is easily shown that the coefficient matrix of (14) is nothing but the well-known "Stiffness Matrix" [2,3] in cases where that term is applicable. We shall refer to (14) as the "standard" finite element formulation.

In the case of finite element analysis, a larger, sparser tableau has been obtained by retaining local variables and generalized coordinates. The familiar process of "Assembly" (i.e., forming the Stiffness Matrix $A^T \frac{\partial\rho}{\partial v} A$) is thus deferred for later handling of the sparse matrix routines. Thus, we shall let an algorithm for ordering the sequence of Gauss Elimination, rather than convention or habit, determine whether or not the stiffness matrix is worth forming. Note that even though $\frac{\partial \rho}{\partial v}$ is block diagonal and A is a very sparse matrix (having exactly 1 non-zero element per row), a substantial amount of additions and data movement are consumed by the matrix operations on both sides of (14).

Before discussing the question of order of elimination in (12) in detail, it is worth mentioning the non-uniqueness of the tableau matrix (12). First, note that we have not mentioned the boundary conditions (4b) and (5). By taking the mandatory integration by parts in (6) and throwing away the integrated part, one automatically gets the "natural" boundary condition (4b). Hence, the very sparse tableau matrix of (12) corresponds to the "all-natural" boundary case $\partial\Omega_D = 0$. If $\partial\Omega_D \neq 0$, then the submatrices A^ℓ and $\partial\rho^\ell/\partial v^\ell$ become even sparser. In fact, if node m of element ℓ lies in $\partial\Omega_D$, one obtains

$$A^\ell_{mn} = 0, \; n = 1,2,\ldots N,$$

$$\rho^\ell_m = \frac{\partial \rho^\ell_m}{\partial v^\ell_j} = \frac{\partial \rho^\ell_j}{\partial v^\ell_m} = 0, \; j \neq m. \tag{15}$$

The effect of the Dirichlet boundary condition (15) is to make (12) quite a bit sparser and to reduce the operations count required to form the stiffness matrix in (14).

As a final point, note that sparse tableaux other than (12) are possible. For example, if the form of (4) admitted a variational principal, the energy function might be minimized by solving (cf. (7) above)

$$\rho^\ell(v^\ell) = 0 \quad , \quad \ell = 1,2,\ldots L \quad , \tag{16a}$$

subject to the linear constraints

$$Cv - b = 0. \tag{16b}$$

In this case, (11) would be replaced by

$$\rho(v) + C^T\lambda = 0 \quad , \tag{17a}$$

$$Cv \qquad = b \quad , \tag{17b}$$

where λ is the Lagrange multiplier of (16b) in the energy functional. The Newton iteration for (17) would be

$$\begin{bmatrix} \frac{\partial \rho(v)}{\partial v} & C^T \\ C & 0 \end{bmatrix} \begin{matrix} \nabla\lambda \\ \\ \nabla v \end{matrix} = \begin{matrix} -(\rho(v) + C^T\lambda) \quad , & (18a) \\ \\ -(Cv - b) . & (18b) \end{matrix}$$

The tableau matrix of (18) is roughly similar to that of (12) except that C would be somewhat more complicated than A. Typically, C would be used to establish the Dirichlet boundary conditions (5) and to establish the interelement continuity class of the global approximations. For example, with this approach, local approximation

$$(v^\ell)^T \phi^\ell = v^\ell_{00} + v^\ell_{10}x + v^\ell_{01}y + v^\ell_{20}x^2 + \ldots \tag{19}$$

in terms of the basic monomials might be employed, whence C might be expressed as a matrix of integers.

4. SPARSE TABLEAU ASSEMBLY VS. NODAL ASSEMBLY

We shall refer to the process of constraining the first 2K pivots in (12) to the diagonal as "NA" (Nodal Assembly). The process of unconstrained pivoting order in (12) will be termed STA (Sparse Tableau Assembly). If the pivots are chosen optimally for sparsity and/or numerical stability, STA should have better operation counts and/or accuracy than NA or, at least, no worse. This is simply because optimization in a subspace cannot be better than optimization in the whole space. Thus, in theory, STA cannot lose to NA in any comparison. Practice, however, is always a very different and more immediately important question. Given non-optimal sparse ordering and solution algorithms, the comparison cannot be resolved a priori. Further, there may be a theorem unknown to the author or, as yet, undiscovered, that shows the NA method currently in vogue to be optimal. For the time being, the best we can do is to compare computed results for STA and NA that result from the application of practical ordering and solution algorithms. We choose for this the algorithms OPTORD and GENSO [1].

OPTORD orders the equations and unknowns of a given tableau according to a locally minimum multiplications count (Markowicz) heuristic. However, OPTORD distinguishes between operands that are +1, -1, constant, or variable in various senses. Thus, multiplication by ± 1 or addition of +1 to -1 are regarded as no-ops, and multiplication of two constants is recognized as a once-only operation. (Note that the Newton iteration (12) may be executed many times for a large problem.) OPTORD also gives weight to relative pivot magnitude in pivot selection; but, in the computer experiments that are reported below, this numerical weight was set to zero so as to concentrate on sparsity.

OPTORD also accounts for sparsity of the RHS and of the unknown vector (cf. [9]) so that all operations count comparisons are to be regarded as overall counts for L/U factorizations plus forward elimination, plus back-substitution. Thus, if OPTORD does obtain an optimal pivoting order, it is for all operations involved in solving (12) and not just for L/U factorization of the Sparse Tableau Matrix of (12).

Note that in (12) the right hand side has only K nonzero entries and these are of nonlinear type. Similarly, only K answers must be computed, e.g., the vector v. Of course, the OPTORD may choose to trade-off L/U operations for back-substitution operations and, profitably, compute the sum of the η's and r's.

OPTORD obeys pivot constraints and will start the ordering with 2K points on command. We submit that this pivot constraint is the only difference between solving (12) and solving (14).

OPTORD breaks ties on a first encounter basis which may be a factor in the apparent suboptimality in the experimental data we describe in the next section.

We now move to operations count comparisons between STA and NA. For each distinct case, OPTORD will be called twice -- once with a constraint of 2K diagonal pivots (NA) and once without a constraint (STA). It should be recognized that since OPTORD conceivably will, but probably won't, get the best possible order, four distinct subcases are possible for each case considered:

NA OPTIMAL,	STA OPTIMAL ,	(20a)
NA OPTIMAL,	STA SUB-OPTIMAL ,	(20b)
NA SUB-OPTIMAL,	STA OPTIMAL ,	(20c)
NA SUB-OPTIMAL,	STA SUB-OPTIMAL .	(20d)

5. COMPUTER EXPERIMENTS -- NA VS. STA

Since our purpose is to measure the practical value of the theoretical dvantage of STA over NA, we have chosen a model problem drawn from the finite element analysis of semiconductor devices [10]. For our present purposes, the main impact of this choice is the form taken by $F(u,\nabla u)$ in (4). Due to the nonlinearity of $F(u,\nabla u)$ in semiconductor and many other applications, the stiffness matrix $A^T \frac{\partial \rho}{\partial v} A$ of (14) is <u>not</u> symmetric. As is common in semiconductor applications, we choose a regular triangularization of a rectangular doman (Ω). That is, we divide the rectangle into NX x NY subrectangles -- each subrectangle forming two triangles. Thus, in the computed results that follow (unless otherwise specified), our problems will have 2xNXxNY elements and 2x(NX+1) x (NY+1) nodes. We shall consider four main variables in our comparisons:

1,2) Number of x and y subdivisions NX and NY (i.e., problem size and grid elongation).

3) Type of Boundary Condition. $\partial\Omega_D = 0$ means all natural boundary conditions. $\partial\Omega_D > 0$ means, as is typical in semiconductor problems, that about half the boundary is Dirichlet; the other half natural.

4) Element Type. $\nabla = 1\text{-}3$ means linear triangular elements with nodes at vertices; $\nabla = 2\text{-}6$ means quadratic elements with nodes at vertices and and midpoints.

The comparison data is compiled in Table I. Note that each case comprises two lines of data. The columns on the left give NX, NY, $\partial\Omega_D$, ∇ (element type), NN (number of nodes or number of equations in [14]). NEL (number of elements),

NEQ (number of equations and unknowns in [12]), and NZ (number of non-zeros in the tableau matrix of [12]). The columns on the right give computed data with the number over the fraction bar giving the operation counts measured in fill-ins, multiplication count, and code length* in 4-byte words for the STA method. The number below the fraction bar for a given case corresponds to the NA method. Addition counts are omitted since they correspond in most cases very closely to multiplication counts. (Exceptions are noted below.)

TABLE I

	Case Description				OPS(STA)/OPS(NA)		
	NX	$\partial\Omega D$	NN	NEQ	FILLS	MULTS	CODE
	NY	∇	NEL	NZ	STA/NA	(≈ADDS)	(WORDS)
1	2	=0	9	121	(275)	(923)	(3729)
	2	2-6	4	528	(579)	(923)	(3841)
2	12	=0	175	1039	4796	21804	
	3	2-6	72	4752	6395	22740	72119
3	6	=0	169	1033	4841	28812	87461
	6	2-6	72	4752	6764	30417	92317
4	12	=0	169	1897	6233	22935	75650
	12	1-3	288	6912	6272	24171	79830
5	9	=0	361	2305	13760	124878	150672
	9	2-6	162	10692	18045	133714	375263
6	9	>0	361	2305	12321	101394	288744
	9	2-6	162	10007	15542	99155	284382

*Actually, the code length indicated gives the entire storage requirement for all NZ elements of the original matrix plus the RHS plus the solution vector plus fill-ins as well as the actual machine solution code that is generated.

6. DATA ANALYSIS -- NA VS. STA

The most striking feature of the data in Table I is that multiplication count and code length ratios (STA/NA) are clustered near unity. In fact,

$$.93 \leq \left\{ R_M \doteq \frac{\text{MULTS (STA)}}{\text{MULTS (NA)}} \right\} \leq 1.02 \tag{21}$$

for the data of Table I. The multiplication ratio R_M fell within the bounds of (21) for about fifty other cases not shown in Table I. For each case, the weights assigned by OPTORD to the variability types "+1,-1, nonlinear" were varied so that the STA and the NA counts were the best available from OPTORD.

It is worth noting that in Case 1 of Table I, $R_M = 1$. (For this case, STA won over NA in addition count 853 to 858.) This data suggests that, in this small problem, OPTORD has found very nearly optimum pivot orders for both STA and NA. The clustering around unity suggests that OPTORD is finding nearly optimum pivoting orders for medium to large problems as well (Cases 5-7).

However, note that for the mixed boundary condition $\partial\Omega_D > 0$ (Case 6), $R_M > 1$. For all mixed boundary condition cases studied (> 10 cases), we have found $R_M \geq 1$. These data suggest that we are encountering mode (20b) of the ordering algorithm OPTORD for which NA is optimal but STA is suboptimal. Our explanation is that the simplifications (15) of the tableau matrix in (14) give OPTORD a sufficiently simple matrix that an optimum ordering can be obtained.

Another implication of the data clustering is that, at least for the small to medium size problems tested, NA is surprisingly effective. Perhaps the reason for this is that the structure of the stiffness matrix $A^T \frac{\partial\rho}{\partial v} A$ of (14) reflects directly the topological structure of the finite element discretization of $\partial\Omega$. In a sense, this structural information is obscured in the larger, sparser structure of the tableau matrix of (12).

However, judgement must be reserved in the case of very large problems. Since as NX and NY $\to \infty$, the effect of $\partial\Omega_D > 0$ must vanish, the data shows that:

$$R_M \to 1 \qquad \text{NX,NY} \to 1 \tag{22a}$$

$$R_M < 1 \qquad \text{NX,NY} >> 1 \tag{22b}$$

The question of whether the advantage of STA over NA becomes appreciable for large problems still, therefore, remains unanswered. It is unfortunate that artificial limitations in the OPTORD-GENSO data structure and in our computing environment have so far prevented us from obtaining data for larger problems (note the rapid growth in non-zero count and code length as NX and NY are increased). Nevertheless, it must not be assumed that these limitations are due to having formed the larger tableau (12). In fact, we have a program that forms the tableau (14) in the conventional manner. Code length and multiplications count for conventional NA exceeded those for the NA cases shown in Table I.

7. CONCLUSIONS

For the small to medium size range of practical, finite element problems studied, the Sparse Tableau approach to finite element assembly (STA) has been shown to be slightly more effective than the conventional "Nodal Assembly" (NA) of the finite element equations. The degrading effect of NA pivot constraint has been shown to be very small in the cases studied. In fact, the multiplication count ratio of STA to NA has remained sufficiently close to unity so that differences from unity could (just) conceivably be explained by nonoptimality of the pivoting order obtained by the program OPTORD*. Thus, the possibility of a theorem proving the optimality of NA has not been eliminated.

On the other hand, (22) suggests that the advantage of STA may become substantial for large problems. Unfortunately, it becomes very difficult to obtain optimal pivoting orders for large matrices. Of course, it would be possible to do nested dissection-like orderings for NA that were at least asymptotically optimal. There is no existing way to carry the nested dissection heuristic over to STA. This might be a profitable avenue for future work.

---

*OPTORD breaks ties on a first-encounter basis. Brayton and Gustavson have shown in an unpublished work that ties must be broken with graph-oriented heuristics before Markowitz-based algorithms like OPTORD can achieve optimum counts on George-like "Nested Dissection Graphs".

In summary, although NA performs surprisingly well, nothing in our data suggests that STA isn't at least as efficient, if not better. Since it can be shown that STA offers many advantages in input flexibility (e.g., heterogeneous element populations) and data management (e.g., code and data buffering), STA might be well recommended to anyone who is considering the programming of a new finite element production code.

References

1) G. D. Hachtel, R. K. Brayton and F. G. Gustavson, (1971), "The Sparse Tableau Approach to Network Analysis and Design," IEEE Trans. on Circuit Theory, VOL. CT-18, No. 1, p.101.

2) J. A. George, (1971), "Computer Implementation of the Finite Element Method," Ph.D. Dissertation, Department of Computer Science, Stanford University.

3) W. G. Strang and G. J. Fix, (1973), "An Analysis of the Finite Element Method," Prentice-Hall, Englewood Cliffs, New Jersey.

4) J. A. George, (1973), "Nested Dissection of a Regular Finite Element Mesh," SIAN J. of Num. Anal., 10, 345-363.

5) B. Speelpenning, (1973), "The Generalized Element Method," private commumnication.

6) G. D. Hachtel, "Extended Applications of the Sparse Tableau Approach-Finite Elements and Least Squares," in "Basic Questions of Design Theory," W.R. Spillers, Ed., North Holland Publishing Company, Amsterdam, 1974.

7) I. Duff, (1975), "On the Solution of Sparse Overdetermined Systems of Equations," in "Sparse Matrix Computations," J. Bunch, Ed., Academic Press, New York.

8) A. Bjorck, (1975), "Methods for Sparse Linear Least Squares Problems," in (Sparse Matrix Computations," J. Bunch, Ed., Academic Press, New York.

9) G. D. Hachtel, (1972), "Vector and Matrix Variability Type in Sparse Matrix Algorithms," in "Sparse Matrices and Their Applications," D. J. Rose and R. A. Willoughby, Editors, Plenum Press, New York, p. 53.

10) G. D. Hachtel, M. Mack and R. O'brien, (1974) "Semiconductor Device Analysis via Finite Elements," Conference Proceedings, Eighth Asilomar Conference on Circuits and Systems, Pacific Grove, California.

A CAPACITANCE MATRIX TECHNIQUE

by

B. L. Buzbee

LOS ALAMOS SCIENTIFIC LABORATORY

Introduction

Modification methods obtain the solution of matrix equation Au = v by solving another equation Bu = w. Usually B differs only slightly from A, so B is naturally thought of as a modification of A. If A is an elliptic finite difference matrix, then B is chosen so that the equation Bu = w can be solved efficiently. The modification w of the right-hand side v must be computed and in order to determine w one must know certain elements of (AB^{-1}). Since these elements are independent of v, this computation is done as a preprocessing phase. Using the results of the preprocessing, the solution of a particular equation Au = v can then be done efficiently.

The capacitance matrix technique is a modification method that is associated with elliptic difference systems and is characterized by modification of rows. Typically, this technique is used to modify a system of elliptic difference equations such that fast Poisson technology is applicable to the modified system. Since fast Poisson techniques are very efficient, one has the possibility of achieving efficiency in the entire capacitance process.

Let p be the number of modified equations. Using results from [1] and [2], Buzbee and Dorr [3] have developed a capacitance matrix technique such that the preprocessing phase on an n by n mesh requires $O(pn^2 + p^3)$ operations. This paper will present a unified development of this technique. First we will develop the capacitance method, then discuss some practical situations where we might use it and summarize known properties of it. Finally we develop the matrix decomposition Poisson solver and exploit it to achieve the aforementioned operation count.

The Capacitance Matrix Technique

Suppose we are given an elliptic finite difference system to solve, say Ax = y where A is N by N and N is "large". Further suppose that A is not amenable to fast Poisson techniques, but if we modify a few rows of it, say p rows where p is on the order of a hundred, then the modified matrix is amenable to fast Poisson techniques. Of course, if we modify the matrix, we must also modify the right hand side y. So given Ax = y, we will change A into B and calculate w from y such that Bx = w. Given A and B, the capacitance technique is one way to determine w. We now develop this technique.

Let

$R = \{1,2,\ldots,N\}$

S = p element subset of R

and

$$A\underset{\sim}{x} = \underset{\sim}{y} \tag{1}$$

where A is N by N. Let e_i denote the ith column of the N by N identity matrix. Let B be a nonsingular matrix such that systems involving it are easily solved and

$$\text{row } j \text{ of } A = \text{row } j \text{ of } B \ \forall \ j \in R \backslash S.$$

Define

$$B\underset{\sim}{g_i} = \underset{\sim}{e_i} \quad \forall \ i \in S$$

$$B\underset{\sim}{\hat{x}} = \underset{\sim}{y},$$

and

$$\underset{\sim}{z} = \underset{\sim}{\hat{x}} + \sum_{S} \beta_i \underset{\sim}{g_i},$$

where the β_i are not yet determined. Then

$$(A\underset{\sim}{z})_j = (B\underset{\sim}{z})_j = y_j \ \forall \ j \in R \backslash S,$$

and thus z satisfies the N−p unmodified equations of (1). All that remains is to determine the β_i so that the other p equations of (1) are also satisfied. For $i \in S$, we require

$$(A\underset{\sim}{z})_i = y_i = (A\underset{\sim}{\hat{x}})_i + \sum_{S} \beta_k (A\underset{\sim}{g_k})_i .$$

This defines a p by p linear system

$$C\underset{\sim}{\beta} = \underset{\sim}{\gamma}.$$

We call this the capacitance system and we call C the capacitance matrix. Also, we partition the algorithm into a preprocessing phase which is independent of y followed by a solution phase for a given right hand side.

Preprocessing:

1. Solve $B\underset{\sim}{g_i} = e_i \quad \forall \ i \in S$

2. Calculate and factor C

Solution:

3. Solve $B\underset{\sim}{\hat{x}} = \underset{\sim}{y}$

4. Solve $C\underset{\sim}{\beta} = \underset{\sim}{\gamma}$

5. Solve $B\underset{\sim}{x} = \underset{\sim}{y} + \sum_{S} \beta_i \underset{\sim}{e_i}$

Now let's look at some situations where we might use this algorithm.

Some Practical Applications

Suppose that we are given Poisson's equation with Dirichlet boundary on the slightly irregular region of Fig. 1. Further suppose

a. that we embed this region in a rectangle as indicated in Fig. 2,

b. that we impose a regular mesh on the rectangle,

and

c. that we use a standard difference equation at all interior mesh points of the rectangle except the p points τ_i in the original region which are adjacent to the irregular boundary, (see Fig. 3).

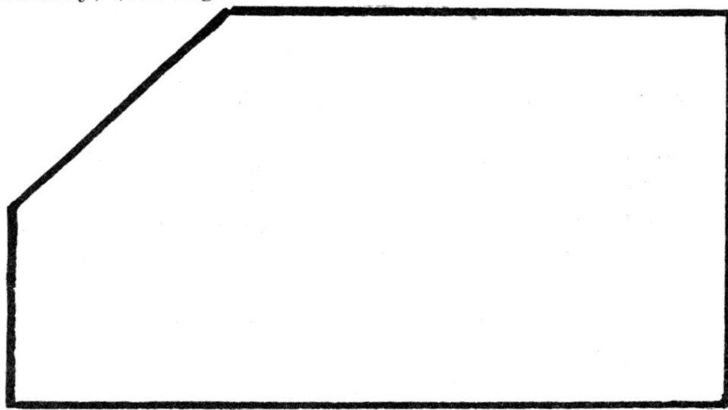

Fig. 1
Initial and slightly irregular region.

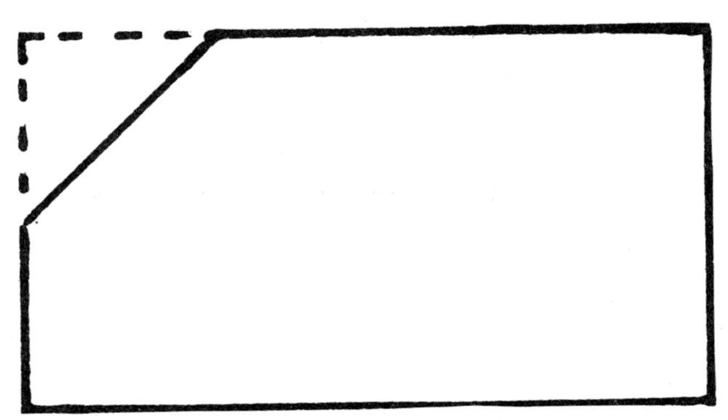

Fig. 2.
Initial region embedded in a rectangle.

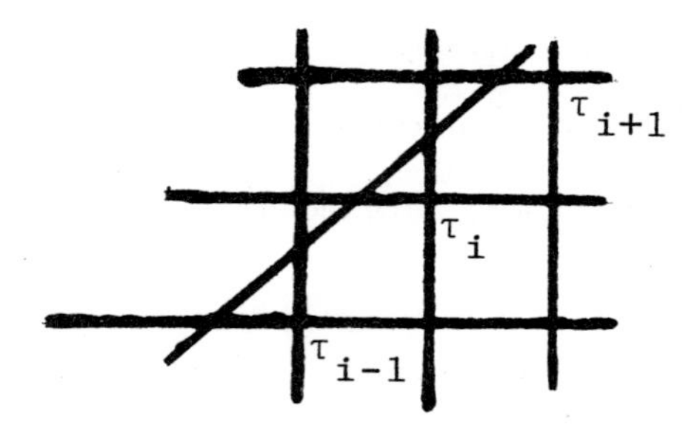

Fig. 3.
Mesh points where the difference equations are modified.

At the points τ_i we will use an appropriate difference equation involving the solution along the curved boundary, thus we get a system Ax = y which is not amenable to fast Poisson technology. So we modify the rows of A corresponding to τ_i such that A is changed to B and B is amenable to fast Poisson solvers, and proceed to apply the capacitance algorithm.

Although the capacitance technique is commonly associated with embedding, we can also use it to split a region. Suppose that we have the L shaped region in Fig. 4

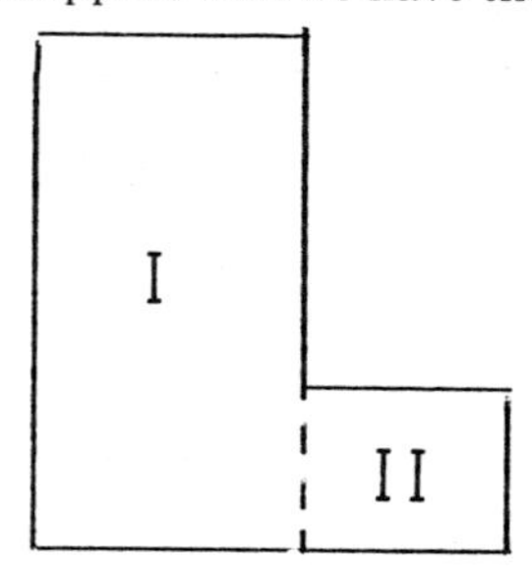

Fig. 4.
Splitting of L-shaped region into two rectangles.

with mesh points ordered along vertical lines. Then A has the form

$$A = \left(\begin{array}{cc|cc} & A_I & J & \emptyset \\ \hline \emptyset & J^T & & A_{II} \end{array}\right)$$

where $A_I(A_{II})$ is a Poisson difference matrix corresponding to region I(II) and J represents the "coupling" between regions I and II. If we take

$$B = \left(\begin{array}{c|c} A_I & \emptyset \quad \emptyset \\ \hline \emptyset \quad J^T & A_{II} \end{array}\right)$$

then systems involving B reduce to two Poisson systems in Regions I and II.

Thus far we have glossed over some important questions which should be asked about the capacitance system. For example, under what conditions is the system solvable? This question is addressed in [2] where the following is shown.

If $\text{Det}(B) \neq O$, then

1. $\text{Det}\ (C) = \dfrac{\text{Det}\ (A)}{\text{Det}\ (B)}$

and

2. if $\text{Det}(A) = 0$, then the capacitance system is consistent.

Thus, the previous algorithm is valid for all cases with $\text{Det}(B) \neq 0$.

Since the assumption of $\text{Det}(B) \neq 0$ is a severe restriction, we now develop the capacitance technique for singular B.

Let

$\text{rank}(B) = N - 1$

$B\underset{\sim}{u} = 0$

$B^t\underset{\sim}{v} = 0$

and assume $\exists\ k \in S \ni$

$\underset{\sim}{v}^t\underset{\sim}{e}_k \neq 0.$

Let

$$B\hat{\underset{\sim}{x}} = \underset{\sim}{y} - \left(\frac{v^t y}{v^t e_k}\right)\underset{\sim}{e}_k$$

$$B\underset{\sim}{g}_i = \underset{\sim}{e}_i - \left(\frac{v^t e_i}{v^t e_k}\right) \underset{\sim}{e}_k \qquad \forall \ i \in S\backslash\{k\}$$

$$\underset{\sim}{g}_k = u$$

and

$$\underset{\sim}{z} = \underset{\sim}{\hat{x}} + \sum_s \beta_j \underset{\sim}{g}_j .$$

Then for $j \in R\backslash S$

$$(A\underset{\sim}{z})_j = (B\underset{\sim}{z}_j) = y_j .$$

We require for $i \in S$

$$(A\underset{\sim}{z})_i = y_i = (A\underset{\sim}{\hat{x}})_i + \sum_s \beta_j (A\underset{\sim}{g}_j)_i .$$

Thus, we again get a p by p capacitance system, and we summarize the algorithm as follows.

Preprocessing:

1. If $\underset{\sim}{u}$ and $\underset{\sim}{v}$ are unknown, then compute them.

2. Calculate $\underset{\sim}{g}_j \ \forall \ j \in S\backslash\{k\}$

3. Calculate and factor C.

Solution:

4. Solve $Bx = y - (v^t y / v^t e_k) e_k$

5. Solve $C\underset{\sim}{\beta} = \underset{\sim}{\gamma}$

6. Solve $B\underset{\sim}{x}' = \underset{\sim}{y} - (\underset{\sim}{v}^t \underset{\sim}{y} / \underset{\sim}{v}^t \underset{\sim}{e}_k) \underset{\sim}{e}_k + \sum_s \beta_i [\underset{\sim}{e}_i - (\underset{\sim}{v}^t \underset{\sim}{e}_i / \underset{\sim}{v}^t \underset{\sim}{e}_k) \underset{\sim}{e}_k]$

7. $\underset{\sim}{x} = \underset{\sim}{x}' + \beta_k \underset{\sim}{u}$.

The assumption that rank (B) = N−1 is legitimate since we are dealing with Poisson problems. Also, if $\mathrm{Det}(A) \neq 0$, then $\mathrm{Det}(C) \neq 0$ and there always exists $k \in s$ such that $v_k \neq 0$ [3].

Each g_i requires the solution of a Poisson system, thus computation of them seems formidable. However, since A is sparse, we do not need all of the components of each g_i to form the elements of C. Rather, we only need those components associated with the nonzero elements of the ith row of A for all i in S. Typically, this is a few hundred, so we now develop the matrix decomposition Poisson solver and show how it can be used to calculate only the required elements of each g_i.

Matrix Decomposition

Consider a finite difference approximation to

$$-\nabla^2 u = f \tag{2}$$

in a rectangle R with $u = g(x,y)$ on the boundary ∂R. We will assume N discretizations in the horizontal direction and M discretizations in the vertical direction. Let $u_{ij} = u(ih,jk)$, approximate (2) by the five-point difference equation, and order the equations along vertical lines.

Let

$$\mathbf{u_i} = \begin{bmatrix} u_{i1} \\ u_{i2} \\ \vdots \\ u_{iM} \end{bmatrix}$$

The difference equations for all N vertical lines can be written

$$\begin{aligned} &Au_1 - u_2 = y_1 \\ &-u_{i-1} + Au_i - u_{i+1} = y_i, \quad i = 2,3,\ldots,N-1, \\ &-u_{N-1} + Au_N = y_N \end{aligned} \tag{3}$$

Since A is symmetric, there exists a unitary matrix V such that

$$\mathbf{V^tAV} = \begin{bmatrix} \lambda_1 & & & 0 \\ & \lambda_2 & & \\ & & \ddots & \\ 0 & & & \lambda_M \end{bmatrix} = \mathbf{D}.$$

Multiplying each equation in (3) by V^T and inserting $I = VV^T$ on the left of each occurrence of A gives

$$(V^TAV)(V^Tu_1) - V^Tu_2 = V^Ty_1$$

$$-V^Tu_{i-1} + (V^TAV)(V^Tu_i) - V^Tu_{i+1} = V^Ty_i, \quad i=2,3,\ldots,N-1,$$

$$(V^Tu_{N-1}) + (V^TAV)(V^Tu_n) = V^Ty_N;$$

or

$$D\tilde{u}_1 - \tilde{u}_2 = \tilde{y}_1$$

$$-\tilde{u}_{i-1} + D\tilde{u}_i - \tilde{u}_{i+1} = \tilde{y}_i, \quad i = 2,3,\ldots,N-1$$

$$-\tilde{u}_{N-1} + D\tilde{u}_N = \tilde{y}_N \tag{4}$$

where

$$\tilde{u}_i = V^Tu_i \text{ and } \tilde{y}_i = V^Ty_i.$$

Now collect all of the equations in (4) in which λ_1 occurs. We obtain

$$\lambda_1\tilde{u}_{11} - \tilde{u}_{21} = \tilde{y}_{11},$$

$$-\tilde{u}_{i-1,1} + \lambda_1\tilde{u}_{i,1} - \tilde{u}_{i+1,1} = \tilde{y}_{i1}, \; i = 2,3,\ldots,N-1,$$

$$-\tilde{u}_{N-1,1} + \lambda_1\tilde{u}_{N1} = \tilde{y}_{N1}$$

This is a N by N tridiagonal linear system; and consideration of each λ_j yields a similar system, call it $\Gamma_j\hat{u}_j = \hat{y}_j$, where

$$\hat{u}_j = \begin{bmatrix} \tilde{u}_{1j} \\ \tilde{u}_{2j} \\ \vdots \\ \tilde{u}_{Nj} \end{bmatrix} \quad \text{and} \quad \hat{y}_j = \begin{bmatrix} y_{1j} \\ y_{2j} \\ \vdots \\ y_{Nj} \end{bmatrix}$$

Solutions of these systems yield all of the components of each $\tilde{u}_i$. Thus, the matrix decomposition Poisson solver is defined by:

1. Calculate V and D.
2. Calculate y_i and $\tilde{y}_i = V^Ty_i, \quad i = 1,2,\ldots,N.$
3. Solve $\Gamma_j\hat{u}_j = \hat{y}_j, \quad j = 1,2,\ldots,M.$
4. Calculate $u_i = V\tilde{u}_i, \quad i = 1,2,\ldots,N.$

The matrix decomposition algorithm has some very nice properties. First, it is easy to implement. If one has a subroutine for eigenanalysis of a tridiagonal system and if one has a subroutine for solving tridiagonal systems, then matrix decomposition can be implemented with about fifty FORTRAN statements. Second, it is applicable to any separable problem on a rectangle [2]. Thus variable mesh sizes and non-Cartesian geometries are tractable. Third, it is highly parallel. Conceptually, all y_{ij} (u_{ij}) can be computed simultaneously. Also, all of the tridiagonal systems can be solved simultaneously. Thus it vectorizes nicely. Finally, the Fast Fourier Transform (FFT) can be used to improve its efficiency if the elements of V are trigonometric functions. As formulated above, it requires $0(NM^2)$ operations. Of course, there are no constraints on N or M. If FFT is used to carry out steps 2 and 4, then it requires O(NMlogM) operations, but there are constraints on M.

We now show how to use this algorithm in the capacitance technique. Recall that we must solve numerous systems of the form

$$B_{gk} = e_k \tag{5}$$

and that step 2 of matrix decomposition is

$$y_i = V^T y_i . \tag{6}$$

Since e_k has only one nonzero element, the associated y_i can be obtained without computation. All tridiagonal systems must be solved. However, since each u_{ij} is an inner product in $\tilde{u}_i$, we can compute only those that we need.By doing this, the preprocessing phase for an n by n mesh becomes $O(pn^2 + p^2)$[3]. Some numerical experiments with this technique are also reported in [3]

References

1. B.L. Buzbee, G.H. Golub, and C.W. Nielson, *On direct methods for solving Poisson's equations, SINUM* 7 (1970), pp. 627-656.

2. B.L. Buzbee, F.W. Dorr, J.A. George, and G.H. Golub, *The direct solution of the discrete Poisson equation on irregular regions, SINUM* 8 (1971), pp. 722-736.

3. B.L. Buzbee and F.W. Dorr,*The direct solution of the biharmonic equation on rectangular regions and the Poisson equation on irregular regions, SINUM* 11 (1974), pp. 753-763.

M-Matrix Theory and Recent Results in Numerical Linear Algebra

Richard S. Varga*
Department of Mathematics
Kent State University
Kent, OH 44242

*Dedicated to Garrett Birkhoff on the occasion of his 65th birthday, January 10, 1976.

§1. Introduction.

That the theory of M-matrices, as introduced by Ostrowski [19] in 1937, provides fundamental tools in the analysis of problems in numerical linear algebra, in particular in the iterative solution of large sparse systems of linear equations, is of course well known to all. What is perhaps surprising is that the theory of M-matrices *continues* to be the underlying theme, whether directly or indirectly, for many *recent* contributions in numerical linear algebra. The aim of this short contribution is sketch how the theory of M-matrices serves as a basis for some recent contributions.

We remark that the material presented here is a condensation of the material to appear in [2], [26], and [27].

*Research supported in part by the Air Force Office of Scientific Research under Grant AFOSR-74-2729, and by the Energy Research and Development Administration (ERDA) under Grant E(11-1)-2075.

§2. Notation and Terminology.

For a positive integer n, let $\langle n\rangle := \{1,2,\cdots,n\}$, let $\mathbb{C}^{n,n}$ denote the collection of all n × n complex matrices $A = [a_{i,j}]$, and let $\mathbb{C}_{\pi}^{n,n}$ denote the subset of matrices of $\mathbb{C}^{n,n}$ with all diagonal elements nonzero. Similarly, let $\mathbb{C}^n$ denote the complex n-dimensional vector space of all column vectors $\underline{v} = [v_1, v_2, \cdots, v_n]^T$, where $v_i \in \mathbb{C}$ for all $i \in \langle n\rangle$. The restriction to real entries or components similarly defines $\mathbb{R}^{n,n}$, $\mathbb{R}_{\pi}^{n,n}$, and $\mathbb{R}^n$.

For any $A \in \mathbb{C}^{n,n}$, set spec $[A] := \{\lambda : \det(A-\lambda I)=0\}$, and call $\rho(A) := \max\{|\lambda| : \lambda \in \text{spec}[A]\}$ the spectral radius of A. Next, for $\underline{v} \in \mathbb{R}^n$, we write $\underline{v} > \underline{0}$ or $\underline{v} \geq \underline{0}$, respectively, if $v_i > 0$ or if $v_i \geq 0$ for all $i \in \langle n\rangle$. Similarly, for $A = [a_{i,j}] \in \mathbb{R}^{n,n}$, we write $A > \mathcal{O}$ or $A \geq \mathcal{O}$, respectively, if $a_{i,j} > 0$ or if $a_{i,j} \geq 0$ for all $i, j \in \langle n\rangle$. Also, given $A = [a_{i,j}] \in \mathbb{C}^{n,n}$, then $|A| \in \mathbb{R}^{n,n}$ is defined by $|A| := [|a_{i,j}|]$. Next, given $A = [a_{i,j}] \in \mathbb{C}^{n,n}$, the set of matrices defined by

$$(2.1) \quad \Omega(A) := \{B = [b_{i,j}] \in \mathbb{C}^{n,n} : |b_{i,j}| = |a_{i,j}|;\ i, j \in \langle n\rangle\}$$

is called the equimodular set of matrices associated with the given matrix A.

For any $A = [a_{i,j}] \in \mathbb{C}_{\pi}^{n,n}$, we can decompose each $B = [b_{i,j}]$ in $\Omega(A)$ into the sum

$$(2.2) \quad B = D(B) - L(B) - U(B),$$

where $D(B) = \text{diag}[b_{1,1}, b_{2,2}, \cdots, b_{n,n}]$ is nonsingular, and where L(B) and U(B) are respectively strictly lower and strictly upper triangular matrices. From the decomposition of (2.2), three familiar iteration matrices can be defined. For any $\omega > 0$,

$$(2.3) \quad J_\omega(B) := \omega(D(B))^{-1}\{L(B)+U(B)\} + (1-\omega)I$$

is the (point) <u>Jacobi overrelaxation iteration matrix</u> for B,

$$(2.4) \quad \mathcal{L}_\omega(B) := [D(B)-\omega L(B)]^{-1}\{(1-\omega)D(B)+\omega U(B)\}$$

is the (point) <u>successive overrelaxation iteration matrix</u> (SOR matrix) for B, and

$$(2.5) \quad S_\omega(B) := [D(B)-\omega U(B)]^{-1}\{(1-\omega)D(B)+\omega L(B)\}[D(B)-\omega L(B)]^{-1} \cdot \{(1-\omega)D(B)+\omega U(B)\}$$

is the (point) <u>symmetric successive overrelaxation iteration matrix</u> (SSOR matrix) for B.

Next, to define M-matrices, consider any $B = [b_{i,j}] \in \mathbb{R}^{n,n}$ with $b_{i,j} \leq 0$ for all $i \neq j$ with $i, j \in \langle n \rangle$. Then, B can be expressed as the difference

$$(2.6) \quad B = \tau_B \cdot I - C(B),$$

where $\tau_B := \max\{b_{i,i} : i \in \langle n \rangle\}$, and where $C(B) = [c_{i,j}] \in \mathbb{R}^{n,n}$, satisfying $C(B) \geq \mathcal{O}$, has its entries given by

$$(2.7) \quad c_{i,i} = \tau_B - b_{i,i} \geq 0; \; c_{i,j} = -b_{i,j}, \; i \neq j; \; i, j \in \langle n \rangle.$$

Following Ostrowski [19], we say that such a matrix B is a <u>nonsingular M-matrix</u> iff $\tau_B > \rho(C(B))$. Next, given any $A = [a_{i,j}] \in \mathbb{C}^{n,n}$, we define, as in [19], its <u>comparison matrix</u> $\mathfrak{M}(A) = [\alpha_{i,j}] \in \mathbb{R}^{n,n}$ by

$$(2.8) \quad \alpha_{i,i} = |a_{i,i}|; \; \alpha_{i,j} = -|a_{i,j}|, \; i \neq j; \; i, j \in \langle n \rangle,$$

and A is defined, again as originally in Ostrowski [19], to be a <u>nonsingular H-matrix</u> iff $\mathfrak{M}(A)$ is a nonsingular M-matrix.

How does one recognize if a given matrix $A \in \mathbb{C}^{n,n}$ is a nonsingular H-matrix, or equivalently, if $\mathfrak{M}(A)$ is a nonsingular M-matrix? Perhaps best known to most readers (cf. [25, p. 83]) is the following: $\mathfrak{M}(A)$ is a nonsingular M-matrix iff $(\mathfrak{M}(A))^{-1} \geq \mathcal{O}$. But this is just one of the many known equivalent conditions for a nonsingular M-matrix which has been studied in the classic and famous papers of Taussky [23], Fan [10], and Fiedler and Pták [11]. This last paper lists <u>thirteen</u> equivalent conditions for a nonsingular M-matrix!

§3. <u>Main Result and Commentary</u>.

In analogy with the previous mentioned papers by Taussky [23], Fan [10], and Fiedler and Pták [11], we collect now some known, as well as some new, equivalent conditions for a given $A \in \mathbb{C}_{\pi}^{n,n}$ to be a nonsingular H-matrix, and then we comment briefly on how these equivalent conditions relate to recent results (back to 1968) in the literature.

<u>Theorem 1</u>. For any $A = [a_{i,j}] \in \mathbb{C}_{\pi}^{n,n}$, $n \geq 2$, the following are equivalent:

i) A is a nonsingular H-matrix, i.e., $\mathfrak{M}(A)$ is a nonsingular M-matrix;

ii) there exists a $\underline{u} \in \mathbb{R}^n$ with $\underline{u} > \underline{0}$ such that $\mathfrak{M}(A)\cdot\underline{u} > \underline{0}$;

iii) $\mathfrak{M}(A)$ is of <u>generalized positive type</u>, i.e., there exists a $\underline{u} \in \mathbb{R}^n$ such that

a) $\underline{u} > \underline{0}$, $\mathfrak{M}(A)\cdot\underline{u} \geq \underline{0}$, and $\{i \in \langle n\rangle : (\mathfrak{M}(A)\cdot\underline{u})_i > 0\}$ is nonempty;

b) for each $i_o \in \langle n\rangle$ with $(\mathfrak{M}(A)\cdot\underline{u})_{i_o} = 0$, there exist indices $i_1, i_2, \cdots, i_r$ in $\langle n\rangle$ with $a_{i_k, i_{k+1}} \neq 0$, $0 \leq k \leq r-1$, such that $(\mathfrak{M}(A)\cdot \underline{u})_{i_r} > 0$;

iv) there exists a $\underline{u} \in \mathbb{R}^n$ with $\underline{u} > \underline{0}$ such that $\mathfrak{M}(A)\cdot\underline{u}) \geq \underline{0}$, and such that $\sum_{j\leq i} \alpha_{i,j}u_j > 0$ for all $i \in \langle n\rangle$, where $\mathfrak{M}(A) := [\alpha_{i,j}]$ (cf. (2.8));

v) there exist respectively lower triangular and upper triangular nonsingular M-matrices L and U such that $\mathfrak{M}(A) = L\cdot U$;

vi) for any $B \in \Omega(A)$, $\rho(J_1(B)) \leq \rho(|J_1(B)|) = \rho(J_1 \mathfrak{M}(A)) < 1$;

vii) for any $B \in \Omega(A)$ and any $0 < \omega < 2/[1+\rho(J_1(B))]$, $\rho(J_\omega(B)) \leq \omega\cdot\rho(J_1(B)) + |1-\omega| < 1$;

viii) for any $B \in \Omega(A)$ and any $0 < \omega < 2/[1+\rho(|J_1(B)|)]$, $\rho(\mathcal{L}_\omega(B)) \leq \omega\cdot\rho(|J_1(B)|) + |1-\omega| < 1$;

ix) for any $B \in \Omega(A)$ and any $0 < \omega < 2/[1+\rho(|J_1(B)|)]$, $\rho(S_\omega(B)) < 1$.

That ii) is equivalent to i) is known, and is due to Fan [10]; that v) is equivalent to i) is known, and is due to Fiedler and Pták [11]. The definition of generalized positive type in iii) is a slight extension of a definition due to Bramble and Hubbard [7]; that iii) is equivalent to i) can be found in Varga [26], Rheinboldt [20], and Moré [17]. That iv) is equivalent to i) is new, and is due to Beauwens [4]. That vi), vii), and viii) are equivalent to i) is essentially new, and can be found in Varga [26]; similarly, that ix) is equivalent to i) is new, and can be found in Alefeld and Varga [2].

We now briefly comment on how certain recent results in the literature (back to 1968) are equivalent to, or weaker than, the listed conditions of Theorem 1. To begin, James and Riha [14] have defined $A = [a_{i,j}] \in \mathbb{C}_{\pi}^{n,n}$ to have generalized column diagonal dominance if there exists a $\underline{u} = [u_1, u_2, \cdots, u_n]^T \in \mathbb{R}^n$ with $\underline{u} > \underline{0}$ such that

$$(3.1) \quad |a_{i,i}|u_i > \sum_{\substack{j \in \langle n \rangle \\ j \neq i}} |a_{i,j}|u_j, \qquad \text{for all } i \in \langle n \rangle.$$

This definition, however, is precisely equivalent to ii) of Theorem 1, i.e., there exists a $\underline{u} \in \mathbb{R}^n$ with $\underline{u} > \underline{0}$ such that $\mathfrak{M}(A) \cdot \underline{u} > \underline{0}$. These authors in essence show [14, Theorem 4], under the added (unnecessary) assumption that A is irreducible, that ii) of Theorem 1 is equivalent to the combined hypotheses that $\rho(J_1 \mathfrak{M}(A)) < 1$ and $\rho(\mathcal{L}_\omega \mathfrak{M}(A)) < 1$ for $0 < \omega \leq 1$, which is weaker than the separate equivalences of ii), vi), and viii) of Theorem 1. Similarly, James [13, Theorem 1] shows that strict irreducible diagonal dominance for A (cf. [25, p. 23]), a stronger assumption than ii) of Theorem 1, implies that $\rho(\mathcal{L}_1(A)) < 1$, which is weaker than viii) of Theorem 1.

Next, given $A = [a_{i,j}] \in \mathbb{C}^{n,n}$, suppose that A is diagonally dominant, i.e., $\mathfrak{M}(A) \cdot \underline{\zeta} \geq \underline{0}$, where $\underline{\zeta} := [1, 1, \cdots, 1]^T$. If A is singular, so that A is not a nonsingular H-matrix, then from i) and iii) of Theorem 1, $\mathfrak{M}(A)$ cannot be of generalized positive type with respect to the vector $\underline{\zeta}$. Without going into detail, we simply remark that negating the property of generalized positive type (with respect to $\underline{\zeta}$) duplicates the main result of Erdelsky [9, Theorem 1].

In defining his Zeilensummenbedingung, Bohl [5, 6] weakens the assumption for generalized positive type matrices in iii) of Theorem 1 by allowing $\underline{u} \in \mathbb{R}^n$ in iii a) to satisfy

$\underline{u} \geq \underline{0}$, but then immediately shows [6, Satz 2.1] that if $\underline{u} \in \mathbb{R}^n$ with $\underline{u} \geq \underline{0}$ satisfies the remaining conditions of iii), then in fact $\underline{u} > \underline{0}$. Consequently, Bohl's Zeilensummen-bedingung and the hypothesis of generalized positive type, iii) of Theorem 1, are equivalent. For $A \in \mathbb{R}^{n,n}$, with $A \geq \mathcal{O}$, Bohl [6, Satz 2.2] shows that i), ii), and iii) of Theorem 1 are equivalent when applied to I-A.

Defining for any $A = [a_{i,j}] \in \mathbb{C}^{n,n}$ its induced operator norm $\|A\|_\infty$ by

$$(3.1) \quad \|A\|_\infty := \max_{i \in \langle n \rangle} \{ \sum_{j \in \langle n \rangle} |a_{i,j}| \},$$

Schäfke [21, Satz 1], improving on a paper by Walter [28], gives six equivalent conditions for an $A = [a_{i,j}] \in \mathbb{R}^{n,n}$ satisfying $A \geq \mathcal{O}$ and $\|A\|_\infty \leq 1$, to have $\rho(A) < 1$. This can be viewed as finding equivalent conditions on A for I-A to be a nonsingular M-matrix. Condition 4 of [21, Satz 1], in fact, reduces to ii) of Theorem 1 in this case.

Next, Kulisch [15, Theorem 1], as a special case, establishes that $\rho(\mathcal{L}_\omega(B)) < 1$ for any $0 < \omega < 2/[1+\rho(|J_1(B)|)]$ and for any B with $\rho(|J_1(B)|) < 1$, and deduces [15, Cor. 1.3] that B, being either strictly or irreducibly diagonally dominant, is sufficient for $\rho(|J_1(B)|) < 1$. This last deduction follows more generally from the equivalence of iii) and vi) of Theorem 1. See also Apostolatos and Kulisch [3].

Continuing, Elsner [8] gives the definition of a verallgemeinerte Zeilensummenkriterium, which turns out to be precisely condition ii) of Theorem 1, applied to the matrix I-A. As consequences of his definition, Elsner in essence shows that ii) implies iii) of Theorem 1, and that ii) implies the convergence of the Gauss-Seidel iterative method, a special case $\omega=1$ of viii) of Theorem 1.

Next, in the iterative solution of nonlinear systems of equations, a number of authors have contributed results which, in the linear case, relate to various parts of Theorem 1. For example, Müller [18] formulates the concept [18, Def. 5] of a chained weakly contractive system which, when applied to the linear matrix equation $(I-A)\underline{x} = \underline{k}$, reduces precisely to iii) of Theorem 1, i.e., $\mathfrak{M}(I-A)$ is of generalized positive type. In the spirit of Theorem 1 and the work of Schäfke [21], Müller [18, Sätze 4, 5, 5a] develops ten consequences and equivalences (when $A \geq \mathcal{O}$) of his chained weakly contractive system in the linear case, as, for example, the equivalence of ii), iii), and vi) of Theorem 1. In a similar vein, Rheinboldt [20, Thm. 4.4] deduces, as a special case of such nonlinear investigations, the equivalence of i), ii), and iii) of Theorem 1, while Moré [17] gives the definition in the linear case of Ω-diagonally dominant matrices, which turns out to be equivalent to the assumption that $\mathfrak{M}(A)$ be of generalized positive type (cf. iii) of Theorem 1) with respect to the vector $\underline{\zeta} = [1,1,\cdots,1]^T$. Moré then shows [17, Thm. 4.7] the equivalence of iii) and i) of Theorem 1.

Continuing our discussion, in Young [29, p. 43], one can deduce the equivalence of i) and vi) of Theorem 1, and it is shown [29, p. 107] that if A is irreducibly diagonally dominant, a stronger hypothesis than iii) of Theorem 1, then $\rho(J_\omega(A)) < 1$ and $\rho(\mathcal{L}_\omega(A)) < 1$ for any $0 < \omega \leq 1$, which are respectively weaker than vii) and viii) of Theorem 1. It is also shown [29, p. 126] that i) of Theorem 1 implies $\rho(\mathcal{L}_\omega(\mathfrak{M}(A))) < 1$ for any $0 < \omega < 2/[1+\rho(J_1(\mathfrak{M}(A)))]$, which is a special case of viii).

Next, it is interesting to note that Beauwens [4], who introduces the condition iv) of Theorem 1, calls this property when $\underline{u} = \underline{\zeta}$, lower semi-strictly diagonal dominance, and shows [4] that this property, coupled with irreducibility (cf. [25, p. 19]), is equivalent with irredicible diagonal dominance. Next, the result of Jacobsen [12] and Meijerink and Van der Vorst [16] is that a nonsingular symmetric M-matrix (which is necessarily positive definite) can be factored as $G\cdot G^T$, where G is a nonsingular triangular M-matrix, which in essence is weaker than the equivalence of i) and v).

Finally, in Shivakumar and Chew [22], one finds as the main result that the special case $\underline{u} = \underline{\zeta}$ of ii) of Theorem 1, implies that A is nonsingular, which is weaker than the equivalence of i) and ii) of Theorem 1.

§4. On Bounding $\|A^{-1}\|_\infty$.

In a recent paper, Varah [24] established

Theorem A. Assume that $A = [a_{i,j}] \in \mathbb{C}^{n,n}$, $n \geq 2$, is strictly diagonally dominant, i.e.,

$$(4.1)\quad \{|a_{i,i}| - \sum_{\substack{j\in\langle n\rangle \\ j\neq i}} |a_{i,j}|\} > 0, \quad \text{for all } i \in \langle n\rangle,$$

and set

$$(4.2)\quad \alpha := \min_{i\in\langle n\rangle} \{|a_{i,i}| - \sum_{\substack{j\in\langle n\rangle \\ j\neq i}} |a_{i,j}|\}.$$

Then (cf. (3.1)),

$$(4.3)\quad \|A^{-1}\|_\infty \leq 1/\alpha,$$

and

Theorem B. Assume that $A = [a_{i,j}] \in \mathbb{C}^{n,n}$ and A^T are both strictly diagonally dominant, and set

$\beta := \min_{i\in\langle n\rangle} \{|a_{i,i}| - \sum_{\substack{j\in\langle n\rangle \\ j\neq i}} |a_{j,i}|\}$. Then, the smallest singular value $\sigma_n(A)$ of A can be bounded below by

(4.4) $\sigma_n(A) := (\|A^{-1}\|_2)^{-1} \geq \sqrt{\alpha\beta}$.

We first remark that Theorem A is known in the literature: see Ahlberg and Nilson [1, p. 96]. Now, using the theory of M-matrices, we show how Theorems A and B can be generalized. First, the assumption in Theorem A that A is strictly diagonally dominant is equivalent to assuming that $\mathfrak{M}(A)\cdot\underline{\zeta} > \underline{0}$, whence ii) of Thoerem 1 is satisfied with the particular vector $\underline{u} = \underline{\zeta} = [1,1,\cdots,1]^T$. Thus, from the equivalence of i) and ii) of Theorem 1, assuming that A is a nonsingular H-matrix implies (after a simple normalization) that the set

(4.5) $U_A := \{\underline{u} \in \mathbb{R}^n: \underline{u} > \underline{0},\ \mathfrak{M}(A)\cdot\underline{u} > \underline{0}, \text{ and } \|\underline{u}\|_\infty = 1\}$

is nonempty. Then, define

(4.6) $f_A(\underline{u}) := \min_{i\in\langle n\rangle} \{(\mathfrak{M}(A)\cdot\underline{u})_i\}$, for any $\underline{u} \in U_A$.

The generalization (cf. [27]) of Theorem A is

Theorem 2. If $A \in \mathbb{C}_\pi^{n,n}$, $n \geq 2$, is a nonsingular H-matrix, then

(4.7) $$\sup_{B\in\Omega_A} \|B^{-1}\|_\infty = \|(\mathfrak{M}(A))^{-1}\|_\infty = \frac{1}{\max\{f_A(\underline{u}): \underline{u} \in U_A\}} .$$

Note that if A, as in Theorem A, is strictly diagonally dominant, then $\underline{\zeta} \in U_A$, and $f_A(\underline{\zeta}) = \alpha$, where α is defined in (4.2). Recalling that A is an element of Ω_A from (2.1), we see that (4.7) of Theorem 2 implies (4.3) of Theorem A. Similarly, the following (cf. [27] then generalizes Theorem B.

Theorem 3. If $A \in \mathbb{C}^{n,n}$, $n \geq 2$, is a nonsingular H-matrix, then (cf. (4.4))

(4.8) $\sigma_n(A) \geq \{f_A(\underline{u})\cdot f_{A^T}(\underline{v})\}^{1/2}$, for any $\underline{u} \in U_A$, $\underline{v} \in U_{A^T}$.

References

1. J. H. Ahlberg and E. N. Nilson, "Convergence properties of the spline fit", J. SIAM 11(1963), 95-104.

2. G. Alefeld and R. S. Varga, "Zur Konvergenz des symmetrischen Relaxationsverfahrens", Numerische Mathematik (to appear).

3. N. Apostolatos and U. Kulisch, "Über die Konvergenz des Relaxationsverfahrens bei nicht-negativen und diagonaldominanten Matrizen", Computing 2(1967),17-24.

4. Robert Beauwens, "Semi-strict diagonal dominance",SIAM J. Numer. Anal. (to appear).

5. Erich Bohl, Monotonie: Lösbarkeit und Numerik bei Operatorgleichungen, Springer Tracts in Natural Philosophy, 25(1974).

6. Erich Bohl, "Über eine Zeilensummenbedingung bei L-Matrizen", Lecture Notes in Mathematics 395(Sammelband der Tagung über Numerische Lösung nichtlinear partieller Differential- und Integrodifferentialgleichungen, in Oberwolfach), Springer, 1974, 247-263.

7. J. H. Bramble and B. E. Hubbard, "On a finite difference analogue of an elliptic boundary value problem which is neither diagonally dominant nor of non-negative type", J. Math. and Phys. 43(1964), 117-132.

8. L. Elsner, "Bemerkungen zum Zeilensummenkriterium", Zeit. Angew. Math. Mech. 49(1966), 211-214.

9. P. J. Erdelsky, "A general theorem on dominant-diagonal matrices", Linear Algebra Appl. 1(1968), 203-209.

10. K. Fan, "Topological proof for certain theorems on matrices with non-negative elements", Monatsh. Math. 62(1958), 219-237.

11. Miroslav Fiedler and Vlastimil Pták, "On matrices with non-positive off-diagonal elements and positive principal minors", Czech. Math. J. 12(87)(1962), 382-400.

12. D. H. Jacobsen, "Factorization of symmetric M-matrices", Linear Algebra and Appl. 9(1974), 275-278.

13. K. R. James, "Convergence of matrix iterations subject to diagonal dominance", SIAM J. Numer. Anal. 10(1973), 478-484.

14. K. R. James and W. Riha, "Convergence criteria for successive overrelaxation", SIAM J. Numer. Anal. 12(1974), 137-143.

15. U. Kulisch, "Über reguläre Zerlegungen von Matrizen und einige Anwendungen", Numer. Math. 11(1968), 444-449.

16. J. A. Meijerink and N. A. Van der Vorst, "Iterative solution of linear systems arising from discrete approximation to partial differential equations", J. Computational Physics (to appear).

17. Jorge J. Moré, "Nonlinear generalizations of matrix diagonal dominance with application to Gauss-Seidel iterations", SIAM J. Numer. Anal. 9(1972), 357-378.

18. Karl Hans Müller, "Zum schwachen Zeilensummenkriterium bei nichtlinearen Gleichungssystemen", Computing 7(1971), 153-171.

19. A. M. Ostrowski, "Über die Determinanten mit überwiegender Haupdiagonale", Comment. Math. Helv. 10(1937), 69-96.

20. Werner C. Rheinboldt, "On M-functions and their application to nonlinear Gauss-Seidel iterations and to network flows", J. Math. Anal. Appl. 32(1970), 274-307.

21. F. W. Schäfke, "Zum Zeilensummenkriterium", Numer. Math. 12(1968), 448-453.

22. P. N. Shivakumar and Kim Ho Chew, "A sufficient condition for nonvanishing of determinants", Proc. Amer. Math. Soc. 43(1974), 63-66.

23. O. Taussky, "A recurring theorem on determinants", Am. Math. Monthly 56(1948), 672-676.

24. J. M. Varah, "A lower bound for the smallest singular value of a matrix", Linear Algebra and Appl. 11(1975), 3-5.

25. Richard S. Varga, Matrix Iterative Analysis, Prentice-Hall, Englewood Cliffs, N.J., 1962.

26. Richard S. Varga, "On recurring theorems on diagonal dominance", Linear Algebra Appl. (Olga Taussky Special Issue), (to appear).

27. R. S. Varga, "On diagonal dominance arguments for bounding $\|A^{-1}\|_\infty$", Linear Algebra Appl. (to appear).

28. Wolfgang Walter, "Bemerkungen zu Iterationsverfahren bei linearen Gleichungssystemem", Numer. Math. 10(1967), 80-85.

29. David M. Young, Iterative Solution of Large Linear Systems, Academic, New York, 1971.

VI

Applications

Sparse Matrix Problems in a Finite Element Open Ocean Model

Joel E. Hirsh
Center for Earth and Planetary Physics

William L. Briggs
Applied Mathematics, Aiken Computation Laboratory

Harvard University, Cambridge, Massachusetts 02138

1. Introduction

The earliest efforts in numerical oceanography were devoted to large scale models that typically encompassed an entire ocean basin. These models furthered the understanding of the general features of ocean circulation, such as the Gulf Stream, but left unresolved the dynamics of sub-basin or mesoscale phenomena. Two such features are the meandering of the Gulf Stream and mid-ocean eddies. Both of these phenomena are highly non-linear, time dependent and almost certainly affect the global circulation.

In an attempt to resolve these mesoscale problems, several models have been constructed (Bretherton and Karweit, 1975; Holland and Lin, 1975). However, these models have the restriction of either periodic boundary conditions or of a closed basin domain. Our model removes these restrictions.

In Section 2, we write the quasigeostrophic equations of motion and boundary conditions for an open-sided ocean. The equations consist of an advection equation for the quasi-geostrophic potential vorticity and a three-dimensional Poisson equation for the streamfunction.

In constructing our model, we have found it desirable to minimize the effects of friction so as not to have simply a spin-down problem. In addition, we would like to consider a domain of arbitrary shape so as to be able to include irregular coastlines or to use arbitrary data sets as boundary conditions. For these reasons, we have chosen a finite element method (Section 3) to solve the time dependent, quasi-geostrophic problem.

The finite element method gives a scheme which conserves kinetic energy, vorticity and enstrophy (mean squared vorticity) which is desirable in a problem with little or no physical dissipation. In addition, these conservation properties can be maintained on arbitrary grids with open boundaries even in the case of higher order schemes (Fix, 1975).

This paper proceeds from a preliminary formulation of the associated two-dimensional problem (Fix, 1975). In that study a method of solution was proposed and its stability, convergence and conservation properties were analyzed. For our present purposes we only assert that these properties may be extended to the three-dimensional problem. Our emphasis in this paper will be upon practical and efficient implementations of numerical solutions. It is first necessary to raise some pre-computational questions which will determine the cost and complexity of the model as a numerical undertaking. The original physical problem and its numerical formulation are posed with a degree of generality that makes any method of solution very time consuming and hence makes any long time integrations very costly. Is it possible to sacrifice some of the generality of the original problem to make its solution feasible computationally? How might the original problem be altered to make a numerical solution less prohibitive? We will discuss the terms of such compromises and show that they lead to efficient models which still retain physical utility and meaning.

In its most complete generality the mid-ocean problem is posed in a domain of arbitrary horizontal shape with the possibility of bottom topography. This would include the situation in which the bottom actually comes to the surface and the side walls vanish. To accommodate this sort of geometry triangular elements in the horizontal were originally proposed. In this way tetrahedral elements could be formed to fit an irregular bottom. Clearly, consideration of a flat bottomed domain is a major simplification. As will be shown, some bottom topography can be included implicitly in the boundary conditions. With this concession a cylindrical domain of arbitrary cross-section is still possible.

The problem is still fully three-dimensional due to the coupling through the vertical elements. The next step (Section 4) in the reduction is to eliminate the vertical coupling in the time dependent advection equation while retaining it in the Poisson equation. The physical implications of this decoupling will be discussed. The computational gain is that now a two-dimensional problem can be solved on each level.

In Section 5 we consider the further constraint of a rectangular domain. In this case it is possible to dispense with triangular elements and use a tensor product basis. Finally by imposing a uniform grid in the horizontal further improvements in efficiency may be realized.

In Section 6, we show some initial results on a barotropic problem and discuss problems that must be solved in order to make efficient use of the finite elements method for ocean modelling.

2. Quasigeostrophic equations and boundary conditions

If x, y, z represent the east, north and vertical directions, the quasigeostrophic equations take the form (Pedlosky, 1964)

$$\frac{D}{Dt}(q+f) = 0 \tag{2.1a}$$

$$\frac{\partial^2\psi}{\partial x^2} + \frac{\partial^2\psi}{\partial y^2} + \frac{\partial}{\partial z}\left(\frac{f_o^{\,2}}{\eta^2(z)}\frac{\partial\psi}{\partial z}\right) = q \tag{2.1b}$$

where

$$\frac{D}{Dt} = \frac{\partial}{\partial t} + u\frac{\partial}{\partial x} + v\frac{\partial}{\partial y}$$

is the total time derivative. The horizontal velocity components u, v in the eastward and northward directions can be defined in terms of the streamfunction, ψ, as

$$u = -\psi_y, \quad v = \psi_x \ .$$

q as defined by eq. (2.1b) is the quasigeostrophic potential vorticity. The Coriolis parameter f is given by $f_o+\beta y$ and $\eta^2(z) = -g\rho_z(z)/\rho_o$ is the Brunt-Vaisala frequency. $\overline{\rho}(z)$ is horizontally averaged vertical density distribution, ρ_o is the mean density and g is gravity. Note that if $\partial/\partial z \doteq 0$, eqs. (2.1) become the barotropic equations studied by Fix (1975). To simplify the notation, let $\sigma(z)=f_o^2/\eta^2(z)$. Eqs. (2.1) can be rewritten as

$$\frac{\partial q}{\partial t} + J(\psi,q) + \beta\psi_x = 0 \tag{2.2a}$$

$$(\nabla^2 + \frac{\partial}{\partial z}(\sigma(z)\frac{\partial}{\partial z}))\psi = q \tag{2.2b}$$

where $\nabla^2 = \partial^2/\partial x^2 + \partial^2/\partial y^2$ and $J(\psi,q)=\psi_x q_y-\psi_y q_x$. Eqs. (2.2) hold for $(x, y, z)\varepsilon\Omega$ and $t>0$.

The initial conditions are simply given by

$$\psi(x,y,z,t=0) = \psi_0(x,y,z) \qquad (x,y,z)\varepsilon\Omega$$

$$q(x,y,z,t=0) = q_0(x,y,z) \qquad (x,y,z)\varepsilon\Omega \ .$$

We now express Γ, the boundary as a union of the top, bottom and lateral walls, i.e.,

$$\Gamma = \Gamma_T \cup \Gamma_B \cup \Gamma_L \ .$$

The lateral boundary conditions will be the Charney-Fjörtoft-

Von Neumann (1950) conditions, i.e.,

$$\psi(x,y,z,t) = \psi_\Gamma(x,y,z,t) \qquad (x,y,z)\varepsilon\Gamma_L,\ t\geq 0 \qquad (2.3a)$$

$$q(x,y,z,t) = q_{IN}(x,y,z,t) \qquad (x,y,z,t)\varepsilon\Gamma_{IN}, t\geq 0 \quad (2.3b)$$

where $\Gamma_L = \Gamma_{IN} \cup \Gamma_{OUT}$. Γ_{IN} is defined to be that portion of Γ_L where the normal velocity given by the tangential derivative of ψ is into the domain, i.e.,

$$\Gamma_{IN}(t) = \{(x,y,z)\varepsilon\Gamma_L | \frac{\partial\psi}{\partial s}(x,y,z,t)>0\}$$

where $\partial/\partial s$ is the tangential derivative in the counterclockwise direction.

The top and bottom boundary conditions come from the condition that there be no flow normal to these boundaries. In terms of ψ, the vertical velocity w can be expressed as $D/Dt(\sigma(z)\partial\psi/\partial z)$ (Bretherton and Karweit, 1975). If $z= -H+B(x,y)$, describes the bottom topography, then the boundary conditions for a rigid lid are

$$\frac{D}{Dt}(\sigma(0)\frac{\partial\psi}{\partial z}) = 0 \qquad (x,y)\varepsilon\Gamma_T, t>0 \qquad (2.4a)$$

$$\frac{D}{Dt}(\sigma(z=-H)\frac{\partial\psi}{\partial z}) = \frac{D}{Dt}B \qquad (x,y,z)\varepsilon\Gamma_B, t>0 . \qquad (2.4b)$$

Note that if $B<<H$, where H is the mean depth, condition (2.4b) can be imposed at $z= -H$. The quantity $\sigma(z)\partial\psi/\partial z$ can be computed as the solution of a two-dimensional partial differential equation where the differential operator D/Dt is the same as in (2.1a). Solving for the top and bottom boundary condition can use much of the same computer code that is used in solving (2.1a).

3. Finite element approximation

Our approximation to the quasigeostrophic equations is obtained by the classical Galerkin (or weighted residuals) method used in conjunction with finite elements. We will formulate the problem for a triangular grid in the horizontal; the solution with a rectangular grid will then appear as a special case. The domain Ω is first divided by K arbitrary levels in the vertical. Each horizontal domain, denoted Ω_k, is then subdivided into triangles. We consider only the case of a flat-bottomed domain with topography in the boundary condition (2.4b). The surface and bottom then correspond to Ω_1 and Ω_K.

In practice, three specific finite elements will be

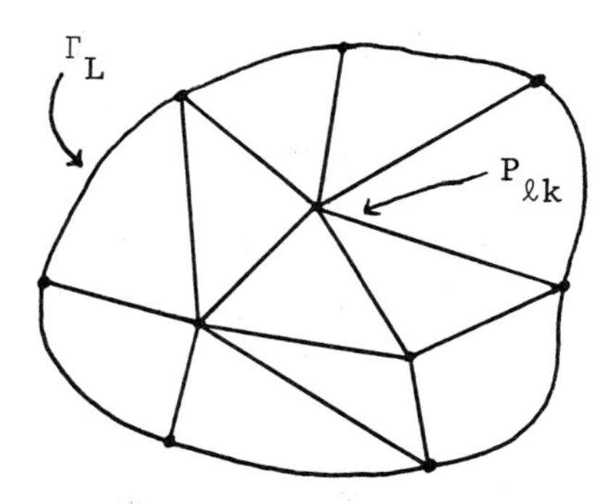

Fig. 1 - Discretization of Ω_k.

considered. However, the formalism that follows applies much more generally. Let $P_{\ell k}=(x_\ell,y_\ell,z_k)$ $1\leq\ell\leq N$, $1\leq k\leq K$ be the points at which the solution is to be approximated. Any function $v^h(x,y,z)$ in the finite element space S^h may be written

$$v^h(x,y,z) = \sum_{i=1}^{N} \sum_{j=1}^{K} v^h(x_i,y_i,z_j)\phi_i(x,y)\Phi_j(z) \quad .$$

The space S^h is then a tensor product of two smaller spaces. In the following discussion $h>0$ will denote an average mesh spacing in the horizontal grid. Quantities associated with finite elements are then parameterized by h. The three elements we will consider are the piecewise linear, quadratic and cubic polynomials.

For the linear elements, the points $p_{\ell k}$ on each horizontal plane correspond to the vertices of the triangles. In the case of the quadratic elements, the $p_{\ell k}$ are the vertices plus the midpoints of the sides of the triangles. For the cubic elements the vertices, the centroid and two equally spaced points on each edge constitute the $p_{\ell k}$.

We now approximate the streamfunction and vorticity in the form

$$\psi^h(x,y,z,t) = \sum_{i=1}^{N} \sum_{j=1}^{K} \psi_{ij}(t)\phi_i(x,y)\Phi_j(z) \tag{3.1a}$$

$$q^h(x,y,z,t) = \sum_{i=1}^{N} \sum_{j=1}^{K} q_{ij}(t)\phi_i(x,y)\Phi_j(z) \quad . \tag{3.1b}$$

The coefficients $\psi_{ij}(t)$ and $q_{ij}(t)$ are to be determined.

Following Fix (1975), we define the inner product as

$$(u,v) = \iiint uv\,dx\,dy\,dz \quad .$$

The quasigeostrophic equations in the integral form are then

$$\left(\frac{\partial q}{\partial t}, v\right) + (J(\psi, q+f), v) = 0 \qquad \forall v \varepsilon L_2(\Omega) \qquad (3.2a)$$

$$(\nabla\psi, \nabla v) + (\sigma\psi_z, v_z) = -(q, v) \qquad \forall v \varepsilon H_o^1(\Omega) \qquad (3.2b)$$

where $H_o^1(\Omega)$ denotes the Sobolev space of functions v satisfying

$$||v||_1 = \{\iiint_\Omega [|\nabla v|^2 + \sigma(z) v_z^2 + v^2] dxdydz\} < \infty$$

and v=0 on Γ_L.*

The Galerkin idea is to use (3.2) with the test functions restricted to the finite element space S^h. Using the approximations to the streamfunction and vorticity given in (3.1) we have the equations

$$\left(\frac{\partial q^h}{\partial t}, \phi_\ell \Phi_k\right) + \left(J(\psi^h, q^h+f), \phi_\ell \Phi_k\right) = 0 \qquad (3.3a)$$

for $\qquad P_{\ell k} \ \varepsilon \ \Omega \cup \Gamma_{OUT}(t) \cup \Gamma_B \cup \Gamma_T \qquad (3.3b)$

and $\qquad (\nabla\psi^h, \nabla\phi_\ell\Phi) + (\sigma(z)\psi_z^h, \phi_\ell\Phi_{k,z}) = -(q^h, \phi_\ell\Phi_k)$

for $\qquad P_{\ell k} \ \varepsilon \ \Omega \cup \Gamma_B \cup \Gamma_T$.

Eq. (3.3a) is a system of non-linear ordinary differential equations in the coefficients $q_{\ell k}(t)$ and $\psi_{\ell k}(t)$. It may be rewritten more simply as

$$\frac{d}{dt} \underset{\sim}{C} q = \underset{\sim}{J}(\psi)(q+f)$$

where the notation $\underset{\sim}{J}(\psi)$ is meant only to indicate the dependence of J on ψ. The symbols q and ψ are perhaps best understood as NxK matrices.

The mass matrix $\underset{\sim}{C}$ is the tensor product $\underset{\sim}{C}^H \otimes \underset{\sim}{C}^V$ where $\underset{\sim}{C}^H$ and $\underset{\sim}{C}^V$ are the inner product matrices for the horizontal and vertical basis functions. The non-linear term on the right hand side is not easily expressed in a compact form. Viewed as an NxK matrix its (ℓ,k) entry is given by

---

* For simplicity we are taking the top and bottom boundary conditions to be simply $\psi_z=0$ which is the natural boundary condition when the equations are in this integral form.

$$\sum_{i,m=1}^{N} \sum_{j,n=1}^{K} \left\{ q_{ij}\psi_{mn}(\Phi_j\Phi_n,\Phi_k) \left(\frac{\partial\phi_i}{\partial x}\frac{\partial\phi_m}{\partial y} - \left(\frac{\partial\phi_i}{\partial y} + \beta\right)\frac{\partial\phi_m}{\partial x} , \phi_\ell \right) \right\} \tag{3.4}$$

The elliptic eq. (3.3b) may be expressed in matrix terms as

$$\underset{\sim}{K}\psi = \underset{\sim}{C}q \ . \tag{3.5}$$

The stiffness matrix K may be considered the tensor product $\underset{\sim}{C}^V \otimes \underset{\sim}{D}^H + \underset{\sim}{C}^H \otimes \underset{\sim}{D}^V$ where $\underset{\sim}{D}^H$ and $\underset{\sim}{D}^V$ have entries $(\nabla\phi_i, \nabla\phi_\ell)$ and $(\sigma(z)\Phi_{j,z}, \Phi_{k,z})$ respectively.

The problem must now be discretized in time also. We have chosen an implicit leap frog method given by

$$\underset{\sim}{C}\left(\frac{q^{(n+1)} - q^{(n-1)}}{2\Delta t}\right) = -\underset{\sim}{J}(\psi^{(n)})\left[\frac{1}{2}\left(q^{(n+1)} + q^{(n-1)}\right) + f\right] \tag{3.6}$$

where Δt is the time step and $q^{(n)}(x,y,z)$ and $\psi^{(n)}(x,y,z)$ approximate $q(x,y,z,n\Delta t)$ and $\psi(x,y,z,n\Delta t)$ respectively. Implicitness (in addition to the implicitness inherent in the finite element method) is required in order to overcome numerical instability which arises from the one-sided differences which appear on outflow boundaries where q is computed. This instability has been studied in the case of linear advection in one dimension (Kreiss, Oliger, 1973). The scheme (3.6) is stable provided that the Courant-Friedrichs-Lewy condition $U\Delta t/h \leq \alpha$ is satisfied where U is the maximum of the fluid speeds and the phase speeds present and α is a constant of $O(1)$.

The leap frog method requires initial conditions at both $t=0$ and $t=\Delta t$. Given the initial conditions

$$q^{(0)} = q_0(x,y,z) \qquad x,y,z \in \Omega \cup \Gamma$$

$$\psi^{(0)} = \psi_0(x,y,z) \qquad x,y,z \in \Omega \cup \Gamma$$

we can compute $q^{(1)}$ and $\psi^{(1)}$ to $O(\Delta t^2)$ by integrating out to $t=\Delta t$ using two different time steps and applying Richardson extrapolation.

4. Method of solution

Eqs. (3.5) and (3.6) present several difficulties in

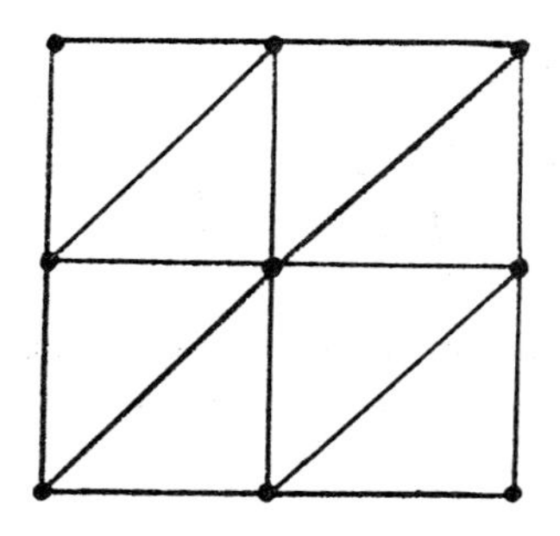

Fig. 2 - Regular triangular grid.

their solution. If a regular triangular grid (see Fig. 2) is used in each horizontal plane with linear elements in the horizontal and vertical, then each unknown is related to as many as twenty other unknowns by these equations. From the outset we are faced the task of inverting two matrices with non-trivial bandwidths. In addition, because of the implicitness and the inflow boundary conditions, not only do the elements of the implicit mass matrix, $\underset{\sim}{C}+\Delta t\ \underset{\sim}{J}(\psi)$, change at each time step but the size and structure of the matrix change as well. Therefore the thought of decomposing it once at the beginning must be abandoned. The Poisson equation, in addition to being three-dimensional, has a variable coefficient and must in general be solved on a highly non-uniform grid in the vertical.

The first numerical solutions of this problem were obtained using a point SOR iteration in the vorticity equation and an ADI algorithm to solve the Poisson equation. This approach involves not only solving eq. (3.6) for the vector $q^{(n+1)}$ but constructing the implicit part of the mass matrix from eq. (3.4). Initial results showed that the amount of computer time required for a 32x32x7 grid would be greater than one minute per time step on the CDC7600. This is totally impractical. Further examination showed that the bulk (>60%) of the effort was involved in computing the implicit part of the mass matrix (3.4).

The inflow-outflow conditions and the non-linearity are inherent features of the problem with which we must contend in any version that is solved. However, we do seek some simplification in the coupling of the problem which does not seem to be as fundamental to the physical problem. The vorticity equation is a statement about horizontal advection. In fact horizontal processes clearly dominate in the physical model. The vertical coupling in the vorticity equation is introduced by the finite element approximation. Therefore as a simplifying assumption with physical plausibility we decouple the vertical levels in the vorticity equation. The result is that a two-dimensional advection problem can be solved on each level rather than solving one three-dimensional problem. The most immediate consequence of this assumption is that the conservation properties, whose proof was so natural in the finite element framework, no longer hold. However, we do know that the horizontal schemes which include all of the advection are conservative.

The vertical coupling in the physical model arises through the Poisson equation and that must be reflected in the discrete problem also. To obtain some simplification while retaining the vertical dependence, we choose to use a lumping process with respect to the mass matrices $\underset{\sim}{C}^V$ and $\underset{\sim}{C}^H$. This leads to a five point scheme on each level together with the usual three point, centered difference in the vertical. With these modifications, we may now consider numerical solutions.

For the sake of discussion we confine our attention to linear elements for the remainder of this section. We consider the Poisson equation first. If a uniform mesh is used in each horizontal plane, the stiffness matrix has the form

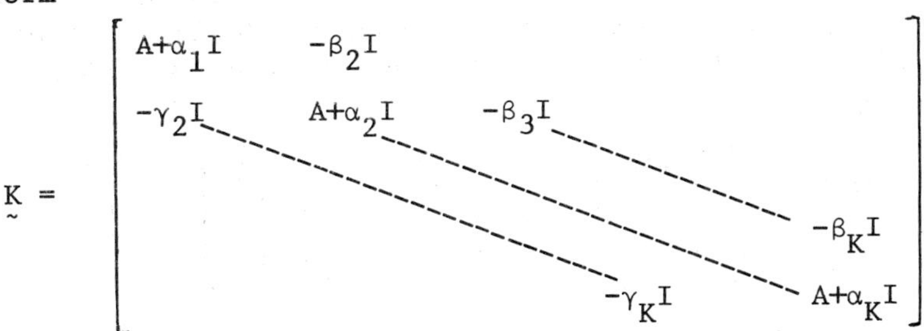

$$\underset{\sim}{K} = \begin{bmatrix} A+\alpha_1 I & -\beta_2 I & & & \\ -\gamma_2 I & A+\alpha_2 I & -\beta_3 I & & \\ & \ddots & \ddots & \ddots & \\ & & & & -\beta_K I \\ & & & -\gamma_K I & A+\alpha_K I \end{bmatrix}$$

where A is NxN and corresponds to the constant coefficient five point scheme on one plane. The α_i, β_i, γ_i arise due to the variable coefficient and the non-uniform grid.

The solution of this linear system of equations is ideally suited to a marching algorithm (Bank; Bank and Rose). On each level a two-dimensional problem can be solved using the generalized marching algorithm. With an appropriate choice of the marching parameter k, this problem can be solved in $O(n^2 \log(n/k))$ operations where we have assumed that $N=n^2$. Because the number of levels is not expected to exceed 8, straight marching is used in the vertical. The separable version of the algorithm must be used to handle the variable coefficient in z. The operation count for the three-dimensional problem is then $O(Kn^2 \log(n/k))$. If a non-uniform grid is used in the horizontal planes, the problem becomes variable coefficient in all three variables and the separable version of the algorithm is used for the horizontal problem also. The operation counts remain the same.

We turn now to the vorticity equation. The decoupling of the levels allows us to consider the equation on one level at a time. We therefore rewrite eq. (3.6) in the form

$$\underset{\sim}{C}q = \underset{\sim}{J}(\psi)q+b \tag{4.1}$$

where b comprises all terms which depend on previous time levels including the Coriolis term. $\underset{\sim}{C}$ is the mass matrix associated with the horizontal elements which was denoted $\underset{\sim}{C}^H$ before. With q and ψ viewed now as N- vectors the kth element of the Jacobian term is

$$\sum_{i,m=1}^{N} q_i\psi_m\left(\frac{\partial\phi_i}{\partial x}\frac{\partial\phi_m}{\partial y} - \frac{\partial\phi_i}{\partial y}\frac{\partial\phi_m}{\partial x}, \phi_k\right).$$

As noted earlier $\underset{\sim}{C}$ and $\underset{\sim}{J}(\psi)$ are both time dependent. In order to isolate the time dependent part of the mass matrix, the unknowns corresponding to the interior of the domain are ordered first; the unknowns on the boundary are ordered last. Eq. (4.1) may then be expressed in the partitioned form

$$\begin{bmatrix} C_I & L^T \\ L & C_B \end{bmatrix}\begin{bmatrix} q_I \\ q_B \end{bmatrix} = \begin{bmatrix} J_I & J_1 \\ J_2 & J_B \end{bmatrix}\begin{bmatrix} q_I \\ q_B \end{bmatrix} + \begin{bmatrix} b_I \\ b_B \end{bmatrix} \tag{4.2}$$

C_I is the mass matrix for the interior points and does not vary in time. It may be decomposed once and stored. The time dependence has been isolated in C_B, corresponding to the boundary points, and L which represents linkage between the interior and the boundary. J is similarly paritioned but still changes in time due to its dependence on ψ.

An iteration of some kind is inevitable. We first write (4.2) in the form

$$\begin{bmatrix} C_I & 0 \\ 0 & C_B \end{bmatrix}\begin{bmatrix} q_I \\ q_B \end{bmatrix} = \begin{bmatrix} J_I & J_1-L^T \\ J_2-L & J_B \end{bmatrix}\begin{bmatrix} q_I \\ q_B \end{bmatrix} + \begin{bmatrix} b_I \\ b_B \end{bmatrix} \tag{4.3}$$

We may then do the following coupled iteration

$$C_I q_I^{(k+1)} = J_I q_I^{(k)} + (J_1 - L^T) q_B^{(k)} + b_I \qquad (4.4a)$$

$$C_B q_B^{(k+1)} = (J_2 - L) q_I^{(k+1)} + J_B q_B^{(k)} + b_B \ . \qquad (4.4b)$$

The first equation (4.4a) involves only a back solution against the decomposed C_I. C_I is symmetric, positive definite and with linear elements is essentially block tridiagonal. The solution of this linear system is suited to recent implementations of the nested dissection algorithm (Rose and Whitten, 1975). Again letting $N=n^2$, this algorithm provides a decomposition in $O(n^3)$ operations with $O(n^2 \log n)$ storage. Each back solution then requires $O(n^2 \log n)$ operations.

In eq. (4.4b), C_B is fundamentally tridiagonal but its changing structure disallows a preliminary decomposition and makes any direct solution difficult. For this reason we choose to use a point SOR iteration on this subproblem. If (4.4b) is written in the form

$$(C_B - J_B) q_B^{(k+1)} = (J_2 - L) q_I^{(k+1)} + b_B \qquad (4.4c)$$

we may view this inner iteration as being against the interior points rather than the non-linear terms.

The solution of the vorticity equation then has been reduced to a nested iteration process. An analysis of the convergence of the complete iteration is not feasible due to the time dependence of the iteration matrices. Both the SOR parameter and the relative number of outer and inner iterations are subject to experiment determination. There is reason for some optimism. Because we are solving a time dependent problem each iteration will begin with a very good guess, namely the solution at the previous time step or even an extrapolated solution. In addition we would expect the number of iterations necessary for convergence to decrease with the time step.

5. Rectangular grids

We now retreat one last order of simplicity and consider rectangular domains in the horizontal. The levels are again decoupled as described in the previous section so that the vorticity equation is solved level by level. The formalism of the previous section still pertains but we may now take advantage of a tensor product basis. The unknowns q and ψ on a given level are approximated by

$$\psi^h(x,y,z,t) = \sum_{i=1}^{m} \sum_{j=1}^{n} \psi_{ij}(z,t)\alpha_i(x)\gamma_j(y) \quad .$$

$$q^h(x,y,z,t) = \sum_{i=1}^{m} \sum_{j=1}^{n} q_{ij}(z,t)\alpha_i(x)\gamma_j(y) \quad .$$

Here we have assumed the horizontal grid to be m x n.

Let us again consider the case of linear elements in both x and y. The bilinear elements give a nine-point scheme at each interior point which is globally second order in the mesh spacing.

The solution of the Poisson equation remains the same. The marching algorithm may be used on either the five or nine-point scheme in the plane. The vertical dependence is treated in exactly the same manner. The additional gains are made in solving the vorticity equation.

We again write the vorticity equation on a given level in the matrix form

$$\underset{\sim}{C}q = \underset{\sim}{J}(\psi)q + b \quad . \tag{5.1}$$

The mass matrix $\underset{\sim}{C}$ is now the tensor product $\underset{\sim}{C}^x \otimes \underset{\sim}{C}^y$ where $\underset{\sim}{C}^x$ and $\underset{\sim}{C}^y$ are the inner product matrices for the x and y elements respectively. If the unknowns q and ψ are now regarded as m x n matrices, the left hand side of (5.1) is simply the matrix product $\underset{\sim}{C}^x q \underset{\sim}{C}^y$ while the (k,ℓ) element of the m x n matrix $J(\psi)q$ may be written

$$\sum_{i,r=1}^{m} \sum_{j,s=1}^{n} \psi_{ij} q_{rs} \{ (\alpha_i'(x)\alpha_r(x),\alpha_k(x))(\gamma_j(y)\gamma_s'(y),\gamma_\ell(y)) - (\alpha_i(x)\alpha_r'(x),\alpha_k(x))((\gamma_j'(y)-\beta)\gamma_s(y),\gamma_\ell(y))\} \quad . \tag{5.2}$$

The inner products are with respect to the single independent variable indicated.

The problem is again partitioned into an interior and a boundary problem. An SOR iteration is used for the boundary problem as before. However, for the interior problem, the mass matrix, denoted $\underset{\sim}{C}_I$ earlier, is now easily inverted. The system

$$\underset{\sim}{C}^x q \underset{\sim}{C}^y = p$$

is solved in the two steps (a) $\underset{\sim}{C}^x v=p$, (b) $q\underset{\sim}{C}^y=v$ or $\underset{\sim}{C}^y q^T=v^T$. $\underset{\sim}{C}^x$ and $\underset{\sim}{C}^y$ are both symmetric tridiagonal matrices that may be decomposed once. The interior problem requires 6mn operations for each iteration. If bicubic elements are used, the same solution requires 16mn operations.

If we restrict our bilinear mesh to be uniform, then at interior points (5.2) is exactly the nine-point Arakawa scheme (Jespersen, 1975) which can be evaluated explicitly rather than by the more general formula (5.2). This represents a tremendous savings since evaluation of the non-linear contribution to the mass matrix (see Table 1) by a formula like (5.2) represents a major portion of the computational effort.

Table 1

Model Run Times (CDC7600)

	Quasigeostrophic Regular Bilinear Elements		Barotropic Triangular Elements
	9 x 9 x 4	17 x 17 x 8	17 x 17
Setting up	12 msec	72 msec	112 msec
Solution for $q^{(n+1)}$	4 outer itr. } 3 SOR itr. } 32 msec	3 outer itr. } 3 SOR itr. } 144 msec	6 SOR itr. } 26 msec
Solution of Poisson eq.	10 msec	80 msec	20 msec
Total time	54 msec	296 msec	158 msec

Some preliminary computer runs have been made with bilinear elements using the method just outlined. An iteration strategy was determined experimentally. On a 9x9x4 grid, three SOR iterations for the solution on the boundary seemed adequate while two or three outer iterations insured convergence to within truncation error. As mentioned earlier the number of iterations will certainly depend not only on the size of the problem but on the time step also.

Actual timing of the method was done to confirm our expectations that the model would be competitive with those presently being used. In Table 1 the results for two grids using bilinear finite elements on a regular grid are shown. These are compared to the results from a barotropic (one-level) code which uses the linear triangular elements on a regular grid but evaluates eq. (5.2) as shown. The advection equation in the barotropic code is solved by point SOR and the Poisson equation is solved using the marching algorithm on the five-point scheme (which is exactly the scheme that the linear triangular elements give on a regular grid).

Note that the eight level bilinear code uses approximately twice the computer time as the barotropic code with the same horizontal resolution. Setting up includes generating right hand sides and determining inflow-outflow boundary conditions. The bulk of the computer time used in the barotropic code is consumed in evaluating the non-linear advection terms. This calculation is replaced in the regular bilinear case by the explicit evaluation of the Arakawa Jacobian.

In practice slightly larger grids will be used. However, it should be possible to keep the model running under one second per time step. At this point no optimization with respect to the marching parameter k has been done since the solution of the Poisson equation is the least costly part of the problem. In addition acceleration techniques might be used to advantage on the iterations if more than three or four outer iterations become necessary.

6. Conclusions and possibilities

We have presented arguments that have reduced a general physical model to successively simpler versions which submit to efficient numerical solution yet still retain some physical value. A combination of fast linear algebra algorithms is used in each case to produce a model which is at least comparable in terms of running time to those currently in use. With these numerical foundations, it is now possible to pose new problems. These will be dictated chiefly by physical considerations.

The bilinear element model can be used to explore a variety of questions. The assumption that the problem can be decoupled in the vertical should be confirmed by a computational test of the conservation laws. Processes which depend upon the vertical stratification of the ocean are believed to be very important. These so-called baroclinic modes can be investigated with this model also.

Initial studies have begun in the barotropic model on a problem of forcing by Gulf Stream meanders. The problem is set in a square domain of side L where the east, west and south walls are rigid. The flow is driven by meanders along the north wall. The boundary conditions are defined as

$$\psi = \begin{cases} \frac{V_o}{k}\,[\sin\,(kx-\omega t)+\sin\omega t]G(t), & y=L;t>0 \\ 0 & x=0,L;y=0;t>0 \end{cases}$$

$$q = -V_o k\,\sin(kx-\omega t)G(t) \qquad (x,y)\varepsilon\Gamma_{IN},t>0$$

where $G(t)=1/2[\tanh(t-t_o/\tau)+1]$ is a smooth function in time which allows the model to be started from rest without introducing a strong singularity at t=0. V_o, k and ω are the velocity amplitude, wavenumber and angular frequency of the forcing. Linear analysis shows that energy propagates to the south if a critical parameter $M=\beta L/2\pi\omega \geq 1$. If M<1, the energy is trapped at the northern wall. These results have been confirmed in the numerical model where V_o=20 km/day, L=4500 km, $k=4\pi/L$. Contour plots of the streamfunction are shown in Figs. 3, 4 for $\omega=2\pi/4$, $2\pi/17$ which correspond to M=0.79, 3.35. These studies will be continued with the bilinear code.

In terms of extensions of these models, there are several directions in which to turn. We have already indicated how triangular elements can be used to consider domains of arbitrary horizontal shape. Perhaps more immediate is the problem of explicit bottom topography. The use of tetrahedral elements could become quite unwieldy, not only computationally but in terms of their assembly. If this approach is taken, it may be possible to include the topographic elements iteratively in much the same manner that the changing boundary points were included in the original problem.

There is another approach to the bottom topography which is appealing because it could be done entirely within the bilinear element model. Letting $z=B(x,y)$ denote the profile of the bottom, the change of vertical coordinate $\zeta=z/B(x,y)$ maps the domain back into a rectangular box. The ensuing changes of variable unavoidably couple the levels in the advection equation making that problem fully three-dimensional again. But with the savings the bilinear element model affords, these complications may well be tolerable.

The power in the finite element method lies in its ability to generate difference schemes (including higher order difference schemes) on an arbitrary, irregular grid in an arbitrary and open-sided domain while maintaining conservation properties like those of the Arakawa scheme. However, our results have shown that the cost of assembling the matrices in a time dependent three-dimensional problem are prohibitively expensive. While work is in progress on finding efficient algorithms for solving the sparse matrix problems that arise in problems of this type, we believe it is important that algorithms be developed for generating the finite element matrices so that the real power in the finite element method can be exploited.

The speculative nature of our final remarks indicates

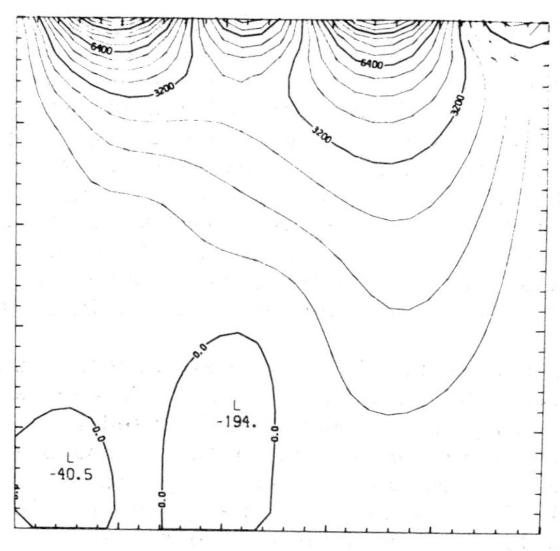

Fig. 3 - Contours of streamfunction ψ for Gulf Stream forcing problem. M=0.79 corresponds to trapped modes.

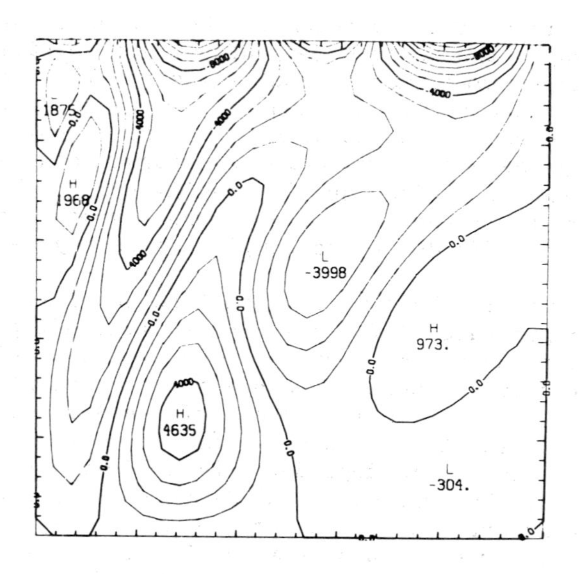

Fig. 4 - Contours of streamfunction ψ for Gulf Stream forcing problem. M=3.35 corresponds to propagating modes.

that the problem is still open in many directions. The physical problems are abundant and will always need accurate numerical solutions at as little cost as possible. Our hope is that numerical methods can keep pace with these demands.

Acknowledgements. This work was sponsored by the National Science Foundation under grant #IDO 71-04213 A04 (JEH) and by the Office of Naval Research under contract number N00014-75C-0243 (WLB). We would like to thank the National Center for Atmospheric Research, sponsored by the National Science Foundation, for providing the computing facilities used in this research. We would also like to thank Professor George Fix of Carnegie-Mellon University and Professor Donald Rose of Harvard University for their many valuable discussions. Acknowledgement is also given to Professor A. R. Robinson of Harvard University for suggesting the problem and for his many discussions.

References

Bank, R.E., 1975: Marching algorithms for elliptic boundary value problems II: Non-constant coefficient case. Submitted to SIAM J. Numerical Analysis.

Bank, R.E., and D.J. Rose, 1975: Marching algorithms for elliptic boundary value problems I: Constant coefficient case. Submitted to SIAM J. Numerical Analysis.

Bretherton, F., and M. Karweit, 1975: Midocean mesoscale modelling. Numerical Models of Ocean Circulation Proceedings, Ocean Affairs Board, National Academy of Sciences.

Charney, J., G. Fjörtoft, and J. Von Neumann, 1950: Numerical integration of the barotropic vorticity equation. Tellus, 2(4), 237-254.

Fix, G.J., 1975: Finite element models for ocean circulation problems. To appear in SIAM J. Appl. Math.

Kreiss, H., and J. Oliger, 1973: Methods for the approximation of time dependent problems. GARP Publication, Series No. 10, World Met. Org. P. 73.

Holland, W.R., and L.B. Lin, 1975: On the generation of mesoscale eddies and their contribution to the oceanic circulation. I. A preliminary numerical experiment. To appear in J. Phys. Oceano.

Jespersen, D.C., 1974: Arakawa's method is a finite element method. J. Comp. Phys., 16, 383-390.

Pedlosky, J., 1964: The stability of currents in the atmosphere and ocean: Part I. J. Atmos. Sci., 21, 201-219.

Rose, D.J., and G. Whitten, 1975: A recursive analysis of dissection strategies. This volume.

CALCULATION OF NORMAL MODES OF OCEANS USING A LANCZOS METHOD

Alan K. Cline
Departments of Computer Sciences and of Mathematics and Center for Numerical Analysis, University of Texas at Austin

Gene H. Golub
Department of Computer Science, Stanford University

George W. Platzman
Department of Geophysical Sciences, University of Chicago

1. Introduction

There are several numerical methods available for the solution of large, sparse, symmetric eigenproblems. Among these are the power method (and its various extensions such as simultaneous iteration), relaxation, and the method of Lanczos. (For an excellent survey of these methods see Stewart [1974].) The first two of these determine only extreme eigenvalues (although it may be possible to transform a problem so that interior eigenvalues are mapped to extremes). The reported use of the Lanczos method has been limited to extreme values although its usefulness may include certain cases where interior values are desired.

In Platzman [1975], the author sought to solve such an eigenproblem to determine normal modes of the Atlantic and Indian Oceans with periods between 8 and 100 hours, those modes likely to be excited by tidal forces. The actual matrix has 959 double eigenvalues (which are inversely proportional to the squares of the periods) with perhaps 650 greater than the region of interest and 275 smaller. In a continuum model, the larger values are unbounded; however, in the particular discrete model, they extend to about 3. The smaller values cluster at the origin in the continuum model, and the discretized version has values smaller than 5×10^{-8}. The region of interest of eigenvalues is only about [.0001, .0245]. Thus the problem truly is to locate interior values; procedures which "peel away" several extreme values, then deflate to expose the internal values, would probably not be useful here.

In section 2, a model of 675 six-degree squares covering the North and South Atlantic and Indian Oceans is described as well as the bathymetry and coastal assumptions. The operator derived from Laplace's "tidal" equations is also introduced along with its discretization corresponding to the finite grid.

In section 3, the Lanczos method for determining eigenvalues is defined and a characterization of it as being optimal (in some sense) is given. The difficulties associated with interior eigenvalues are discussed.

The modification of the Lanczos algorithm to determine such interior values is presented in section 4. Results of the computations are also presented. Error bounds for the computed periods are given showing the values to be quite acceptable since these errors are far exceeded by the expected discretization errors associated with the model. Maps of several of the modes are presented; however, the physical interpretation of the modes is found in Platzman [1975]. The description here is brief.

In section 5, an alternative method for the refinement of approximations to interior eigenvalues is presented. This method applies a variant of the Lanczos algorithm for the solution of sparse symmetric indefinite systems to an inverse iteration type of eigenvalue improvement.

2. The Model and Operator

The model used in Platzman [1975] contained 675 squares of length 6° longitude and width 6° Mercator latitude. (One degree of Mercator latitude is a map distance equal to one degree of longitude on a Mercator projection.) A particular square is included in the model if over half of the corresponding region on the earth is Atlantic or Indian Ocean. Truncations are made at several sea-arcs to exclude the Arctic and Pacific Oceans. The largest of these are a longitudinal slice at 150°E from Australia to Antarctica to exclude the South Pacific and another at 66°W, the Drake Passage south of South America. Three other minor sea-arcs in the North exclude Baffin Bay and the Greenland and Norwegian Seas.

Bathymetry (bottom configuration) is introduced by using average depth values associated with each square. Figure 1 is a contour map of the bathymetry in the region of interest. The boundary of the region is shown in solid (coastal) and dotted (sea-arc) lines. Using the variables ζ, the surface elevation, and u and v, the eastward and northward components of velocity, Laplace's tidal equations for unforced motion without frictional dissipation can be written as

$$\frac{\partial a}{\partial t} = i\mathcal{L}a$$

where

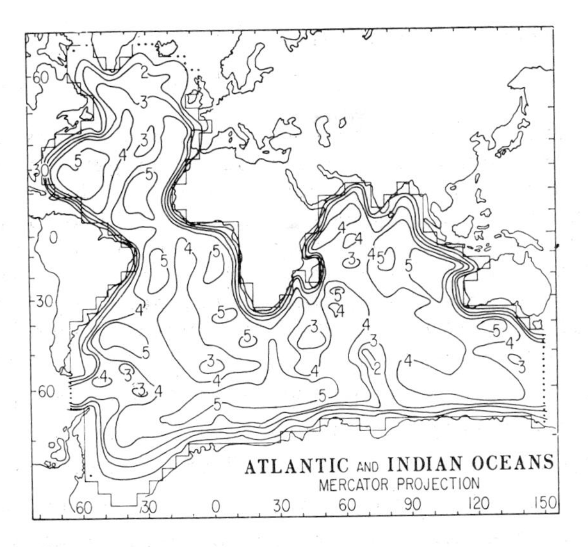

Fig. 1. The Atlantic and Indian Oceans resolved on a grid of 6° Mercator squares. Bathymetry (km) is drawn from grid-square averages. Dotted segments of the boundary show locations of ports.

$$a \equiv \begin{pmatrix} \zeta \\ u \\ v \end{pmatrix} \quad \text{and} \quad \mathcal{L} \equiv i \begin{pmatrix} 0 & \frac{\partial}{\partial x} h & \frac{\partial}{\partial y} h \\ g \frac{\partial}{\partial x} & 0 & -f \\ g \frac{\partial}{\partial y} & f & 0 \end{pmatrix},$$

h is the depth, f is the Coriolis parameter (a function of latitude), g is gravitational acceleration (assumed constant), and the x- and y-directions correspond to east and north, respectively. Adiabatic boundary conditions prohibiting the flux of energy across the boundary of the region are imposed. These can be met by assuming the surface elevation is zero (used on some sea-arc sections of the boundary) or by assuming the normal component of velocity is zero (used on other sea-arc sections and on coastal sections). The boundary is thus totally reflecting with respect to wave energy, and since there is no frictional dissipation in the model, the normal modes must be purely oscillatory. If we seek them of the simple-harmonic form, $\bar{a}e^{i\sigma t}$, where $\bar{a}$ is complex-valued and dependent on spatial variables only, then we have the eigenvalue problem

$$\mathcal{L}\bar{a} = \sigma\bar{a}.$$

With inner product defined by

$$\langle a,a'\rangle = \int_S (g\zeta^*\zeta' + hu^*u' + hv^*v')\,dS ,$$

(where S represents the oceanic region under consideration and * indicates complex conjugation) the operator $\mathcal{L}$ is self-adjoint. This is a consequence of Green's identity and the adiabatic boundary conditions. In fact, the operator $i\mathcal{L}$ is skew-symmetric; thus we may expect the eigenvalues of $\mathcal{L}$ to be not only real but to occur as + and - pairs. We may deal solely with the positive semidefinite, self-adjoint operator $\mathcal{L}^2$, if we please, since if σ^2 is a double eigenvalue with orthogonal eigenfunctions b_1 and b_2 of $\mathcal{L}^2$, then σ and $-\sigma$ are eigenvalues of $\mathcal{L}$ with associated eigenfunctions $b_1 + ib_2$ and $b_1 - ib_2$, respectively. Furthermore, if b_1 is known, b_2 is easily determined as $-i/\sigma\mathcal{L}b_1$. (A "normal mode" is the real part of $(b_1+ib_2)e^{i\sigma t}$ or $(b_1-ib_2)e^{-i\sigma t}$.) Although methods (for example, the Lanczos method) for dealing with self-adjoint eigenproblems can easily be modified to handle skew-symmetric ones, a good deal of simplicity results if we consider the self-adjoint operator $\mathcal{L}^2$ rather than $i\mathcal{L}$.

In our discretization, the value of ζ is determined in the center of each grid square, the velocity component u on the midpoint of the east and west edges of each square, and the component v on the midpoint of the north and south edges of each square. This leads to 675 ζ-values, 729 u-values (621 internal, 95 on coasts, 13 on sea-arcs), and 727 v-values (623 internal, 99 on coasts, 5 on sea-arcs). The operator $\mathcal{L}$ is discretized by replacing the partial derivatives by centered, divided differences. After having removed the boundary constraints, an order 1919 eigenvalue problem results. One eigensolution is trivial ($\sigma = 0$, $\zeta \equiv 1$, $u,v \equiv 0$); the other 1918 form the 959 pairs noted above.

Theoretical considerations of the continuous problem lead us to expect both very large and very small eigenvalues in magnitude. An oscillating basin such as this should have an unbounded sequence of eigenvalues corresponding to "gravity modes" and a second sequence of eigenvalues clustered at zero corresponding to "rotational modes." Gravity modes are those induced by disturbances of the mass field (here, the free surface) and exist without the Coriolis force. Rotational modes are those induced primarily by deformations of the potential vorticity f/h, and exist without gravity. In our model, we may expect approximations to some of each, however. The effect of the discretization is not only to perturb some of the values but, in fact, to filter out the extremes (i.e., the high-frequency gravity and low-frequency rotational modes). We are, however, at present only interested in the range of modes with periods between 8 and 100 hours, those most likely to be excited by semidiurnal and diurnal tidal forces.

Upon discretization the inner product defined above becomes

$$\langle a,a' \rangle = \hat{x}^T D^2 \hat{x}' = x^T x'$$

where the column $\hat{x}$ is the discretized a and D^2 is a positive diagonal matrix whose elements consist of g and the discretized h. With the use of $x = D\hat{x}$, the discretized inner product leads to a Euclidian norm. Similarly, if $\hat{A}\hat{x} = \lambda\hat{x}$ is the discretized form of $\mathcal{L}^2 a = \sigma^2 a$ (with $\lambda = \sigma^2$), then $Ax = \lambda x$ with $A = D\hat{A}D^{-1}$. For the purpose of the following discussion, we assume we are given the symmetrix matrix A in "operator" form. That is, a procedure is available whereby Az can be computed from any vector z without storing the elements of A.

3. The Lanczos Method and Internal Eigenvalues

The Lanczos algorithm is described many places in the scientific literature (cf. Lanczos [1950]). It is not our intention here to develop the algorithm fully and all its properties but simply to define it and give one characterization.

We seek eigenvalues λ and eigenvectors x satisfying $Ax = \lambda x$. The algorithm initiates with a given vector q_1, then generates additional vectors $q_2, q_3, \ldots$ as well as scalar sequences $\alpha_1, \alpha_2, \ldots$ and $\beta_2, \beta_3, \ldots$ from

$$\alpha_k = q_k^T A q_k$$

$$\beta_{k+1} = \| (A-\alpha_k) q_k - \beta_k q_{k-1} \|$$

and, provided $\beta_{k+1} \neq 0$,

$$q_{k+1} = \frac{1}{\beta_{k+1}} [(A - \alpha_k) q_k - \beta_k q_{k-1}]$$

(where for $k = 1$, $\beta_1 q_0$ is replaced by zero and $\|\cdot\|$ indicates

Euclidean norm).

If we let Q_k be the matrix whose columns are $q_1,\ldots,q_k$ and T_k be the k x k tridiagonal matrix with principal diagonal α_1, $\ldots,\alpha_k$ and sub- and super-diagonals $\beta_2,\ldots,\beta_k$, we may represent the recurrence as

$$AQ_k = Q_kT_k + \beta_{k+1}q_{k+1}e_k^T,$$

where e_k^T is a k-vector with all components zero except the last which is one.

The vectors $\{q_k\}$ can be shown to be orthonormal, thus $Q_k^TQ_k = I$. If any $\beta_{k+1} = 0$, the algorithm stops with

$$AQ_k = Q_kT_k,$$

thus

$$Q_k^TAQ_k = T_k,$$

and the eigenvalues of T_k are eigenvalues of A. In this case, if y is an eigenvector of T_k, Q_ky is an eigenvector of A. If additional eigenvalues of A not found in T_k are to be determined, it is necessary to restart the algorithm with a new q_1' orthogonal to the original set $q_1,\ldots,q_k$. Since the number of orthogonal vectors $\{q_k\}$ cannot exceed the order of the matrix (call it n), theoretically some $\beta_{k+1} = 0$ for a $k \leq n-1$. When finite-precision arithmetic is used, the orthogonality property suffers severely. In fact, it can occur that even for values of k far exceeding n, no β_{k+1} may be small. This does not necessarily imply the algorithm loses its usefulness. This situation is discussed in Paige [1971].

It is the case that very good approximations to eigenvalues of A may be obtained by stopping the procedure prematurely (i.e., before any β_{k+1} is small). For any scalar λ' and normalized vector x', the quantity $\|Ax' - \lambda'x'\|$ bounds the error between λ' and some eigenvalue λ of A. If λ' is an eigenvalue of some T_k and y is the associated eigenvector, then with $x' = y$, we have

$$\begin{aligned} |\lambda-\lambda'| \le \|Ax' - \lambda'x'\| &= \|AQ_k y - \lambda' Q_k y\| \\ &= \|AQ_k y - Q_k T_k y\| \\ &= \|\beta_{k+1} q_{k+1} e_k^T y\| \\ &= |\beta_{k+1}| \cdot |y_k|. \end{aligned}$$

Thus the eigenvalue error is bounded by $|\beta_{k+1}| \cdot |y_k|$, and if y_k, the last component of the vector y, is small, then the bound may be sufficiently small even though β_{k+1} may only be of moderate size. This analysis indicates that the more accurate approximations to eigenvalues may be those eigenvalues of T_k whose associated eigenvectors have rapidly dwindling components.

We now turn to another characterization of the eigenvalues of T_k and their relation to those of A. This derives from the Courant-Fisher theorem which states that λ_j, the jth largest eigenvalue of A, satisfies

$$\lambda_j = \min_{p_1,\ldots,p_{j-1}} \ \max_{\substack{x \perp p_i \\ i=1,\ldots,j-1}} \frac{x^T A x}{x^T x} .$$

Furthermore, the minimum in this expression is assumed if and only if $p_1,\ldots,p_{j-1}$ span an eigenspace of A corresponding to eigenvalues $\lambda_1,\ldots,\lambda_{j-1}$. Let λ_j' denote the jth largest eigenvalue of T_k (for some fixed k) and let P denote the space spanned by polynomials of degree k-1 or less in the matrix A applied to q_1; then λ_j' satisfies

$$\lambda_j' = \min_{p_1,\ldots,p_{j-1} \in P} \ \max_{\substack{x \perp p_i \\ i=1,\ldots,j-1 \\ x \in P}} \frac{x^T A x}{x^T x} .$$

To see this we first notice that the vectors $q_1,\ldots,q_k$ (which are the columns of Q_k) form an orthonormal basis for P. Furthermore, applying the Courant-Fisher theorem to T_k we have

$$\lambda_j' = \min_{r_1,\ldots,r_{i-1}} \max_{\substack{y \perp r_i \\ i=1,\ldots,j-1}} \frac{y^T T_k y}{y^T y} .$$

But since $T_k = Q_k^T A Q_k$ and $Q_k^T Q_k = I$,

$$\lambda_j' = \min_{r_1,\ldots,r_{j-1}} \max_{\substack{y \perp r_i \\ i=1,\ldots,j-1}} \frac{(Q_k y)^T A (Q_k y)}{(Q_k y)^T (Q_k y)}$$

$$= \min_{r_1,\ldots,r_{j-1}} \max_{\substack{x \perp Q_k r_i \\ i=1,\ldots,j-1 \\ x \in P}} \frac{x^T A x}{x^T x} ,$$

which follows by letting $x = Q_k y$, which implies $x \in P$ and observing that $y \perp r_i$ if and only if $x \perp Q_k r_i$. The original characterization is then proved by letting the p_i's (which must be in P) be related to r_i's through the relation $p_i = Q_k r_i$.

It can be shown that analogous to the interlacing property for the sets of eigenvalues of principal submatrices of symmetric matrices, the set of eigenvalues associated with various T_k also interlace and that $\lambda_j' \le \lambda_j$. Similarly, the jth smallest eigenvalue of A cannot exceed the jth smallest of any T_k. From the characterization we see the important effect of the subspace P in the approximating power of λ_j' to λ_j. If P were an eigenspace associated with values $\lambda_1,\ldots,\lambda_{j-1}$, then $\lambda_j' = \lambda_j$; however, if P is far from such a space, λ_j' may be much different from λ_j. Furthermore, since as j increases, the characterization depends increasingly on the space P (since x and j-1 vectors p_i are to be selected from

P), we might expect better behavior from the larger eigenvalues than from smaller. However, since the eigenvalues can be given an analogous max min characterization in which the influence of P increases as the size of the values, we find that the extreme eigenvalues (large and small) of T_k are less influenced by the space P than the interior ones, and thus we might expect better approximations there and poorer ones in the interior.

The preceding analysis is hardly rigorous, although it can be made so. Obviously the space P itself is critical: if P were an eigenspace associated with interior eigenvalues, then accurate interior values would be obtained from T_k. A first treatment of the eigenvalue convergence properties was given by Kaniel [1966]. A more thorough one is in Paige [1971]. Experience has indicated, however, that although the Lanczos process may produce quite accurate extreme eigenvalues with very few iterative steps, the interior eigenvalues are very difficult to obtain.

4. Eigenvector Refinement and Computational Results

An approach employed by the third author resulted in accurate determination of many of the interior eigenvalues and associated eigenvectors for the problem presented in the previous section. This approach begins with an initial approximation to the eigenvalues from a T_k. In the particular problem of order 1919, since all eigenvalues are double but one (which is zero), it should be theoretically possible to obtain the eigenvalues of A using T_{959}. In fact, due to the effect of finite-precision arithmetic, many of the eigenvalues of A are not displayed in T_{959} or even T_{1918}. In the region of interest [.0001, .0245], 107 eigenvalues are found from T_{1918}. Many of these are found multiply, however, and after forming equivalence classes of nearly coincident values, only 37 distinct eigenvalues can be considered as approximations to those of A. As described above, corresponding eigenvectors of A can be obtained from those of T_{1918}.

Before proceeding with the discussion of the eigenvector refinement process, it should be commented that an eigenvalue λ of some T_k may be a poor approximation to any eigenvalue of A. A computation of the corresponding eigenvector x, and the residual norm $\|Ax - \lambda x\|$ provides a bound on the error, however. Nevertheless, even an accurate eigenvalue of A may be a poor approximation to any actual eigenvalue of a continuous operator of which A may be a discrete representation. To estimate this error it may be necessary to compute residual norms with more accurate discretizations.

Returning now to the problem of refining approximate eigenvectors, we will briefly describe the inverse-power method used to obtain the results listed below in Table 1. Given x, a normalized approximate eigenvector, and $\lambda = x^T Ax$, its associated Rayleigh quotient, solve $(A - \lambda I)z = x$ for z. This leads to the problem of solving a large, linear system. We begin as before by generating a sequence of Lanczos vectors $q_1, q_2, \ldots, q_{k+1}$ and corresponding tridiagonal T_k, except that now we take the initial vector $q_1 = x$. These steps can again be stated compactly as

$$AQ_k = Q_k T_k + \beta_{k+1} q_{k+1} e_k^T.$$

If we now seek a solution z as a linear combination of the Lanczos vectors, hence of the form $z = Q_k y$, then y is the solution of $(T_k - \lambda I)y = e_1$. Postmultiplying the preceding equation by y, we obtain

$$(A - \lambda I)z = x + \beta_{k+1} y_k q_{k+1}.$$

TABLE 1

Mode	Eigenvalue (non-dimensional)	Residual (non-dimensional)	Period (hours)	Error Bound (hours)
		GRAVITY MODES		
1	3.559(-4)	1.3(-8)	66.61	1.2(-3)
2	9.001(-4)	6.4(-12)	41.88	1.5(-7)
3	1.823(-3)	1.7(-11)	29.43	1.3(-7)
4	2.867(-3)	2.7(-11)	23.47	1.1(-7)
5	3.618(-3)	3.6(-11)	20.89	1.0(-7)
6	4.711(-3)	4.5(-11)	18.31	8.8(-8)
7	5.327(-3)	4.6(-11)	17.22	7.4(-8)
8	6.795(-3)	6.3(-11)	15.24	7.1(-8)
9	7.656(-3)	6.5(-11)	14.36	6.1(-8)
10	8.333(-3)	7.1(-11)	13.77	5.9(-8)
11	9.661(-3)	2.4(-12)	12.78	1.6(-9)
12	1.067(-2)	9.7(-11)	12.16	5.5(-8)
13	1.226(-2)	9.3(-12)	11.35	4.3(-9)
14	1.292(-2)	7.8(-13)	11.05	3.3(-10)
15	1.340(-2)	1.4(-11)	10.86	5.9(-9)
16	1.405(-2)	5.1(-13)	10.60	1.9(-10)
17	1.467(-2)	2.9(-13)	10.38	1.0(-10)
18	1.621(-2)	9.4(-13)	9.869	2.9(-10)
19	1.707(-2)	7.6(-12)	9.619	2.1(-9)
20	1.867(-2)	1.1(-11)	9.198	2.8(-9)
21	1.928(-2)	2.6(-12)	9.049	6.1(-10)
22	2.047(-2)	1.4(-13)	8.784	2.7(-11)
23	2.130(-2)	2.3(-12)	8.611	4.7(-10)
24	2.201(-2)	1.6(-12)	8.471	3.1(-10)
25	2.315(-2)	1.8(-11)	8.260	3.2(-9)
26	2.440(-2)	1.4(-11)	8.044	2.3(-8)
		ROTATIONAL MODES		
1	8.041(-4)	6.8(-12)	44.31	1.9(-7)
2	7.675(-4)	6.3(-12)	45.36	1.9(-7)
3	6.634(-4)	5.4(-12)	48.79	2.0(-7)
4	5.229(-4)	3.5(-11)	54.95	1.9(-6)
5	4.333(-4)	1.6(-9)	60.36	1.1(-4)
6	3.046(-4)	1.6(-9)	72.00	1.9(-4)
7	2.645(-4)	6.7(-10)	77.26	9.7(-5)
8	2.416(-4)	4.0(-10)	80.85	6.8(-5)
9	2.235(-4)	9.2(-10)	84.04	1.7(-4)
10	2.143(-4)	9.7(-9)	85.84	1.9(-3)
11	1.612(-4)	1.2(-6)	98.97	3.5(-1)

This gives $|\beta_{k+1}y_k|$ as the norm of the residual $(A-\lambda I)z - x$, and termination of the Lanczos sequence (i.e., the terminal value of k) could be based on this bound. As explained below, the procedure actually followed terminated on the basis of an estimate of the eigenvalue residual norm $\|(A-\lambda I)z\|/\|z\|$.

The solution of $(T_k-\lambda I)y = e_1$ is accomplished by writing this system as a three-term recurrence relation. Two vectors y' and y'' with starting conditions $y_1' = 1$, $y_2' = 0$, and $y_1'' = 0$, $y_2'' = 1$ are developed by using the recurrence relations in the forward direction. They satisfy

$$(T_k-\lambda I)y' = -\beta_{k+1}y_{k+1}'e_k$$

$$(T_k-\lambda I)y'' = \beta_2 e_1 - \beta_{k+1}y_{k+1}''e_k$$

and consequently the linear combination

$$y = \frac{1}{\beta_2}(y'' - (y_{k+1}''/y_{k+1}')y')$$

satisfies $(T_k-\lambda I)y = e_1$. Wilkinson [1965] found the recurrence method of solving tridiagonal systems of equations for eigenvectors to be unstable. When the residual is large in its pth component, the typically unstable case is one in which an eigenvector of T_k is sought whose p^{th} component is small. In our case, however, since the Lanczos sequence was started with x, a good approximation to an eigenvector of A, the eigenvector y of T_k that we seek is a small perturbation of e_1, and since the residual is itself e_1, the conditions for instability of the recursion described by Wilkinson are absent.

The Lanczos sequence in this refinement procedure was terminated by means of the following estimate of the final residual norm

$$\|(A-\lambda I)z\|/\|z\| = |\beta_2/y_{k+1}''|(1 + O(\theta^2)),$$

where θ is the norm of the difference between the approximate eigenvalue x and the corresponding exact eigenvector. Since x is the starting vector of the sequence, the initial residual norm $\|(A-\lambda I)x\|$ is simply $|\beta_2|$, so the preceding estimate shows that the amount of refinement obtained is controlled by the size of $|y''_{k+1}|$. In general we found that y''_k oscillates with increasing amplitude as k increases. The sequence was terminated when the final residual norm fell below a preassigned bound, or failing this, when a preassigned number of steps was exceeded. Note that explicit construction of y can be deferred until after the decision to terminate is made.

Using this procedure, 37 eigenvectors were refined and the residuals listed in Table 1 were obtained. This table also gives the period of each mode (inversely proportional to the square root of the eigenvalue) and an error bound for the period. Notice the deterioration in the bounds for the longer periods (smaller eigenvalues).

Maps of two of the modes are shown in Figure 2. These are the fundamental gravity mode of the region, which has a period of 66.6 hr., and the second gravity mode of 41.9 hr. period, which exhibits a quarter-wave resonance of the North Atlantic. Other maps, as well as a discussion of the physical characteristics of the modes, are found in Platzman [1975].

5. An Alternative Approach to Eigenvector Refinement

In the preceding section, one method was presented for the refinement of approximate eigenvectors. The approach of this section is very similar in that we seek to solve the inverse iteration equation using a linear combination of the Lanczos vectors generated with the approximate eigenvector as the initial vector. It differs from that approach, though, in being a minor modification of a sparse linear equation-solving

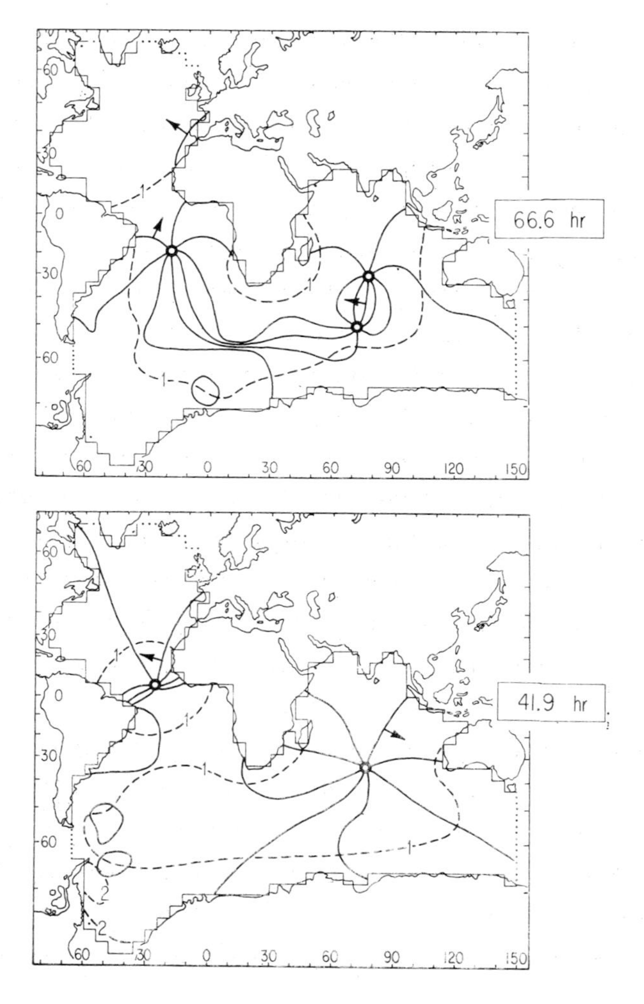

Fig. 2. Phase (solid lines) and amplitude (broken lines) of surface elevation of the first two gravity modes of the Atlantic/Indian Oceans. Upper: fundamental mode; lower: quarter-wave of the North Atlantic. Phase isogons are at intervals of 45° and an arrow in the direction of propagation is attached to the line of zero phase. (The arbitrary additive constant in the phase distribution is fixed by assigning zero phase to the grid point nearest to Lisbon.) Amplitude is normalized to an rms of 1 for the whole domain and co-amplitude lines are drawn at intervals of 1.

method, the Paige-Saunders algorithm [1973], which does not require retention or recomputation of the Lanczos vectors.

The Paige-Saunders algorithm can be directly applied to any symmetric, linear system. The process takes linear combinations of the Lanczos vectors as approximations to the solution of the system. The convergence of the algorithm "MINSYM" (which minimizes the residual of the system over the space spanned by the Lanczos vectors) can be bounded by the quantity $1/|\tau_{k-1}(0)|$, where τ_{k-1} is the $k-1^{st}$-order Chebychev polynomial on the set of eigenvalues of the matrix (i.e., the monic polynomial of order k-1 whose maximum absolute value on the set is minimized). Unfortunately, as eigenvalues get closer to zero this bound (which can be met) indicates slower and slower convergence. If μ is the smallest eigenvalue in magnitude and K the norm of the matrix, the residual norm after k Lanczos steps can be as much as $(1-2\mu/K)^k$.

After an inverse iteration equation has been solved and a new eigenvector approximation obtained, we would compute a new eigenvalue approximation as the Rayleigh quotient and iterate. The dilemma in using the Paige-Saunders approach for inverse iteration is that as our eigenvalue approximation improves, the convergence behavior of the solution algorithm deteriorates. In this section, we present a deflation that removes this effect.

As in the previous section, the problem of solving $(A-\lambda I)z = x$ with the Lanczos vectors reduces to solving $(T_k-\lambda I)y = e_1$. With λ the Rayleigh quotient corresponding to x, the matrix $(T_k-\lambda I)$ has the special form

$$T_k - \lambda I = \beta_2 \begin{bmatrix} 0 & \bar{e}_1^T \\ \bar{e}_1 & \bar{T}_k \end{bmatrix}$$

(the bar indicates a projection into dimension k-1). The quantity β_2 is exactly $\|(A-\lambda I)x\|$, the residual corresponding to the approximate eigenvalue λ and eigenvector x. It is easy to see that the eigenvalues of $\beta_2\bar{T}_k$ are $O(\beta_2^2)$ perturbations of all of those of $T_k-\lambda I$ excluding the one near zero.

Now write $y = y_1\begin{bmatrix}1\\ \bar{y}\end{bmatrix}$ and use the above partitioning of $T_k-\lambda I$ to expand the equation $(T_k-\lambda I)y = e_1$ into the scalar and vector parts

$$\beta_2 y_1 \bar{y}_1 = 1,$$
$$\bar{T}_k\bar{y} = -e_1.$$

The second equation can be solved for $\bar{y}$ and then the scalar equation for y_1. The vector y would then be entirely determined and

$$z = Q_k y = y_1(q_1 + \bar{Q}_k\bar{y})$$
$$= y_1(q_1 + \bar{z}),$$

where $\bar{Q}_k$ is the matrix with columns $[q_2,\ldots,q_k]$ and $\bar{z} = \bar{Q}_k\bar{y}$. However, in the solution of an inverse iteration equation for an eigenvector, a non-zero scalar factor is unimportant; hence, we may accept $q_1 + \bar{z}$ as the solution. This avoids the computation of y_1, something not produced by the Paige-Saunders algorithm. Since $x = q_1$ and $(A - \lambda I)q_1 = \beta_2 q_2$, the equation $(A - \lambda I)z = x$ can be transformed into

$$(A - \lambda I)\bar{z} = -\beta_2 q_2 - \frac{1}{y_1} q_1.$$

Owing to the orthogonality of the q's, this is equivalent to

$$B\bar{z} = -\beta_2 q_2$$

where $B = (I - q_1q_1^T)(A - \lambda I)(I - q_1q_1^T)$. We can solve $B\bar{z} = -\beta_2 q_2$ directly for $\bar{z}$ by using either of the Paige-Saunders algorithms

"SYMLQ" or "MINRES". Now one eigenvalue of B is zero (with eigenvector q_1); the others are small perturbations of those of $A - \lambda I$. Because $\bar{z}$ is orthogonal to q_1, the convergence of these algorithms is governed by the non-zero eigenvalues of B--in other words, by the portion of the spectrum of $A-\lambda I$ bounded away from zero--and hence does not deteriorate with increasing accuracy of λ.

The approach has undergone initial experimental testing with successful results. Other methods for using a Lanczos sequence to determine interior eigenvalues have been explored by Mr. John G. Lewis.

6. Acknowledgements

The work of A. K. Cline was in part supported by NASA Grant NGR 47-102-001 while the author was in residence at ICASE, NASA Langley Research Center. The work of G. H. Golub was in part supported by NSF and ERDA grants. The work of G. W. Platzman was supported by the Oceanography Section of the National Science Foundation under NSF Grant GA-15995.

References

1. Kaniel, S., Estimates for some computational techniques in linear algebra, Math. Comp. 20 (1966), 369-378.

2. Lanczos, C., An iteration method for the solution of the eigenvalue problem of linear differential and integral operators, J. Res. Nat. Bur. Standards 45 (1950), 255-282.

3. Paige, C. C., The computation of eigenvalues and eigenvectors of very large sparse matrices, Ph.D. thesis, London University (1971).

4. Paige, C. C., and Saunders, M. A., Solution of sparse indefinite systems of equations and least squares problems, Report STAN-CS-73-399, Stanford University (1973).

5. Platzman, G. W., Normal modes of the Atlantic and Indian Oceans, J. Phys. Oceanography 5 (1975), 201-221.

6. Stewart, G. W., The numerical treatment of large eigenvalue problems, Proceedings of IFIP Congress 1974, Mathematical Aspects of Information Processing Sec. 3, Stockholm (1974), 666-672.

7. Wilkinson, J. H., The Algebraic Eigenvalue Problem, Oxford Univ. Press, London (1965).

APPLICATION OF SPARSE MATRIX TECHNIQUES TO RESERVOIR SIMULATION

P. T. Woo[1], S. C. Eisenstat[2], M. H. Schultz[2], A. H. Sherman[3]

(1) Chevron Oil Field Research Company, La Habra, California, 90631; (2) Department of Computer Science, Yale University, New Haven, Connecticut, 06520; and (3) Department of Computer Science, University of Illinois, Urbana, Illinois, 61801.

INTRODUCTION:

The purpose of this paper is to show how sparse Gaussian elimination is applied to the numerical simulation of petroleum reservoirs. Our emphasis will be on this particular application, somewhat in the style of [10], rather than on theoretical or implementation questions, which are treated in other papers in these proceedings, [3] and [4]. In particular, we shall present the work and computing-time requirements of sparse Gaussian elimination for some typical problems of reservoir simulation.

In many reservoir simulation problems, we seek the solution of a system of nonlinear parabolic partial differential equations describing multiphase flow in two or three space dimensions. In this paper, we restrict our attention to two-dimensional problems.* The most common technique is to approximate the domain by a rectilinear mesh or grid and to approximate the partial differential equations by five-point difference equations together with suitable linearizations. The result is a sequence of systems of linear equations

$$A x = b \tag{1}$$

where entries in the matrix A and the vector b vary from time step to time step. A simulation problem may involve up to a thousand or more such systems.

In reservoir simulation, systems of the form (1) have usually been solved with iterative rather than elimination methods. This was thought to save both time and storage. However, selecting an efficient iterative method, optimal acceleration parameters, and a good stopping criterion is

*See [8] and [10] for computational results for coarse three-dimensional grid problems. Sparse Gaussian elimination for fine three-dimensional grids is prohibitively expensive, cf. [2] and [3].

sometimes difficult and expensive. Moreover, recently we have found that in some situations, iterative methods will not converge to an acceptable solution within a reasonable number of iterations because of the increasing complexity of simulation problems.

Classical elimination methods, though inefficient in terms of both speed and storage, have avoided these difficulties and have always yielded satisfactory solutions. Recently, sparse matrix techniques have greatly improved the computing speed and storage efficiency of elimination methods, and they have become an important tool in reservoir simulation.

The computing mesh in reservoir simulation is customarily numbered row by row or column by column (the grid row ordering). Depending on the derivation of the difference equations, the system (1) can be in one of the two forms described below.

TYPE 1: In this case, x represents unknown reservoir pressures at the grid points. A is a diagonally dominant band matrix which is usually nonsymmetric, although its incidence matrix is symmetric. As an example, a 3 x 4 grid and a corresponding matrix A are shown in Figure 1.

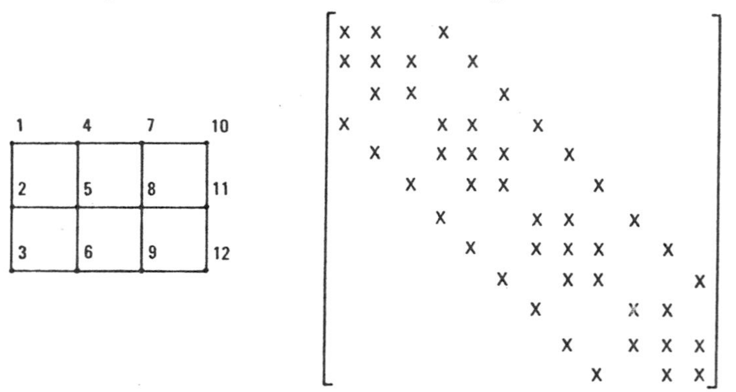

FIGURE 1: TYPE 1 MATRIX FROM A 3 x 4 GRID

TYPE 2: In this case, x represents unknown reservoir pressures and/or fluid saturations (volumetric fractions) at the mesh points. If the unknowns at each grid point are numbered consecutively, the elements of A cluster in blocks as in Figure 2. Some of the matrix entries within each block may be zero. Both A and its incidence matrix are usually nonsymmetric, and A is not always diagonally dominant, although pivoting is not required for numerical stability. This type of matrix will be referred to as a block matrix.

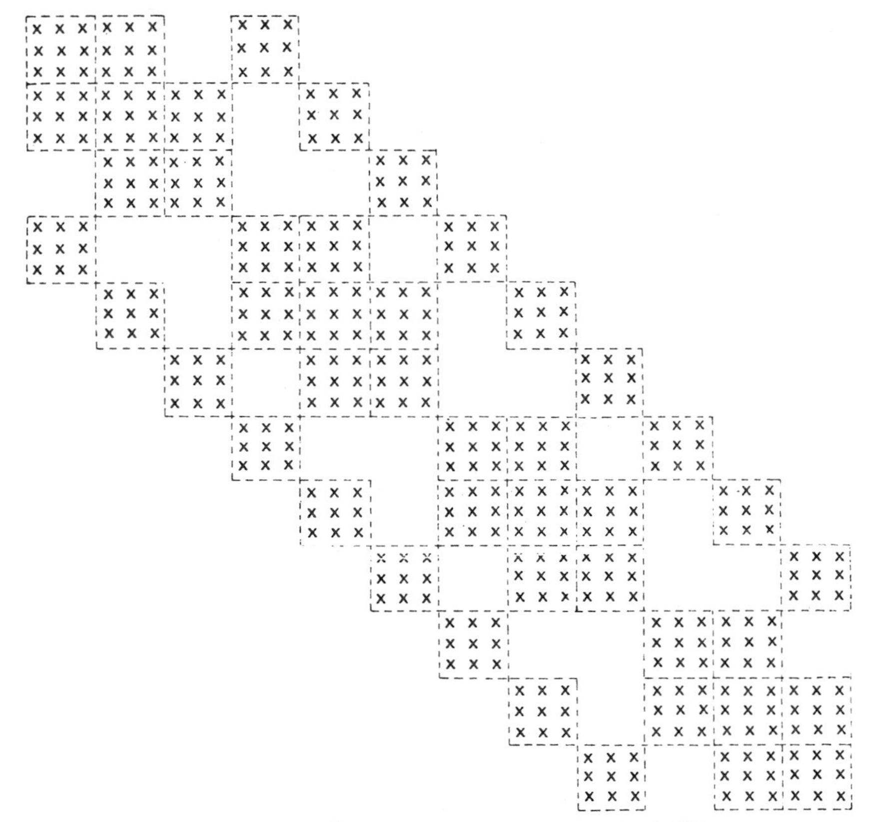

FIGURE 2: TYPE 2 (BLOCK) MATRIX FROM A 3 X 4 GRID

The application of sparse matrix techniques to reservoir simulation consists of two steps. First, we renumber or reorder the grid to minimize the fill-in during elimination and to minimize the number of arithmetic operations involving nonzeroes. This reordering is equivalent to a symmetric permutation of the matrix A of the form PAP^t where P is the permutation matrix corresponding to the reordering of the grid. Second, we solve the permuted system using Gaussian elimination where we store and operate on only the nonzeroes of A plus the new nonzeroes that occur during elimination. In our experience with reservoir simulation, the grids are relatively coarse because of the lack of detailed geological data describing the reservoir or because of the need for only "engineering" estimates of the unknowns. Hence, we will assume that in-core storage is not a critical issue and will emphasize an implementation designed for maximum efficiency with respect to computing time rather than computer storage. For a discussion of related implementations designed to optimize storage rather than speed, see [4]. We will first describe the ordering techniques and second the sparse matrix algorithms that we have found effective in reservoir simulation.

ORDERING SCHEMES:

For Type 1 matrices it has been our experience that the alternate diagonal (AD) and the minimum degree (MD) orderings described below yield good results. The application of the alternate diagonal ordering to reservoir simulation was first reported in [8]. Instead of numbering a two-dimensional grid row by row, it numbers the grid along alternate diagonals. An example of this ordering and the corresponding matrix are shown in Figure 3.

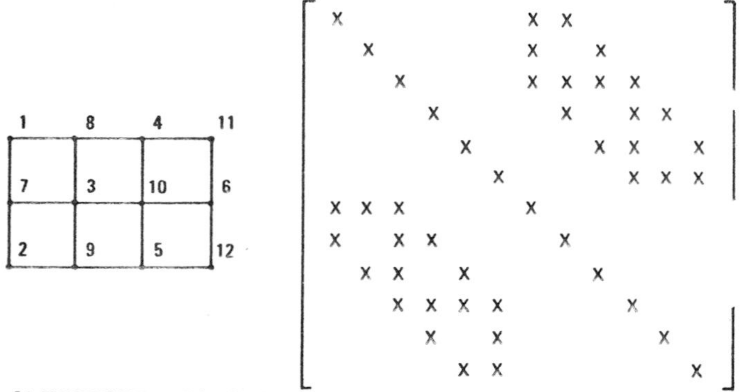

FIGURE 3: ALTERNATE DIAGONAL ORDERING AND ITS TYPE 1 MATRIX

The application of the minimum degree (MD) ordering scheme to reservoir simulation was discussed in [10]. The general idea is that at any stage of the elimination procedure, we select as pivot row that row which has the fewest nonzero off-diagonal elements. In case of a tie, we select any row from the tie. In the minimum degree ordering, we can start with a matrix which corresponds to either the grid row ordering or the alternate diagonal ordering. (The only effect is to modify the breaking of ties.) The latter sometimes results in as many as 15% fewer multiplications than the former.

For Type 2 matrices, we have found that it is convenient and effective to number the grid or permute the matrix blocks according to one of the schemes above. The relative positions within each block are unchanged.

WORK REQUIREMENT:

We define the work requirement as the number of multiplications and divisions required to solve the system (1) by sparse elimination. For Type 1 linear systems, the work requirement of the alternate diagonal ordering was given in [8]. For large two-dimensional grids with dimensions I and J ($I \geq J$), the work requirement with the AD ordering is

$$W_{AD} \approx \frac{IJ^3}{2} - \frac{J^4}{4} \tag{2}$$

while for large I x I squares, it is

$$W_{AD} \approx \frac{I^4}{4} \tag{3}$$

The AD ordering is not asymptotically optimal, since the work required with an optimal ordering scheme is $O(I^3)$ for two-dimensional square grids, cf. [6]. However, in practice the AD ordering works fairly well.

We do not know of any formulas for calculating the work required with the minimum degree ordering. However, the observed work requirement for several typical problems will be given later. This ordering seems to work as well as or better than the alternate diagonal ordering.

The work requirements with various orderings for some rectangular grids are given in Table 1. GR, AD, and MD designate respectively the grid row, alternate diagonal, and minimum degree orderings. We observe from this table that reordering the grid substantially reduces the work requirement, with the greatest reduction for square grids. Reordering is less effective in reducing the work requirement of elongated rectangular grids, though such grids require considerably less work than square grids with the same number of grid points.

Table 2 shows the work requirement of three two-dimensional grids from actual reservoir simulation problems. These grids are characterized by irregular boundaries as shown in Figures 4, 5, and 6. Clearly reordering is also effective in reducing the work requirement for such grids. In simulation problem 3, the AD ordering reduces the work requirement by a factor of about 4, and the MD ordering reduces it by a factor of about 10.

For Type 2 systems, suppose there are k unknowns per mesh point. The simplest way of applying sparse matrix techniques to the system (1) is to assume that each block of

TABLE 1 - WORK REQUIREMENTS FOR TYPE 1 EQUATIONS

Grid Dimensions	Ordering	Row Algorithm
10 x 10	GR	11,696
	AD	5,504
	MD	4.936
30 x 30	GR	855,096
	AD	261,104
	MD	189,276
50 x 50	GR	6,458,496
	AD	1,809,504
	MD	1,149,772
10 x 20	GR	24,896
	AD	13,504
	MD	13,562
20 x 60	GR	541,796
	AD	255,604
	MD	232,372

TABLE 2 - WORK REQUIREMENTS FOR TYPE 1 EQUATIONS SIMULATION PROBLEMS 1, 2, AND 3

Sim. Prob.	Grid Dims	No.of Eqns.	Work Requirement GR	AD	MD
1	8x69	390	20,726	15,218	16,110
2	23x37	507	174,974	62,430	51,766
3	55x72	2,347	~7,200,000	2,037,432	709,442

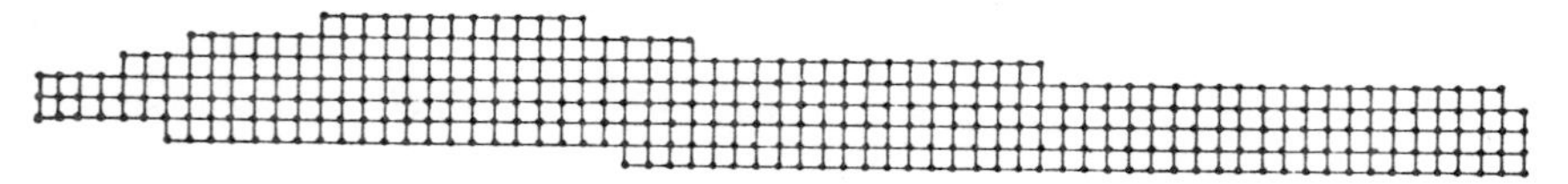

FIGURE 4: SIMULATION PROBLEM NO. 1, 8 x 69, 390 EQUATIONS

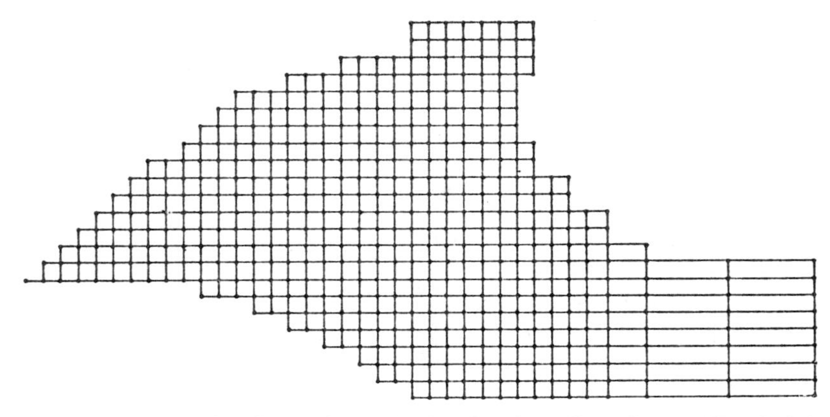

FIGURE 5: SIMULATION PROBLEM NO. 2, 23x37 GRID, 507 EQUATIONS

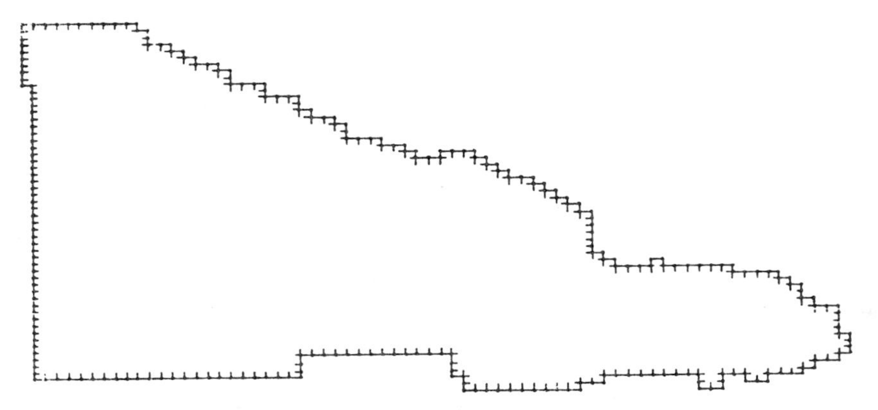

FIGURE 6: SIMULATION PROBLEM NO.3, 55x72 GRID, 2347 EQUATIONS

matrix A has dimensions k x k and contains k^2 nonzeroes. The work requirement for a Type 2 matrix in this simple situation is then approximately equal to k^3 times that of the Type 1 matrix based on the same ordering.

As an example, we give the work requirement for a block matrix A arising from reservoir simulation problem No. 4 with a 7 x 9 grid as shown in Figure 7. There are three unknowns per mesh point resulting in a system with 189 equations. If each block of A is assumed to be dense, then A has 2,547 nonzeroes and the work requirement is 91,640 for the GR ordering; 52,290 for the AD ordering; and 51,894 for the MD ordering (cf. Table 3). Actually

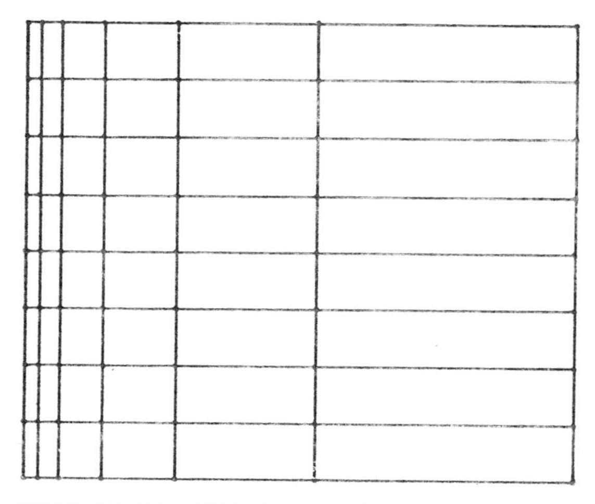

FIGURE 7: SIMULATION PROBLEM NO. 4, 7x9 GRID, 189 EQUATIONS

TABLE 3 - WORK REQUIREMENTS FOR SIMULATION PROBLEM 4, 7 x 9 GRID, 189 EQUATIONS

Nonzeros in A	Grid Row	Alternate Diagonal	Minimum Degree
2,547	91,640	52,290	51,894
1,061	15,213	11,611	17,767

each block of A may contain fewer than nine nonzeroes. In this example, A has only 1,061 actual nonzeroes at the start of the simulation and the number gradually increases to 2,547 nonzeroes in the limit, as the number of time steps becomes very large. The work requirement corresponding to 1,061 nonzeroes in A is 15,213 for the GR ordering, 11,611 for the AD ordering, and 17,767 for the MD ordering (minimum degree with respect to the grid). By using a good grid ordering and taking into account all the zeroes in A, the work requirement is substantially reduced from 91,640 to 11,611. Thus, it is advantageous to account for the change in the number of nonzeroes in A during simulation, when an efficient code is available to do so.

SPARSE MATRIX CODES:

The basic sparse matrix coding techniques have been described in [1] and [7]. We have found the symbolic and numeric (SYMFAC-NUMFAC) factorization technique to be extremely efficient in solving the linear equations in reservoir simulation. A subroutine SYMFAC is used to generate pointers to the nonzeroes of the triangular factors L and U of the LU factorization of A. Given these pointers, a subroutine NUMFAC is used to factor A (or to convert it to U). A version of SYMFAC-NUMFAC that is particularly well suited for reservoir simulation has been developed at Yale University, cf. [5].

For Type 1 linear systems, we apply SYMFAC once per simulation grid and NUMFAC once each time step. In the case of Type 2 linear systems there are two options. First, we can assume that the matrix blocks (see Figure 2) contain only nonzeroes. (The incidence matrix is symmetric as a result.) This allows us to apply SYMFAC once per grid and NUMFAC once per time step as before. Second, we can take into account the zeroes within the matrix blocks at each time step. Then we must apply both SYMFAC and NUMFAC at every time step. The relative advantage of the two options depends on the relative quantities of zeroes and nonzeroes in the matrix blocks and the relative efficiency of the SYMFAC and NUMFAC subroutines. As a general rule, it would pay to assume that all the matrix blocks are dense for the first time step just to find out the cost of NUMFAC for this case; at subsequent time steps, one would take into account zeroes within the matrix blocks until the combined cost of SYMFAC and NUMFAC exceeded the cost of NUMFAC assuming the dense blocks.

We have made a series of computer timing studies to determine the CPU time requirements of the Yale code when applied to reservoir simulation problems. The computer used for the timing studies was an IBM 370/158 with virtual storage. The timing runs were made without background jobs to avoid interference. The programs were compiled by using the IBM FORTRAN IV Level H-Extended Compiler, with OPT=2. Selected single precision floating point instruction speeds for the 370/158 are given in Table 4.

TABLE 4 - SELECTED AVERAGE FLOATING POINT INSTRUCTION SPEEDS, MICROSECONDS

Instruction	IBM 370/158
Add	2.0
Load	.70
Multiply	2.0
Store	0.88

Table 5 gives the CPU time requirement of the Yale codes

TABLE 5 - CPU TIME REQUIREMENT OF SYMFAC AND NUMFAC

			SYMFAC	NUMFAC	
Grid	Nonzeroes of U	Work	Sec	Sec	Micro-sec Multiply
30x30	8,915	189,276	.62	1.8	9.5
50x50	33,961	1,149,772	2.2	10.	9.1
20x60	11,701	232,372	.78	2.2	9.6
55x72	26,514	709,442	1.8	6.6	9.3
7x9	913*	11,611	.10	.14	12.
7x9	2,664	51,894	.23	.49	9.4

*Nonzeroes of L is 1,355

for several typical cases of Type 1 and Type 2 matrices. In each case, we used the best of the three grid orderings. Note that except for the smallest system, the Yale code requires about 9.5 microseconds per multiplication for NUMFAC, while a band algorithm using the grid row ordering requires about 7.2 microseconds per multiplication. The band algorithm, which does not do any pivoting, is the most efficient of the codes we tested in terms of CPU time per mutliplication. However, overall the Yale NUMFAC code is considerably faster. For example, it is approximately eight times faster in the case of the 55 x 72 grid in simulation problem No. 3. This favorable comparison is due to the reduced number of multiplications (a factor of about 10).

The computer time required to reorder the grid is unimportant relative to the time required to perform the numeric elimination, since in most cases the grid is reordered only once and the system (1) is solved many times in a simulation study. Typically, the computer time to reorder the grid is 0.5 to 4.0 times that required for one numeric elimination.

Storage requirements for sparse elimination are discussed in [4]. It should be noted that the storage required increases very rapidly with increasing grid dimensions. See Table 5 for the actual number of nonzeroes in U for the test problems.

Since we are solving a time-dependent problem and are primarily concerned with speed and not storage, we use a SYMFAC subroutine which returns pointers to the nonzeroes of

L and U and a NUMFAC subroutine which returns the numeric values of U only. If we were more concerned about storage (because of hardware limitations), we could use the SYMFAC subroutine described in [9] which returns the "compressed" pointers for the nonzeroes of L and U. Or going one step further, we could use the TRKSLV subroutine described in [4], which combines the SYMFAC and NUMFAC subroutines and requires storage for only the "compressed" pointers and numeric values for the nonzeroes of U.

CONCLUSIONS:

1. Sparse Gaussian elimination is a powerful engineering tool that can be used to solve economically many systems of linear equations arising in reservoir simulation.
2. Reordering significantly reduces the number of multiplitions required for elimination. For a Type 1 matrix, when a two-dimensional grid is nearly square, the minimum degree ordering is significantly better than the alternate diagonal ordering. The improvement is further enhanced if the grid is irregularly shaped.
3. Sparse Gaussian elimination can be considerably faster than band elimination algorithms. For a 55 x 72 grid with 2347 equations, the Yale code reduces computing time by a factor of 8 over a band elimination algorithm.

ACKNOWLEDGMENT:

This work was partially supported by ONR grant N0014-67-A-0097-0016.

REFERENCES:

[1] A. Chang, "Application of Sparse Matrix Methods in Electric Power Analysis", SPARSE MATRIX PROCEEDINGS, edited by R. A. Willoughby, IBM, 113-122, 1968.

[2] S. C. Eisenstat, "Complexity Bounds for Gaussian Elimination".

[3] S. C. Eisenstat, M. H. Schultz, and A. H. Sherman, "Applications of an Element Model for Gaussian Elimination", This proceedings.

[4] S. C. Eisenstat, M. H. Schultz, and A. H. Sherman, "Considerations in the Design of Software for Sparse Gaussian Elimination", This proceedings.

[5] S. C. Eisenstat, M. H. Schultz, and A. H. Sherman, "A User's Guide to the Yale Sparse Matrix Package".

[6] J. A. George, "Nested Dissection of a Regular Finite Element Mesh", SIAM J. NUMER. ANAL. 10: 345-363, 1973.

[7] F. G. Gustavson, "Some Basic Techniques for Solving Sparse Systems of Linear Equations", SPARSE MATRICES AND THEIR APPLICATIONS, edited by D. J. Rose and R. A. Willoughby, Plenum Press, New York, 41-52, 1972.

[8] H. S. Price and K. H. Coats, "Direct Methods in Reservoir Simulation", SOC. OF PET. ENGRS. JL., 14: 295-308, 1974.

[9] A. H. Sherman, "On the Efficient Solution of Sparse Systems of Linear and Nonlinear Equations", PhD dissertation, Department of Computer Science, Yale, 1975.

[10] P.T. Woo, S. J. Roberts, F. G. Gustavson, "Application of Sparse Matrix Techniques in Reservoir Simulation", SPE 4544, 48th Annual Fall Meeting of the Soc. of Pet. Engrs., Las Vegas, Nevada, 1973.

On the Origins and Numerical Solution of Some Sparse Nonlinear Systems

T. A. Porsching

This paper is dedicated to Garrett Birkhoff

1. Introduction.

Our intention in this paper is to present some source problems which give rise to sparse nonlinear systems of algebraic equations, and to discuss some numerical methods for their solution. In the next section we consider a class of nonlinear diffusion equations and an associated discretization which takes the form of a sparse nonlinear system. This system has the agreeable property of admitting a unique solution which may be determined by a globally convergent iterative method.

The last section of the paper contains a problem involving heated networks. Its solution is seen to be equivalent to finding a Brouwer fixed point. In this connection we describe a new method due to Kellogg, Li and Yorke for the computation of such fixed points.

2. Nonlinear Diffusion.

Many linear and nonlinear algebraic systems arise as discrete analogs of partial differential equations. In this section we will consider some special types of elliptic boundary value problems, namely two dimensional quasilinear diffusion equations of the form

$$-\nabla \cdot p(x,y,u)\nabla u + q(x,y,u) = 0, \tag{2.1}$$

where p and q are given smooth functions of x, y, and u. Equation (2.1) is required to hold on some region Ω of the x,y plane, while the Dirichlet boundary condition

(2.2) $$u(x,y) = g(x,y)$$

is imposed on $\partial\Omega$, the boundary of Ω. The function $g(x,y)$ is, of course, assumed to be known.

The literature of the natural sciences is replete with equations of the type (2.1). Among other things they have been used to describe the diffusion of odor [22, pg73] as well as the flow of heat [4, pg10], electric current [7, pg79], neutrons [5, pg115], and water [15, pg68].

Without going into existence and uniqueness questions about the solution $u(x,y)$ of (2.1) and (2.2) we will derive a set of nonlinear difference equations which define a discrete approximation of u. We remark that elliptic differential and difference equations both more and less general than those considered here have been analyzed by a number of authors, e.g., [1]-[3], [10]-[12], [14], [16], [21], [24]. The paper of Meyer [14] is particularly relevant since it too treats the solution of a set of nonlinear difference equations (not the present ones) derived from (2.1). We derive the present equations because they provide an important example of a sparse nonlinear system which may be solved by a globally convergent iterative method.

Assume for simplicity that Ω is a union of squares on which we impose a compatible square mesh whose spacing is h. We designate the mesh points by (x_i,y_j) and the mesh box side midpoints by $(x_{i+1/2},y_j)$ etc. (See figure 1). For mesh points $(x_i,y_i) \notin \partial\Omega$ we integrate (2.1) over the subregion Ω_{ij} shown in figure 1 and apply the divergence theorem to obtain

(2.3) $$-(I_{i+1/2,j} + I_{i,j+1/2} + I_{i-1/2,j} + I_{i,j-1/2}) + \iint_{\Omega_{ij}} q(x,y,u)\, dx\, dy = 0,$$

where

$$(2.4)\qquad I_{i+1/2,j} = \int_{y_j-h/2}^{y_j+h/2} p(x_{i+1/2},y,u(x_{i+1/2},y)) \frac{\partial u}{\partial x}(x_{i+1/2},y)dy,$$

and so forth.

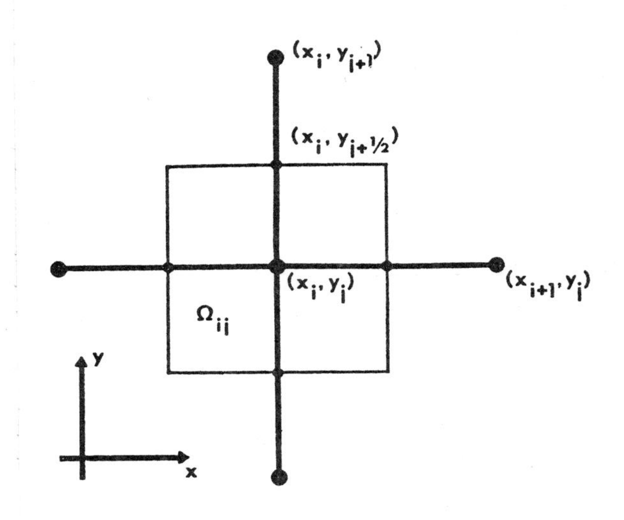

Figure 1. Subregion Ω_{ij}.

Since p is a continuous function of u for each $(x,y) \in \Omega$, we can define a Kirchhoff transformation [4,p11],

$$(2.5)\qquad J_{i+1/2,j}(u) = \int_0^u p(x_{i+1/2},y_j,s)ds,$$

from which it follows that

$$\frac{\partial J_{i+1/2,j}(u)}{\partial x} = p(x_{i+1/2},y_j,u)\frac{\partial u}{\partial x}.$$

This suggests approximating the integral (2.4) by $h\frac{\partial J_{i+1/2,j}}{\partial x}(u(x_{i+1/2},y_j))$, which in turn leads quite naturally to the further approximation

$$J_{i+1/2,j}(u_{i+1,j}) - J_{i+1/2,j}(u_{ij}),$$

where $u_{ij} \equiv u(x_i,y_j)$ etc. . An analogous treatment of the remaining integrals in (2.3) then yields in its place the following (nonlinear) difference equation for the approximation U_{ij}:

(2.6)
$$\begin{aligned}
&[J_{i+1/2,j}(U_{ij}) - J_{i+1/2,j}(U_{i+1,j})] \\
&+ [J_{i-1/2,j}(U_{ij}) - J_{i-1/2,j}(U_{i-1,j})] \\
&+ [J_{i,j+1/2}(U_{ij}) - J_{i,j+1/2}(U_{i,j+1})] \\
&+ [J_{i,j-1/2}(U_{ij}) - J_{i,j-1/2}(U_{i,j-1})] \\
&+ h^2\, q_{ij}(U_{ij}) = 0,
\end{aligned}$$

where $q_{ij}(u) \equiv q(x_i,y_j,u)$.

An equation of the type (2.6) is to hold at each mesh point $(x_i,y_j) \not\in \partial\Omega$. For $(x_i,y_j) \in \partial\Omega$ we use the obvious requirement that

$$U_{ij} = g(x_i,y_j).$$

Although we have given a hueristic derivation of (2.6), a straightforward but tedious calculation shows that (2.6) is a consistent approximation of (2.1). In fact if $u \in C^4(\Omega)$ and $p \in C^3[\Omega\times(-\infty,\infty)]$, and if we divide (2.6) by h^2 and write the result as

(2.7) $$M_h(U) = 0,$$

then

(2.8) $$M_h(u) = 0(h^2).$$

That is, the local truncation error is of second order.

Turning now to the nonlinear system (2.6), we number the mesh points, unknowns, and equations in consecutive order from 1 to (say) n, proceeding through the mesh from left to right, bottom to top. If we relabel the unknowns as $z_1, \ldots, z_n$, and denote the left side of the kth equation (2.6) by $f_k(z_1, \ldots, z_n)$, then our nonlinear system becomes

$$F(z) = 0, \tag{2.9}$$

where $F = (f_1, \ldots, f_n)^T$ and $z = (z_1, \ldots, z_n)^T$. System (2.9) is sparse since each equation depends on at most 5 unknowns. Moreover, its <u>structure matrix</u> $\Lambda = [\lambda_{k\ell}]$, where

$$\lambda_{k\ell} = \begin{cases} 1 \text{ if } f_k \text{ depends on } z_\ell \\ 0 \text{ otherwise} \end{cases}, \quad k, \ell = 1, \ldots, n,$$

has a half bandwidth which does not exceed max b_j where b_j is the number of interior mesh points on the jth horizontal mesh line.

To be assured that (2.9) has a unique solution <u>for any</u> $h > 0$, we make the following assumptions for $(x,y,u) \in \Omega\times(-\infty,\infty)$: i) equation (2.1) is uniformly elliptic, i.e., $p \geq p_0 > 0$ for some constant p_0; ii) $\frac{\partial q}{\partial u} \geq 0$. These assumptions are satisfied for many physically reasonable problems. Using the results of [19] (see also [18]), it can now be shown that F is a continuous, surjective M-function. From this follows the unique solvability of (2.9) as well as the <u>global</u> convergence to that solution of the iterates $\{z^k\}$ generated by the nonlinear Gauss-Seidel method

(2.10)

Solve $f_i(z_1^{k+1},\ldots, z_{i-1}^{k+1}, s, z_{i+1}^{k},\ldots, z_n^{k}) = 0$ for s

Set $z_i^{k+1} = s.$

Since the structure matrix of F is two cyclic and consistently ordered, it should be possible to accelerate the basic Gauss-Seidel method (2.10) by overrelaxation. See for example [6]. On the other hand, as pointed out by Rheinboldt [20], both the convergence rates and numerical stability of these methods will most likely deteriorate with increasing n.

A simplification results when p is independent of x and y. For in this case we can write (2.5) as

$$v = J(u) \equiv \int_0^u p(s)ds. \tag{2.11}$$

Hence, under the previous assumptions, (2.11) implicitly defines u as a function of v, say $u = J^{-1}(v)$. Letting $V_{ij} = J(U_{ij})$, we see that (2.6) becomes

$$4V_{ij} - (V_{i+1,j} + V_{i-1,j} + V_{i,j+1} + V_{i,j-1}) + h^2 q_{ij}(J^{-1}(V_{ij})) = 0. \tag{2.12}$$

As before, system (2.12) may be written in the vector form (2.9), and F is again a surjective M-function. Now however,

$$F(z) = Az + \phi(z), \tag{2.13}$$

where A is an M-matrix and ϕ is a diagonal function. Thus the structive of F-and hence the implementation of (2.10)-is greatly simplified. Indeed, if also q is independent of u, then we can solve the nonlinear system (2.9) by first solving the linear system $Az = -\phi$ and then the n one-dimensional nonlinear equations $J(U_{ij}) = V_{ij} \equiv z_k$.

3. Fixed Points.

Solution of a nonlinear system is sometimes equivalent to finding a fixed point of an associated continuous mapping G which takes a compact, convex subset, $D \subset R^n$, into itself. That is we are required to find $x \varepsilon D$ such that

$$x = G(x). \tag{3.1}$$

In this case the Brouwer Theorem [16, pg161] guarantees the existence of the fixed point, but does not show how to compute it. Recently a number of algorithms have been developed to find approximate Brouwer fixed points [13]. However, since these algorithms are combinatorial in nature, they do not take advantage of the possible sparseness of (3.1)[1].

In this section we shall describe a new method due to Kellogg, Li and Yorke [8], [9] for computing Brouwer fixed points which preserves any sparseness inherent in (3.1). By way of introduction, we present a problem involving heated networks which generates a sparse nonlinear system equivalent to a system of the form (3.1). As this example shows, it is an unfortunate fact that the sparseness of the original system need not be inherited by its fixed point equivalent[2].

---

[1] In the original economic applications given by Scarf [23], this is not a consideration since these nonlinear systems are by their very nature not sparse.

[2] This is also true of certain fixed point problems arising from discretizations of nonlinear elliptic boundary value problems. See for example [10], [24].

We consider a connected network of pipes as shown in figure 2. To each pipe we arbitrarily assign a direction of positive flow by placing an arrow on the pipe pointing in that direction.

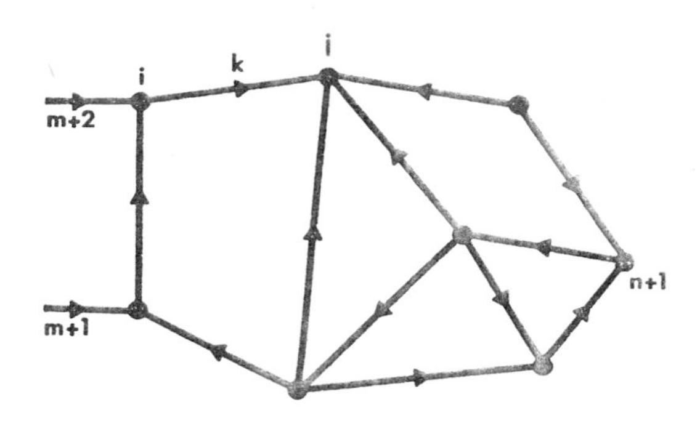

Figure 2. A Piping Network.

Let the junctions (nodes) and pipes (links) of the network be numbered from 1 to n+1 and 1 to m+ν respectively. Fluid is assumed to enter the network through links m+1,..., m+ν at known flow rates $w^*_{m+1},\dots, w^*_{m+\nu}$ and enthalpies $h^*_{m+1},\dots, h^*_{m+\nu}$ (enthalpy may be regarded as a measure of thermal energy). It leaves the network at node n+1 where a reference pressure of 0 is given. The problem is to determine the flow rate and enthalpy distribution in the network links.

To formulate the problem mathematically, let $A = [a_{ik}]$ be the n×m node-link incidence matrix[3] whose rows are associated with nodes 1,..., n. Then the conservation of fluid at these nodes may be expressed by the equation

[3] $a_{ik} = -1$ or $+1$ as the direction of positive flow in link k is toward or away from node i, and $a_{ik} = 0$ if link k is not incident upon node i.

(3.2) $$Aw = b,$$

where $w = (w_1, \ldots, w_m)^T$ is the vector of unknown flow rates and b is a known vector containing the boundary flow rates $\{w_k^*\}$.

If it is assumed that the enthalpy of the fluid in any link is that of its upstream node, than it suffices to determine the enthalpies at the nodes. We obtain a relation involving these enthalpies by utilizing the conservation of the fluid's thermal energy at the nodes. Thus, we write

(3.3) $$B(w)h = q,$$

where $h = (h_1, \ldots, h_{n+1})^T$ is the vector of unknown enthalpies and $B(w) = [b_{ij}(w)]$ is a $(n+1) \times (n+1)$ matrix with entries

$$b_{ij} = \begin{cases} \left. \begin{array}{l} -|w_k| \text{ if link k connects nodes i and j,} \\ \text{and the actual direction of flow in link} \\ \text{k is toward node i.} \end{array} \right\}, \; i \neq j, \\ 0 \quad \text{otherwise.} \end{cases}$$

$$b_{ii} = -\sum_{j \neq i} b_{ij},$$

and q is a known vector containing the thermal energies $\{w_k^* \, h_k^*\}$ of the boundary flows.

To complete the mathematical specification of the problem we assume that the flow rate in any link is related to the pressures and enthalpies at the link's extremities by an equation of the form (Laminar Flow Law)

(3.4) $$w_k = d_k(p_i - p_j) + c_k \, \rho\left(\frac{h_i + h_j}{2}\right), \; k = 1, \ldots, m,$$

where p_i and p_j are pressures, c_k and d_k are constants, $d_k > 0$, and ρ is the fluid density. The subscripts i,j and k are related as shown in figure 2. In this application ρ is assumed to be only a function of enthalpy. That is, the equation of state holds at a constant system pressure. We remark that it is also possible to treat a more general form of (3.4) in which the first term on the right is replaced by a continuous, isotone, surjective function of the pressure drop.

If we write (3.4) in vector form, we obtain

$$w = DA^T p + r(h), \tag{3.5}$$

where $p = (p_1, \ldots, p_n)^T$, $D = \text{diag}(d_k)$ and

$r(h) = (\ldots, c_k\, \rho(\frac{h_i + h_j}{2}), \ldots)^T$. Equations (3.2), (3.3) and (3.5) are sufficient in principle to determine the $2n + m + 1$ unknowns p, h and w.

The nonlinear system (3.2), (3.3), (3.5) is sparse if A is sparse, that is, if each node has only a few links incident upon it. Indeed, it is not difficult to see that upper bounds for the number of unknowns in (3.2), (3.3) and (3.5) are respectively s, $2s + 1$ and 5 where s is the maximum number of links incident upon any node.

The pressures are easily eliminated from this system. From (3.2) and (3.5) we obtain

$$ADA^T p = b - A\, r(h).$$

Since A has full row rank, we can solve this equation for p and substitute in (3.5). Thus

$$w = DA^T(ADA^T)^{-1}b + [I - DA^T(ADA^T)^{-1}A]\, r(h) \equiv g(h). \tag{3.6}$$

Now let $[\alpha,\beta]$ be a compact interval and suppose that $\rho : R^1 \to [\alpha,\beta]$ is continuous. Moreover, assume that for any choice of densities in $[\alpha,\beta]$ no flow vector given by (3.6) produces a stagnant node (i.e., a node with only zero flow rates in the links incident upon it.) Then (3.3) may be uniquely solved for h, i.e., $h = B^{-1}(w)q$, so that by (3.6),

$$h = B^{-1}(g(h))q \equiv G(h). \tag{3.7}$$

It can now be shown that the continuous function G takes some sufficiently large hypercube into itself. Therefore, (3.7) has a fixed point which, along with the flow rates given by (3.6), solves the heated network problem. However it is clear that, unlike the original system, G is not sparse.

The method of Kellogg, Li, and Yorke mentioned earlier applies to the fixed point problem (3.1) when $G \in C^2(D)$, and the boundary ∂D is piecewise smooth and contains no fixed points of G. Let C denote the set of fixed points of G (C is not void by the Brouwer Theorem). The idea of Kellogg et al is to construct a mapping $H : D - C \to \partial D$ by defining H(x) to be the point intersection of ∂D with the ray which emanates from G(x) and passes through x (figure 3). Since $x \notin C$, this mapping is well defined. The main result of [9] is that for almost all $x^0 \in \partial D$, there is a smooth curve, x(t), $0 \le t < T$, such that $H(x(t)) = x^0$. Furthermore, this curve goes from x^0 to within an arbitrary distance from the set of fixed points C.

For computational purposes, x(t) must be more precisely characterized. Also, it would be advantageous to know when a particular $x^0 \varepsilon \partial D$ had such a curve associated with it, i.e., was a so called "regular value". This latter consideration pales however if we note that the probability of choosing a non-regular point $x^0 \varepsilon D$ is zero!

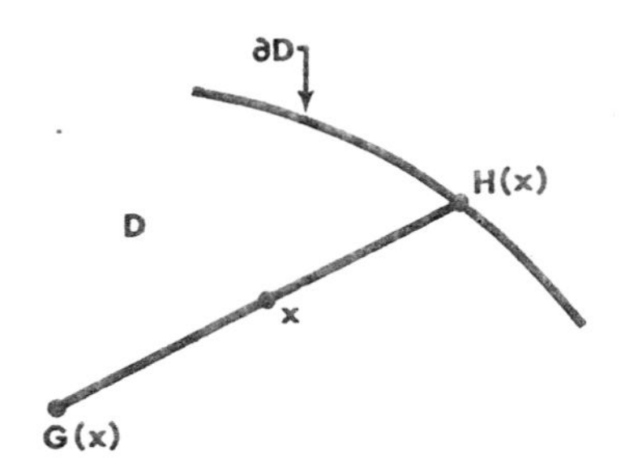

Figure 3. The mapping H(x).

Clearly H(x) is of the form

$$H(x) = (1 - \mu(x))x + \mu(x)\, G(x),$$

where $\mu : D - C \rightarrow (-\infty, 0]$. It is shown in [9] that if for $x \varepsilon D$ the eigenvalues of $G'(x)$ are excluded from the interval $(1,\infty)$, then any x^0 on a C^1 portion of ∂D is a regular value. Moreover, the associated curve solves the initial value problem

$$(3.8)\qquad \begin{aligned} &[G'(x) - (1 - \frac{1}{\mu(x)})I]\dot{x} = -(G(x) - x), \\ &x(0) = x^0. \end{aligned}$$

Each numerical integration procedure for (3.8) may be regarded as a numerical method for computing a fixed point of (3.1). For example, Euler's method with step length h^k is

(3.9)

$$x^{k+1} = x^k - h^k \, [G'(x^k) - (1 - \frac{1}{\mu(x^k)})I]^{-1} \, (G(x^k) - x^k).$$

As x(t) approaches a fixed point, it is easy to see that $\mu(x(t)) \to -\infty$. Therefore, in a small neighborhood of such a point, (3.9) becomes a damped Newton's method applied to $G(x) - x = 0$.

REFERENCES

1. L. Bers, "On mildly nonlinear partial difference equations of elliptic type." J. Res. Nat. Bur. Standards Sect. B, 51 (1953), 229-236.

2. G. Birkhoff, "The Numerical Solution of Elliptic Equations," CBMS Regional Conf. Ser. in Appl. Math., Vol. 1, SIAM, Philadelphia, 1971.

3. J. Bramble and B. Hubbard, "A theorem on error estimation for finite difference analogues of the Dirichlet problem for elliptic equations," Contr. Diff. Eq. 2 (1962), 319-340.

4. H. Carslaw and J. Jaeger, "Conduction of Heat in Solids," Oxford Univ. Press, London, 1959.

5. S. Glasstone and A. Sesonske, "Nuclear Reactor Engineering," Von Nostrand, Princeton New Jersey, 1963.

6. L. Hageman and T. Porsching, "Aspects of nonlinear block successive overrelaxation," SIAM J. Numer. Anal., 12 (1975), 316-335.

7. O. Kellogg, "Foundations of Potential Theory," Dover, New York, 1953.

8. R. Kellogg, T. Li and J. Yorke, "A Method of Continuation for Calculating a Brouwer Fixed Point," to appear in Computing Fixed Points with Applications, S. Karamadian, ed. Academic Press, New York.

9. ____________________, "A constructive proof of the Brouwer fixed point theorem and computational results," University of Maryland IFDAM Technical Note BN-810, Jan., 1975.

10. G. McAllister, "Some nonlinear elliptic partial differential equations and difference equations," SIAM J. Appl. Math., 12 (1964), 772-777.

11. _____________, "Quasilinear uniformly elliptic partial differential equations and difference equations," SIAM J. Numer. Anal., 3 (1966), 13-33.

12. _____________, "Difference methods for a nonlinear elliptic system of partial differential equations," Quart. Appl. Math., 23 (1966), 355-360.

13. O. Merrill, "A summary of techniques for computing fixed points of continuous mappings," in Mathematical Topics in Economic Theory and Computation, R. Day and S. Robinson eds., SIAM, Philadelphia, 1972.

14. G. Meyer, "The numerical solution of quasilinear elliptic equations" in Numerical Solution of Systems of Nonlinear Algebraic Equations, G. Byrne and C. Hall ed. Academic Press, New York, 1973.

15. L. Milne-Thomson, "Theoretical Hydrodynamics," Macmillan, New York, 1955.

16. J. Ortega and W. Rheinboldt, "Monotone iterations for nonlinear equations with application to Gauss-Seidel methods," SIAM J. Numer. Anal., 4 (1967), 171-190.

17. _________________, "Iterative Solution of Nonlinear Equations in Several Variables," Academic Press, New York, 1970.

18. T. Porsching, "Jacobi and Gauss-Seidel methods for nonlinear network problems," SIAM J. Numer. Anal. 6 (1969), 437-449.

19. W. Rheinboldt, "On M-functions and their application to nonlinear Gauss-Seidel iterations and to network flows," J. Math. Anal. and Appl., 32 (1970), 274-307.

20. ____________, "On the solution of large, sparse sets of nonlinear equations," University of Maryland Technical Report TR-324, Sept., 1974.

21. ____________, "On the solution of some nonlinear equations arising in the application of finite element methods," University of Maryland Technical Report TR-362, March, 1975

22. S. Rubinow, "Mathematical Problems in the Biological Sciences," CBMS Regional Conf. Ser. in Appl. Math., Vol. 10, SIAM, Philadelphia, 1973.

23. H. Scarf, "The approximation of fixed points of a continuous mapping," SIAM J. Appl. Math., 15 (1967), 1328-1343.

24. R. Stepleman, "Difference analogues of quasi-linear elliptic Dirichlet problems with mixed derivatives," Math. of Comp., 25 (1971), 257-269.

B 7
C 8
D 9
E 0
F 1
G 2
H 3
I 4
J 5